# KATE BROWN

# Manual for Survival

*A Chernobyl Guide to the Future*

PENGUIN BOOKS

## PENGUIN BOOKS

UK | USA | Canada | Ireland | Australia
India | New Zealand | South Africa

Penguin Books is part of the Penguin Random House group of companies
whose addresses can be found at global.penguinrandomhouse.com.

First published in the United States of America by W. W. Norton & Company, Inc. 2019
First published in Great Britain by Allen Lane 2019
Published in Penguin Books 2020
001

Printed and bound in Great Britain by Clays Ltd, Elcograf S.p.A.

A CIP catalogue record for this book is available from the British Library

ISBN: 978-0-141-98854-2

www.greenpenguin.co.uk

# MANUAL FOR SURVIVAL

'Magisterial . . . Kate Brown sets out to uncover
Chernobyl's true medical and environmental effects . . . an
awe-inspiring journey' *Economist*

'Exemplary . . . chilling reading . . . written with skill
and passion . . . Brown is an indomitable researcher'
Luke Harding, *Observer*

'Brown's page-turner skilfully weaves an original narrative on
the long-term medical effects of the Chernobyl disaster . . . Her
capacity to immerse herself and pick up on nuances brings
these stories from factory workers, technicians, doctors
and villagers alive' *Nature*

'Vital work, making a convincing case for the catastrophic
long-term medical and ecological effects of the disaster'
Tobie Mathew, *Literary Review*

'With bountiful, devastating detail, Brown describes how doctors,
scientists, and journalists – mainly in Ukraine and Belarus – went to
great lengths and took substantial risks to collect information . . . One
of the most alarming – though also eerily beautiful – aspects of
Brown's book is her description of the way radioactive material moves
through organisms, ecosystems, and human society'
Sophie Pinkham, *New York Review of Books*

'A troubling book, passionately written and deeply researched
over 10 years in former Soviet, Ukrainian and Belarusian
archives. Brown has interviewed doctors who have treated poisoned
patients, and farmers whose sheep produce irradiated wool and
contaminated milk. She has talked to scientists who have consistently
lied for years about radiation levels, and to others who have risked
their careers to unearth the truth . . . What is new in Brown's book
are the assertions she makes about western countries, and how, in an
"unholy alliance" with the former eastern bloc, they have hidden their
contribution to global radiation levels . . . Her conclusion is chilling'
Victor Sebestyen, *Sunday Times*

'Kate Brown has spent years trying to uncover the truth . . . Brown
brings home the effects of Chernobyl on individuals and communities.
In doing so, it uncovers a deeper and more disturbing tale'
Philip Ball, *New Statesman*

'A magnificent monograph that stands out among the multiple books on Chernobyl simply because it tells us the truth – the whole unadulterated truth – about one of the worst disasters in history. As such, it may itself be regarded as a survival manual of sorts. And a guide to the future, too' *Engineering and Technology*

'*Manual For Survival* is a remarkable book, distinguished by Kate Brown's rare combination of skills: formidable archival history, investigative research, and vivid storytelling. There are parts of this book that grip with the force of a thriller – but again and again, the plot is proved true' Robert Macfarlane, author of *The Lost Words*

'This deftly written, impassioned, courageous book should make the world think twice about what's at stake when we unleash nuclear reactions' Alan Weisman, author of *The World Without Us*

'Kate Brown presents a convincing challenge to the official narrative of the Chernobyl disaster. Deeply reported and elegantly written, *Manual for Survival* is chilling' Elizabeth Kolbert, author of *The Sixth Extinction: An Unnatural History*

'Kate Brown has blown the lid off the 1986 Chernobyl nuclear disaster and decades of official efforts to suppress its grim truths. Disturbing in its conclusions, destined to incite controversy, *Manual for Survival* is first-rate historical sleuthing' J.R. McNeill, author of *The Great Acceleration*

ABOUT THE AUTHOR

Kate Brown is the author of *A Biography of No Place*, which won the George Louis Beer Prize from the American Historical Association for the best book in International History, and *Plutopia*, which won seven awards, including the Dunning and Beveridge prizes from the American Historical Association for the best book in American history. She is the first historian of the Soviet Union to be nominated to the honorary Society of American Historians, and her research has been funded by the American Academy in Berlin and by Carnegie and Guggenheim fellowships. She teaches environmental and nuclear history at the Massachusetts Institute of Technology, Baltimore County, and lives in Washington, DC.

*For Marjoleine*

# CONTENTS

# Manual for Survival

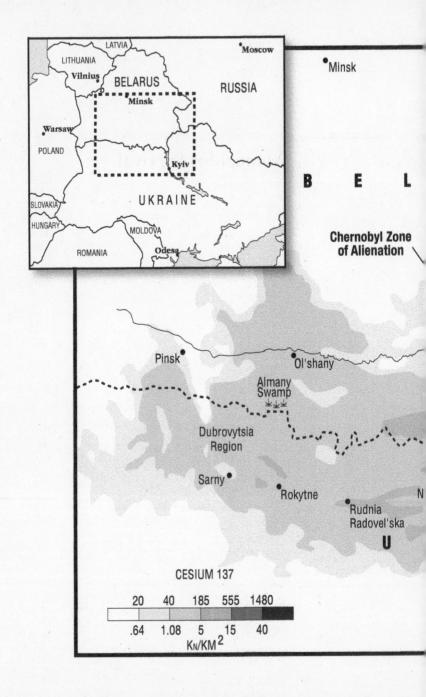

LATVIA

LITHUANIA

Vilnius

BELARUS

Minsk

RUSSIA

Moscow

Warsaw

POLAND

Kyiv

UKRAINE

SLOVAKIA

HUNGARY

MOLDOVA

ROMANIA

Odesa

Minsk

B E L

Chernobyl Zone
of Alienation

Pinsk

Ol'shany

Almany
Swamp

Dubrovytsia
Region

Sarny

Rokytne

Rudnia
Radovel'ska

N

U

CESIUM 137

| 20 | 40 | 185 | 555 | 1480 |
|----|----|-----|-----|------|
| .64 | 1.08 | 5 | 15 | 40 |

$K_N/KM^2$

# INTRODUCTION

# The Survivor's Manual

Three months after the Chernobyl accident in August 1986, the Ukrainian Ministry of Health issued five thousand copies of a pamphlet addressed to "residents of communities exposed to radioactive fallout from the Chernobyl atomic station." The pamphlet, speaking directly to the reader ("you"), begins with assurances.

> Dear Comrades!
> Since the accident at the Chernobyl power plant, there has been a detailed analysis of the radioactivity of the food and territory of your population point. The results show that living and working in your village will cause no harm to adults or children. The main portion of radioactivity has decayed. You have no reason to limit your consumption of local agricultural produce.

If villagers persisted in reading beyond the first page, they found that the confident tone trails off and pivots in contradiction:

> Please follow these guidelines:
> Do not include in your diet berries and mushrooms gathered this year.
> Children should not enter the forest beyond the village.
> Limit fresh greens. Do not consume local meat and milk.

Wash down homes regularly.
Remove topsoil from the garden and bury it in specially
prepared graves far from the village.
Better to give up the milk cow and keep pigs instead.[1]

The pamphlet is actually a survival manual, one that is unique in human history. Earlier nuclear accidents had left people living on territory contaminated with radioactive fallout, but never before Chernobyl had a state been forced to admit publicly to the problem and issue a manual with instructions on how to live in a new, postnuclear reality.

As I worked on this story, I watched TV documentaries and read books on Chernobyl. They tend to have a similar plot development. The clock counts down the seconds, as operators in the control room make decisions that can never be undone. Piercing alarms give way to the persistently creepy ticking of radiation meters. The focus turns to broad-shouldered, Slavic-handsome men who are gruffly unconcerned about their well-being. In front of the smoldering reactor, they smoke cigarettes, crush them out, and get on with the job of saving the world from this new, radioactive protagonist. The drama then shifts to hospital wards where the same men have been reduced to skeletons of rotting flesh. Just when you have had enough of blackened skin and intestinal damage, the narrator comes out with a just-kidding moment, asserting that commentators have long exaggerated the Chernobyl accident.

A journalist troops into the forests of the Chernobyl Zone of Alienation, a thirty-kilometer (km) zone around the plant depopulated in the weeks after the accident. The journalist points to a bird and a tree and pronounces the Zone to be thriving! Through cloying music, a voice-over says that, although Chernobyl was the worst disaster in nuclear history, the consequences were minimal. Only fifty-four men died from acute radioactive poisoning and a few thousand children had a nonfatal thyroid cancer, which is easy enough to cure. The soothing qualities of these made-for-TV narratives work like magic dust. The scary features of nuclear accidents disappear, so too the questions they raise. These narratives draw you in for the high-tech, human drama, while leaving you feeling hopeful about the future and (most importantly) grateful it didn't happen to you. By focusing on the seconds before the blasts and then on the safely contained radioactive remnants in the sarcophagus, most histories of Chernobyl eclipse the accident itself.

Only fifty-four deaths? Is that all? I checked websites of UN agencies and found a range of thirty-one to fifty-four fatalities. In 2005, the UN Chernobyl Forum predicted from 2,000 to 9,000 future cancer deaths from Chernobyl radiation. Responding to the forum, Greenpeace gave much higher numbers: 200,000 people had already died and there would be 93,000 fatal cancers in the future.[2] A decade later, the controversy surrounding Chernobyl consequences has not been resolved. We learn that birds in the Chernobyl Zone are dying from mutations, but then journalists inform us that wolves and caribou are repopulating the Zone. The public is left at a scientific stalemate. The mainstream media tend to report the most conservative numbers—thirty-one to fifty-four people dead. The assertion is that the final death toll will never be known.[3]

Why don't we know more? For decades, scientists around the world called for a large-scale, long-term epidemiological study of Chernobyl's consequences.[4] That study never came together. Why? Was there intention behind the confusion over Chernobyl damage? Within the Grand Canyon–sized gap between the UN and Greenpeace estimates of fatalities there exists a great deal of uncertainty. My aim in this book is to come to a more certain number describing the damage the accident caused and a clearer grasp of the medical and environmental effects of the disaster.

Without a better understanding of Chernobyl's consequences, humans get stuck in an eternal video loop, the same scene playing over and over. After the Fukushima accident in 2011, scientists told the public they had no certain knowledge of the effects of low-dose exposures of radiation to human beings. They asked citizens for patience, for ten to twenty years, while they studied this new catastrophe, as if it were the first. They cautioned the public against undue anxiety. They speculated and stonewalled as if they did not recognize they were reproducing the playbook of Soviet officials twenty-five years before them. And that leads to the pivotal question: Why, after Chernobyl, do societies carry on much as they did before Chernobyl?

I have other questions: What is life like when ecosystems and organisms, among them humans, mingle with technological waste and become inseparable? What does it take to get on with the art of living after the kind of thorough social-environmental-military sacking that the communities around the Chernobyl Zone of Alienation experienced in the twentieth century? Chernobyl, I learned, was not the first disaster to strike the territory. Before the Chernobyl region became synonymous with nuclear disaster, it was a front line in two world wars, one conventional war, a civil war, the Holocaust, plus two famines and three political purges, after which it became home to a Cold War bombing range. That makes the greater Chernobyl zones, where people continue to live, good places to investigate the outer limits of human endurance in the age of the Anthropocene, the epoch when humans became the force driving planetary change.

These are the questions that inspired my journeys around and into the Chernobyl Zone. I started my search for answers in the central archives of former Soviet republics and found reports of widespread health problems caused by exposure to Chernobyl fallout. Wanting to be sure, I went to provincial archives and tracked health statistics down to the county level. Everywhere I went I found evidence that Chernobyl radiation caused a public health disaster in the contaminated lands. Even the KGB reported this story. Soviet leaders banned media discussion of Chernobyl's consequences, so the records I found were classified "for office use only." Finally, in 1989, Soviet leaders lifted the media blackout, and news of serious health problems reached the Soviet and international press. Learning of their exposures, angry protesters demanded aid to relocate from contaminated territory. Panicking over the rising costs, leaders in Moscow called United Nations agencies for help. Two UN agencies provided assessments that backed up Soviet leaders' assertions that doses were too low to cause health problems.

I followed this drama through archives in Vienna, Geneva, Paris, Washington, Florence, and Amsterdam to see how international agencies took over managing public assessments of Chernobyl damage after the collapse of the Union of Soviet Socialist Republics (USSR). I found, sadly, a great deal of ignorance amid an orchestrated effort to minimize what was billed as the world's largest nuclear disaster. International diplomats stonewalled and blocked Chernobyl research because leaders of the big nuclear powers had already exposed millions of people to dangerous radioactive isotopes during the Cold War in producing and testing nuclear weapons. Awakening to this fact in the 1990s, Americans and Europeans were taking their governments to court. In this global context, Chernobyl wasn't the largest nuclear emergency in human history. It was just a waving red flag pointing to other disasters hidden by Cold War national security regimes.

In total over the course of four years and with the help of two research assistants, I worked in twenty-seven archives in the former USSR, Europe, and the United States. I filed freedom of information requests and asked that records be declassified. Often I was the first researcher to view the files. I focused on the major players—the Soviet government, the United Nations, Greenpeace International, and the most powerful UN sponsor, the U.S. government. Wanting to make sure the astonishing story I uncovered in the archives was correct, I looked for ways to cross-check the documents. I interviewed three dozen people—scientists, doctors, and civilians who became specialists in nuclear disaster from their experience living with its consequences. I visited factories, institutes, forests, and bogs in contaminated territories. I followed foresters, biologists, and residents around the Chernobyl Zone and attended scientific conferences to learn ways to read damage from contamination on the landscape.

Because of a Soviet ban on Chernobyl records and due to the usual archival hold of records for twenty or thirty years after an event, many documents about the disaster have only recently been brought to the light of day. Until now, Chernobyl histories have relied on eyewitness accounts and unconfirmed rumors. Writing this book, I vowed that I wasn't going to fall for every maudlin story or get dragged into children's clinics to look at sick kids who may or may not be sick because of Chernobyl. I set out to substantiate every claim, cross-reference it, and use the archives as my guide. Historians are captivated by archives because they help us return to the scene of the crime. It matters what my subjects say now, but it matters even more what they said and did thirty years ago.

On April 26, 1986, reactor No. 4 at the large and growing Chernobyl Nuclear Power Plant in northern Ukraine, a republic of the Soviet Union, exploded. Photojournalist Igor Kostin risked his life to take photos of men in lead aprons, shoulders down, rushing like linebackers to extinguish the radioactive inferno.[5] Kostin's black-and-white images do not show the men's spooky pallor. High doses of radiation cause spasms in surface capillaries of the skin so that faces look strangely white, as if powdered for the stage. Soviet leaders did not warn people to stay indoors during the emergency. Photos of families in Kyiv enjoying the sunny May Day holiday a week after the accident now appear cruelly sardonic. (The name Kyiv, Ukraine's capital city, is called Kiev in English.) The day before the holiday, catching city leaders by surprise, radiation levels spiked suddenly in Kyiv to 30 μSv/hr (microsieverts per hour), which was more than a hundred times higher than preaccident background levels.[6]

The festivities went ahead as planned in Kyiv on orders from Moscow. The parade lasted all day, as flank after flank of schoolchildren, keeping step to the music of brass horns, marched past the tribunal. They carried portraits of leaders they were taught to emulate and trust. At the end of the day, the kids struggled for breath. Their faces showed unusual purpled sunburns. In the following week, mild-mannered Ukrainian Minister of Health Anatoly Romanenko was pushed onto the podium to make public statements about the accident. He announced that the levels of radioactivity in Kyiv were going down, but he didn't say where the radioactive isotopes were going.

Physicists will tell you that energy can be neither created nor destroyed. The newsreels of the May holiday did not record the actions of two and a half million lungs, inhaling and exhaling, working like a giant organic filter. Half of the radioactive substances Kyivans inhaled their bodies retained. Plants and trees in the lovely, tree-lined city scrubbed the air of ionizing radiation. When the leaves fell later that autumn, they needed to be treated as radioactive waste. Such is nature's stunning efficiency at absorbing bursts of radioactivity after a nuclear explosion.

To be fair, Health Minister Romanenko did not know what happened to the radionuclides that blanketed his home city. He had no training in radiation medicine. Only one young doctor in the Ministry of Health knew something. She had taken a short course in nuclear emergencies and she quickly became the in-house specialist. She explained to other doctors and party leaders the differences among a roentgen, a rem, and a becquerel and between radiation in beta form and gamma form.[7] The accident caught public health and civil defense officials off guard because nuclear physicists had been telling the public for years that nuclear power was perfectly safe, while a classified wing of the Ministry of Health furtively dealt with hazardous nuclear accidents that plagued Soviet nuclear enterprises. Having fooled themselves, public health officials were left with little training or skills to handle the nuclear disaster.

The accident drew hundreds, then thousands, and finally hundreds of thousands of people into a three-dimensional space surrounding the disaster site. Helicopter pilots navigated overhead, dropping 2,400 tons of sand, lead, and boron on the reactor to try to snuff out smoldering embers. One helicopter clipped a crane and crashed, killing four men. Soldiers took turns racing onto the roof of reactor No. 3 to shovel off the graphite innards of the blown reactor. Miners tunneled ninety feet under the melted core to build a protective wall. Construction workers created dams to hold back the radioactive Pripyat River. Suspecting sabotage, KGB investigators rifled through filing cabinets, computer records, and the minds of survivors dying on their hospital beds.[8] On April 27, army officers escorted 44,500 residents from the atomic city of Pripyat. In the next two weeks they resettled 75,000 more people from a surrounding thirty-kilometer belt, which was renamed the "Zone of Alienation."

Radiation monitors and medical personnel trailed the Red Army trying to assess the damage. Conscripts ripped up asphalt, washed down buildings, and removed topsoil with the plan to reinhabit the abandoned communities. But every time the wind changed, it dusted the landscape with more fallout, and the boys would have to clean all over again.[9] Whoever said Soviet leaders were not capitalists was wrong.

Like business leaders elsewhere, they emphasized production over safety. Rather than securing the disaster site and closing it to let the most powerful radioactive isotopes decay for several months or years, they fast-tracked a plan to return the Chernobyl plant to full production as soon as possible.

Soviet journalists wrote about Chernobyl as a story of brave, selfless "liquidators" (cleanup workers) who fought the radioactive fires. Archival records show that not everyone behaved honorably. The KGB looked for several thousand plant employees and soldiers who left their duties and fled. Burglars snuck into abandoned Pripyat to steal rugs, motorcycles, and furniture to sell the radioactive possessions elsewhere.[10]

On May 6, Soviet officials broadcast to the world they had put out the raging fire in the reactor core. "The danger is over," they announced. That was not true. The fire continued until the graphite burned down on its own. Classified records show that radioactive gases poured from the disaster site for another week, spiking on May 11.[11] Soviet officials estimated that 3–6 percent of the core vaporized into the air and dropped about 50 million curies of fallout on the surrounding environment. A later study conducted after the USSR collapsed estimated that at least 29 percent of the fuel burned up in the fire for a total closer to 200 million curies of radioactivity dispersed into the environment. The releases were comparable to several very large nuclear warheads.[12]

As the awful magnitude of the disaster became clear in the months following the explosions, Soviet officials wrote more and more guides for citizens living in the wake of the catastrophe. They typed up survivor manuals for doctors treating Chernobyl-exposed patients, for farmers working on radioactive farms, for agronomists and food processors turning radioactive produce into consumer goods, for manufacturers of woolens, textiles, and leather, and for public relations specialists dealing with an anxious public. Dozens of manuals were issued specially for the disaster in thousands of copies. Unfortunately, Soviet survivor manuals were hamstrung by what writers could not say. I would like to provide a better guide to survive nuclear disaster, one that uses the Chernobyl archives to bring together all the players—the operators, doctors, farmers, and radiation monitors—to animate lessons learned from the isotopes, soil, wind, rain, dust, milk, meat, and the soft, porous bodies that took it all in.

When I started working on this book, I was no stranger to nuclear disaster zones or to northern Ukraine where the Chernobyl disaster took place. In 1987, I went to the USSR for the first time to study in the city formerly known as Leningrad (St. Petersburg). That was a year after the Chernobyl accident, but at the time I didn't pay much attention to rumors of radioactive food. Young and living in a miserable Soviet dormitory, I was mostly fixated on not being hungry myself. In the 1990s, I worked in Moscow, studied in Krakow, Poland, to the west of Ukraine, and researched my first book in the archives in Kyiv and Zhytomyr, all the while oblivious to the radionuclides—which I now know from charts in the archives were swirling about me. I heard something about health problems in Zhytomyr and noticed nuclear plant operators picketing in Kyiv, but I did not dwell on them. I had other interests and, like most Westerners in the USSR at the time, I thought Soviet activists were exaggerating the effects of the Chernobyl accident. I was a classic Western traveler in Eastern Europe, confident in the superiority of my society, sure of the natural, beneficial qualities of democracy and capitalism, and suspicious of Soviet truths in whatever form they took. These assumptions often made me, like many Westerners breaching the Iron Curtain, a poor listener and myopic observer. On the trips I took researching this book I tried to be more observant.

The disaster involved millions of people and required a complex sequence of actions. The book's first section deals with the players who responded immediately to assess and "liquidate" the effects of spilling radiation. Part 2 dwells on the people left behind in contaminated zones who continued producing and consuming despite the blanket of radioactive fallout enveloping them. Part 3 explores the ecology and history of the Pripyat Marshes, where the Chernobyl Nuclear Power Plant was located. The following section about politics focuses on Soviet leaders who both classified Chernobyl and used the tragedy to discredit their rivals. Part 5 concentrates on medical discoveries made by Soviet researchers. Part 6 tracks how Chernobyl came to be managed by international agencies as the Soviet Union came down in a cloud of dust. The concluding segment follows the survival artists who figured out how to carry on in an altered landscape.

Accidents happen. They are supposed to have a concluding chapter where humans learn a lesson or two. Calamities with no perceptible end make it harder to draw conclusions. A general lesson I learned from the Chernobyl disaster is that technology promoted as infallible sometimes fails and there is, as yet, no good guide for societies struggling with large-scale technological and environmental disasters. Reactors, many of which are working long past their expiration dates, are most often built in economically strapped, rural communities where people are grateful for the jobs the plant provides. If a reactor or nuclear bomb factory is shuttered because of an accident or planned obsolescence, the immediate territory is abandoned, a cyclone fence goes up, and the radioactive brownfield becomes a nature preserve, but one with strange regulations posted at the entrance to the park: "No dogs. Do not step off the gravel paths. Do not pick up any masonry object."[13] The fencing and designation "nature reserve" normalize disaster, soothe, and reassure like the 1986 Soviet-issue survival manual that begins "Dear Comrades."

Maybe nuclear power is, as advocates say, the best option to reduce carbon emissions and supply energy for a growing global population. And maybe nuclear weapons, which are the genesis of nuclear power, are the best way to defend against "rogue" nations. Perhaps there is no other way. If that is so, then I set out to travel around the Chernobyl Zone of Alienation with my eyes wide open, trying to understand how human life changes in the post-apocalyptic shadow. I made this journey because I don't want to be one of those duped comrades who found out too late that the survival manual contained a pack of lies.

# PART I

———— // ————

# THE ACCIDENT

# Liquidators
## at Hospital No. 6

Doctor Angelina Gus'kova slept with a phone next to her bed. At 2:30 a.m. on April 26, it rang. Chernobyl was on the line. The voice of the dispatcher, scratchy and fading, was desperate for help to treat firemen who were falling ill after battling the inferno at the reactor. Gus'kova was the director of the clinic for radiation medicine at the high-security Hospital No. 6 in Moscow. "The call came an hour after the accident. I was probably the first person in Moscow to hear about it," Gus'kova told a reporter in 2015.[1] The medic at the Chernobyl plant described patients who were nauseous and weak, with reddened skin, and one who was already vomiting. For Gus'kova the diagnosis was easy: "Typical signs of acute radiation sickness."

Gus'kova had seen cases like this before. No one in the world had treated more patients with radiation illness than Gus'kova. At age sixty-two, she was at the peak of a successful career that few knew about because her resume was kept secret in locked filing cabinets. Her biography is typical for her generation. She was born in a small mining town in Stalinist Siberia, where the meagerness of daily life taught resilience and selfless patriotism. From a family of doctors, Gus'kova naturally went to medical school. On graduation, she was handed the kind of unenviable job assignment at a closed military "post box" that was the fate of many Soviet provincials. In 1949, without complaint she moved into a cramped dorm room with four beds for seven women at a secret nuclear weapons installation in Siberia.

Gus'kova worked for ten years at the Mayak Plutonium Plant at a time when no one knew much about radiation and health. Patients would appear at her clinic with symptoms that could spell anything from flu to meningitis to tuberculosis. Radiation damage has no stand-alone symptoms. It causes a body to feel bad in many familiar ways. Because of security regulations, Gus'kova didn't know and could not ask if the prisoners, soldiers, and employees she treated had been exposed to radiation. With no other options, she and her colleagues studied their patients' bodies carefully. Gus'kova, trained as a neurologist, drew on a long tradition in Russian science, starting with Pavlov and his dogs, of looking to the central nervous system for signs of health problems. The doctors reasoned that toxins would first appear as damage to the vulnerable central nervous system before other organs.[2] They learned to detect the effects of radioactivity on nerves at very low doses. They also noticed chromosomal breakages in the quickly reproducing bone marrow cells of exposed patients. They figured out how to estimate doses of radioactivity their patients received from the extent of cell damage. They performed autopsies. Turning bones to ash, they used a gamma-ray spectrometer to detect and measure radioactivity lodged in the body.[3]

As they worked, Gus'kova and her colleagues used their patients' bodies as biological barometers. They came to guess their patients' dose from exterior symptoms and changes in blood cells. In 1953, she coauthored a book called *Radiation Sickness in Man*. For twenty years, the publication appeared only in classified editions, circulating in restricted libraries. Little of this information was shared abroad because Soviet security officials considered radiation medicine an important Cold War secret for surviving nuclear war.[4] In 1957, Gus'kova was promoted to a Moscow institute where her male colleagues disparaged her as a country hick. She switched to a job treating radiologists who had been overexposed by medical X-ray machines. She saved many patients but could not restore to health one young man who loved to scare women by painting his lips, fingers, and nose with glow-in-the-dark radium powder.

In the 1970s the first civilian nuclear power plants started up in the USSR. Gus'kova took over as the director of the radiation medicine clinic at Hospital No. 6. She approached a deputy minister in the Third

Department, the secret radiation medicine division of the Ministry of Health. The Third Department was one of those ghostly Soviet entities that had no address on its letterhead—just a postal number in Moscow, as if this gray, bureaucratic emanation floated above the city. It was set up in the 1950s to manage victims of accidents at Soviet nuclear weapons installations. The Third Department existed as a classified realm unto itself. The Soviet minister of health and his assistants had no idea what the Third Department did, even though it was part of the Ministry of Health.

Gus'kova appealed to the Third Department to publish a pamphlet for doctors with instructions for treating radiation victims. She figured that as civilian nuclear power plants spread across the Soviet Union, there could be mishaps. The deputy minister of health was enraged when he saw it. "You are planning an accident!" he shouted, tossing her manuscript back at her feet. Due to the Third Department's secrecy, civilian public health officials were little prepared for nuclear accidents when they occurred. In the following years, Gus'kova treated hundreds of workers who were exposed in accidents that were kept secret. At least twenty people died, quietly sacrificed to the peaceful atom.[5] Working on hundreds of patients suffering from radiation exposure over three decades, Gus'kova developed a compendium of knowledge on radiation medicine that had no equivalent in the world.

That knowledge became critical on April 26, 1986. At 1:23:48, a Saturday morning, seventeen employees of the Chernobyl Nuclear Power Plant were on shift. In carrying out a routine experiment, they turned off the reactor's emergency SCRAM system, which was, in any case, too slow to prevent an accident.[6] As the operators finished the test, they planned to take the reactor off-line for several weeks of routine maintenance. But on shutdown, the chain reaction in the reactor core went "critical," meaning operators no longer controlled it. The reactor's power surged. The operators remembered how the thick concrete walls wobbled, plaster rained down, and the lights went out. They heard a human-sounding moan as the reactor bolted and then popped.[7] The blast tossed up a concrete lid, the size of a cruise ship, flipping it over to expose the molten-hot core inside. A few seconds later, a more powerful second explosion sent a geyser of radioactive gases into the splendor of the Ukrainian night.[8] Plant worker Sasha Yuvchenko felt the thudding concussions and looked up from the machine hall to see nothing but sky. He watched a blue stream of ionizing radiation careening toward the heavens. "I remember," he later reflected, "thinking how beautiful it was."[9]

Unable to grasp the fact that the core had exploded, Anatoly Diatlov, the deputy chief engineer, sent two subordinates from the control room to the reactor hall (and inadvertently to their deaths). Fire brigades from the neighboring atomic city of Pripyat responded to the alarm. They sped toward the fire that emitted a strange cerulean glow and found a spookily quiet disaster site. Alarm bells and phones had failed. Scattered embers of burning graphite outlined the constellations of the disaster. Water seeped everywhere. With radiation counters maxed out, the men went to work without respirators or equipment for fighting hot fires at long distances. Six firemen climbed to the roof of the still functioning reactor No. 3. They picked their way around glowing shards of graphite and pieces of the blasted machine hall to aim their hoses at tongues of flame shooting hundreds of meters high from the hole where reactor No. 4 had been.[10] The mission was to make sure the fire did not spread to reactor No. 3 and cause a second meltdown. The men battled the fire until they lost consciousness and their comrades carried them down. Others took their place. An ambulance from Pripyat carted away the firemen and operators who had begun to vomit. Pripyat doctors, who had no special training in radiation injuries, were quickly overwhelmed; they called Moscow and got Gus'kova on the line.

Gus'kova told the provincial doctors that she would treat these patients at her Moscow clinic. She had her staff clear the ward. That Saturday evening, 148 men arrived, shivering in hospital gowns, their clothing dumped as radioactive waste.[11] The patients were assigned rooms based on their exposures. The most radioactive went to the top floor, where they were cordoned off in plastic tents to protect them from infection. The staff knew to keep their distance from the men whose bodies were contaminated with dangerously high levels of radioactivity. Over the next few weeks, Soviet officials publicly numbered 207 accident victims who checked into Gus'kova's clinic.

Most of the men started out feeling okay. They were tired, but they could walk around the ward, smoke, and talk over the accident with their mates. Just when the young men started to believe they would soon go home, they took to their beds and stayed there. They suffered from infections, nausea, hair loss, hacking coughs, diarrhea, fevers, and, later, intestinal bleeding. Their lungs filled with fluid. Their raw skin blistered, ulcerated, and blackened like burnt toast before sloughing off. The men found it increasingly difficult to think and communicate.[12] Some slipped in and out of consciousness. These were the outward signs. Gus'kova knew that a great deal more was going on inside her patients' bodies. Radioactive energy bouncing inside organisms causes what radiologist Dr. Karl Morgan described as a "madman loose in the library."[13] Ionizing energy alters and kills cells. Even cells not directly touched by radiation get damaged in ways that knock off-kilter cell communications that govern how cells reproduce and function. Radioactive energy causes the ungluing of strands of DNA, complicating cell repair. Damaged cells cause synaptic interactions of neurons to falter.[14] Radiation, in short, causes the body to fail on many levels from the inside out.

As in the past, KGB officials did not allow Gus'kova and her staff to learn the extent of their patients' exposures.[15] Judging doses from symptoms and blood work, Gus'kova directed her staff to treat the men suffering from acute radiation sickness with nutrients, vitamins, blood and platelet transfusions, antibiotics, chelating agents (which bind to toxic metal ions so they can be excreted from the body), and gamma globulin to boost patients' immune systems. As the days passed, the symptoms multiplied fearsomely. Firemen with large doses, over 6 Sv, experienced a massive cell death that caused many organs to fail or cease to function at all.[16] In the gastrointestinal tract, rapidly dividing crypt cells failed to replace cells of the villi, which are tiny brooms in the small intestine that absorb nutrients and provide a barrier between fecal matter and the body's other organs. As the worn-down villi slumped, the men suffered from malnutrition. Bacteria flooded their organs, causing sepsis. Radiation demolished the men's bone marrow cells, which normally produce billions of blood cells each day. Without them, the patients became severely anemic, bled spontaneously, and had no means to fight infections that ran rampant. After two weeks, the men with higher doses, over 9 Sv, died from burns, gastrointestinal damage, and central nervous system and liver failure. None of this was pretty. Acute whole-body exposure causes the organs to collapse all at once, in concert.[17]

With Chernobyl, the volume of patients was alarming, but Gus'kova had seen cases like this before and she was confident in the expertise of her staff. So it was unnerving when, a week after the explosion, an American doctor showed up in her restricted hospital to help. Robert Gale was a leukemia specialist from the University of California (UCLA). Through his California networks, Gale was acquainted with Armand Hammer, an American millionaire who met Lenin in 1921 and made his fortune in the twenties and thirties trading with the internationally ostracized Soviets. Hearing about the accident, Gale contacted Hammer and told him he wanted a pipeline to Mikhail Gorbachev so that he could donate his expertise for the emergency.[18] Hammer greased the wheels with $600,000 in medical supplies, and Gorbachev, who had initially turned down all other Western pledges of aid, invited Gale soon after the accident to Gus'kova's clinic at Hospital No. 6.

Gale wanted to perform bone marrow transplants on patients who had a chance of survival. Gus'kova was doubtful. She knew that bodies under assault from radiation had trouble withstanding invasive procedures like bone marrow transplants, and her team did them rarely. Gale insisted. He persuaded three more Western doctors to join him in Moscow. They arrived a few days later with heavy crates of supplies. In the 1980s, Soviet medicine suffered from a lack of investment. Gus'kova's special clinic was well funded by Soviet standards, but to the Americans it was "run-down" and sweltering for lack of air conditioners with rats crawling through basement passages.[19] The Americans were not used to these conditions. A lab hood started smoking, burned up, and never worked again. A centrifuge needed to separate blood cells broke down. Soviet lab techs spent hours tediously counting cells on slides, whereas American automated blood-cell counters did the work in twenty seconds.[20]

Gale brought with him a new miracle drug—genetically engineered molecules for bone marrow transplants. He hoped they might restore the zapped bone marrow of Chernobyl firemen. There was just one hitch. The drug had never been tested on humans. Gale was working with the Swiss pharmaceutical company Sandoz, which hoped to market GM-CSF as the drug to stockpile for nuclear emergencies. Imagine the volume of sales if large countries purchased GM-CSF. Testing the drug was a problem because nuclear emergencies were rare. Chernobyl offered a terrific opportunity. Gale proposed to experiment with the new drug on Chernobyl firemen and operators.

Soviet leaders were dubious about using their heroes as lab mice for capitalist pharmaceuticals. "We don't want to become their test site," a commissioner remarked.[21] To prove that GM-CSF was safe, Gale and Soviet hematologist Andrei Vorobiev took the drug themselves. They injected into their veins ten times the maximum dose tolerated by experimental monkeys.

After his shot of GM-CSF, Gale, feeling fine, took off across Moscow for dinner at Spasso House, home of the American ambassador, Arthur Hartman. Over appetizers in the spacious nineteenth-century villa, Gale received a phone call. Vorobiev was dying! He rushed back to the hospital to find the doctor in the coronary care unit, pale and suffering from severe chest pains. Gale surmised that the pain was caused by a buildup of granulocytes in the sternum. That was good news. Producing granulocytes, which fight off bacterial and fungal infections, was just what he hoped the drug would do. After an anxious night, Vorobiev recovered, and Gale got the green light to use the drug on Chernobyl patients.[22]

Gale did not speak publicly about this drug test on human subjects for several decades. Doctors are not supposed to experiment on their patients, not without drawing up a protocol, getting it approved, and receiving patients' written permission. Whilst Gale would argue that he was always acting in the best interests of patients, some might view him as a bit of a medical cowboy. A year before, in 1985, U.S. federal regulators had severely reprimanded Gale as the principal investigator on a research project in which researchers experimentally performed bone marrow transplants on children terminally ill with cancer without the approval of a faculty committee responsible for protecting the rights of the patients.[23] Regulators determined Gale violated several international protocols, including the Nuremberg Laws. Gale maintained he had done nothing wrong. Over the phone he justified to me the tests as "best practices" at the time. Bone marrow transfusions, he pointed out, are now standard practice for children with leukemia.[24]

Trying to understand better how doctors work with human subjects, I contacted an expert on protections for human subjects, Dr. Michael Carome. "The fact that the experimental interventions tested by Dr. Gale eventually became the standard of care," Carome commented in a later email exchange, "does not vindicate him for conducting unethical research without the appropriate informed consent of the human subjects or the review and approval of an institutional review board."[25]

In addition to the use of medical subjects, the Soviet leadership gave Gale unusual access to the secretive bubble they had created around Chernobyl. Invited to Kyiv, he toured the restricted hospital where doctors were treating Chernobyl patients. He flew over the smoking reactor in a helicopter. He asked questions about the number of evacuees, the level of radiation exposure, and future health damage. His KGB handlers gave him "approved answers."[26] Gale was fortunate. He had rare firsthand information. Because of his presence at Gus'kova's hospital, Soviet officials released to the public some news on the firemen. "We will provide information about the number and condition of patients in Hospital No. 6," Moscow Chernobyl commissioners resolved, "because American specialists are working there."[27]

U.S. officials in the USSR were having trouble getting the most basic information about the disaster. With no invitations, American diplomats drove to Ukraine in their private vehicles. Along the way, they measured radiation in the air and soils. Four times KGB agents caught American diplomats trying to enter the Zone of Alienation. Twice they stopped diplomats who were scooping up soil and stuffing it into plastic baggies. The agents "neutralized" these actions. KGB agents also tailed a CBS film crew as they skated around a Kyiv farmers' market with Geiger counters or tried, unsuccessfully, to visit evacuees.[28]

KGB agents eavesdropped on journalists who reported on the phone to their editors that Kyiv was functioning as it always did. "Crowds stroll the streets and enjoy life," a correspondent told his chief. The agents followed reporters to the margins of the city where they shot footage of empty streets and buses at the end of the run so, KGB agents surmised, the buses would have few passengers. The grimmer the pictures, the more foreign photographers snapped their shutters. Western journalists also tried to contact known Ukrainian nationalists, who were sure to complain about Soviet malfeasance. Faced with this unwanted attention, the KGB was forced to "localize" six especially troublesome correspondents.[29]

KGB agents were incensed that international tourists fled Ukraine because of the sensational, negative media coverage, while alien spies, they believed, were streaming in. They conjectured that the special services of foreign governments (the CIA, the West German BND, the French DGSE) were sending agents to Kyiv pretending to be journalists. They listed correspondents from respected news services—NBC, the *New York Times*, *Der Spiegel*, *Süddeutsche Zeitung*, and *Le Figaro*. I don't know what to think when I read intelligence reports, either from the United States or the USSR. I don't see a lot of footnoting in them, which makes it hard to fact-check what could just be the whimsy of a self-promoting or paranoid agent. I came across a file in which a KGB agent reported that a man I know personally was working for the CIA. I found that hard to believe but, as the CIA has not yet opened its archives, I have no way to check that allegation.

In Kyiv after the accident, foreign journalists poked microphones in the faces of nervous Kyivans and asked them if they were unhappy that Soviet officials had waited two days to tell them about the blown reactor. Reporters wanted to know how people felt about buying possibly radioactive food in farmers' markets. A KGB agent planted in the crowd easily dealt with these amateur attempts to "spread panic." He fired back questions about the overblown Western reporting of the accident. That, indeed, was embarrassing. Western reporters conceded that they did have a problem with a "lack of trustworthy material." Major U.S. networks showed video footage of the Chernobyl plant burning. The shots were actually of a fire at a cement factory in Trieste, Italy.[30] The network's apologies sounded lame because the Italian hills clearly visible in the background are hard to mistake for northern Ukraine, which is as flat as a blini. The UPI news service stated that two thousand people were dead from Chernobyl radiation. It soon turned out that the Soviet fatality count in early May of two people was true. Good news in the USSR made for an uninspiring news day in the West. In New York, the Soviet foreign press attaché quipped about the Chernobyl reporting, "I get the feeling the American press is not happy there are so few victims."[31]

Soviet officials could have helped the situation by relaying more information about the global disaster. Even KGB agents were critical of the cheerful Soviet coverage of the accident. An agent wrote in his daily report that TV news featuring collective farm directors near Chernobyl happily prattling about harvest targets had little utility. "The question still stands," the KGB officer wrote, "whether those harvests will be edible. If the soil isn't contaminated, then they should state it directly. Even better would be to avoid dubious messages."[32]

All this was Cold War politics conducted with a dull predictability. For commentators in the West, the more dreadful the disaster appeared, the more the Soviet "Evil Empire" looked worthy of U.S. president Ronald Reagan's moniker. But the Chernobyl accident was not just a Soviet drama. Citizens in Europe and North America, who were already worried about nuclear weapons, saw the accident as proof that civilian nuclear power was just as threatening. Officials at organizations promoting nuclear power, such as the UN International Atomic Energy Agency (IAEA) and the U.S. Department of Energy (DOE), grew increasingly nervous. They had worked for years to make careful distinctions between reactors created to produce nuclear bombs and reactors for "peaceful" production of electricity and isotopes for medical uses. A civilian reactor that also manufactured plutonium for bomb cores and accidentally blew up (like a bomb) greatly muddied these distinctions.

U.S. scientists working for the Department of Energy jumped in to make predictions. They initially estimated that 24,000 people would die from Chernobyl-induced cancers. When this number enflamed public anxieties, they quickly downsized the death count to 5,100 fatalities.[33] In August 1986, an IAEA official rushed to dispel public anxiety. "Chernobyl," Morris Rosen noted, "shows us that even in a catastrophic accident, we are not talking about unreasonable deaths."[34]

The international scramble in the summer of 1986 to control the message on the Chernobyl disaster in a context of secrecy and suspicion started a pattern where public spokespersons in the USSR and abroad made unsubstantiated claims, which later had to be corrected. As time passed, doubt and skepticism became a major by-product of the accident. The blown reactor contaminated not only the soil and air but also the political atmosphere and public faith in science. The half-life of that contaminant has yet to be determined.

GUS'KOVA WASN'T INVOLVED in the public relations side of the catastrophe. In the weeks after the disaster, she spent long hours at the clinic treating patients. Gradually she started to have apprehensions about Gale. The lean, sun-tanned, forty-year-old doctor had an extravagant personality and clearly didn't mind taking risks. He skydived in the Mojave Desert, went to war zones in Beirut, and walked barefoot around Kyiv.[35] In Moscow, Gus'kova suspected that Gale was experimenting with her patients. The American cancer doctors were not specialists in radiation medicine. Gale pursued what she believed was the mistaken idea that a good way to save patients with acute radiation poisoning was with headline-catching and risky bone marrow transplants and experimental drugs.[36] Gale and his UCLA colleague Paul Terasaki advised Soviet doctors on nineteen transplants at Hospital No. 6 on patients considered to have a chance of survival.[37] Six men given fetal liver transplants died soon after surgery.[38] The men who had undergone bone marrow transplants were in distress. And Chernobyl victims did not respond as hoped to the experimental drug GM-CSF.[39]

Gale stayed two weeks on that first trip. Before he left, he held a news conference in a packed hall of journalists hungry for information on the accident. Gale spoke of "battlefield conditions" and noted there would be more deaths among the first responders. He added that there could be long-term health problems for thousands of people who lived near the plant.[40] He and Armand Hammer emphasized that Soviet doctors were doing an excellent job and that Hospital No. 6 met "the highest international standards." The men sought to deflect attention from Chernobyl and focus it on the greater threat of nuclear war, a point that Gorbachev stressed in his remarks on the accident. After the press conference, Gorbachev met Gale and thanked him for his help in winning the global media over to support Soviet relief efforts. Because new censorship regulations, formalized in late July, muffled Gus'kova and her colleagues from speaking about Chernobyl, they did not appear at the press conference or in media reports. Gale dominated the podium.[41]

In Western newscasts it appeared as if the American doctor was at the helm directing the emergency medical response. It looked as if he knew more about radiation medicine than his Soviet counterparts, despite the fact that several visiting Western scientists expressed their admiration of Soviet doctors' mastery of radiation medicine. Gale's American colleagues noted how "unusually capable" Soviet doctors were at estimating dosage simply by studying a patient's vital signs and commented on their impressive range of treatments unknown in the West.[42] Most of Gus'kova's patients who had potentially fatal doses of 6 Sv survived for at least several more years.[43] This high survival rate is testimony to her team's clinical experience.

Gale's patients did not fare as well. Within three months, all but one had expired.[44] Soviet doctors charged that some men, who would have lived, died because of Gale's bone marrow transplants.[45] Gale later acknowledged that Gus'kova's "sophisticated treatments" made his more risky procedure unnecessary.[46] A UN review ten years later pronounced Gale's experimental treatments to be more harmful than beneficial.[47]

Volunteering to help with Chernobyl patients enabled Gale to participate in drug tests on human subjects without onerous American regulations.[48] Gale's KGB handlers benefited too. They used him in their public relations management of the disaster. Fed misinformation, Gale relayed it to information-starved foreign correspondents, who believed without question the testimony of the altruistic American doctor. He walked around Kyiv with his children, repeating the Soviet line that the danger had passed, at a time when Ukrainian health officials were sending children out of Kyiv because of the health risks. He praised the orderliness of the Soviet cleanup when it was in great disarray. He also helped to focus attention on the severely exposed firemen in Hospital No. 6. The dying firemen diverted the media away from a much larger drama taking place elsewhere around the new Chernobyl Zone of Alienation to which Gale had no access.

After Chernobyl, Gale became known as a world expert in radiation medicine and a go-to commentator for nuclear emergencies. In 1987, he appeared in Goiania, Brazil, where a thousand curies of radioactive cesium salt killed four people and hospitalized two hundred others. He showed up at the 1987 Armenian earthquake site, where fault lines came dangerously near a nuclear power plant. He made appearances in Japan for two nuclear disasters, including the 2011 event when a tsunami demolished the Fukushima Daiichi plant. He continued to experiment with GM-CSF in these foreign locations. The U.S. Food and Drug Administration approved the drug in 2015.

After his work in Moscow, Gale continued to advocate for Chernobyl victims. In the 1980s he set up a joint program to study 100,000 Chernobyl-exposed people for long-term damage, but that project never went anywhere.[49] In 1990, he appealed to the international community to support Chernobyl relief.[50] But over the years, his statements about the dangers of radiation exposure softened. In 1994, he testified as an expert witness in court against Rung C. Tang, a bedridden Nuclear Regulatory Commission inspector who claimed her leukemia was due to "radioactive fleas" issuing from the San Onofre nuclear power plant in southern California. Other plant workers later sued when they also suffered from myeloid leukemia, one of the known causes of which is radiation exposure.[51] After the Fukushima accident, Gale wrote editorials to stem panic about the nuclear disaster.[52]

The encounter between Soviet doctors and Gale's team of Western physicians revealed how the Cold War had incubated two distinct silos of knowledge on radiation medicine. Soviet doctors, because of their unfortunate surfeit of exposed patients, had a great deal more expertise than their Western counterparts. In 1986, that message did not come across. Instead, Gale stood as a symbol of the superiority, ingenuity, and initiative of Western, capitalist medicine over the supposed inherent shortcomings of socialist medicine. That false impression persisted and deepened in the coming years.

# Evacuees

Before it was abandoned, Nadia Shevchenko lived in the city of Pripyat in a five-story building now swallowed by forest. She started her family there. "I loved it," she remembered. "Rose bushes along the paths. The town was surrounded by forest and lakes, a sandy beach along the river. Friends. Pripyat had everything."

For several years I met Nadia each summer in the town in northern Ukraine where she resettled after the accident. We became friends. She liked to sing and dance and would break into song just about anywhere. I enjoyed her sunny outlook, though her life as an evacuee from the Chernobyl disaster hadn't been that sunny. One day, Nadia told me again about the events in Pripyat thirty-six hours after the accident. This time the story came out with a surprising vehemence. "I worked at the Chernobyl plant in the ventilation department. On April 27, the police told us to pack a few things to leave for three days. I took a bag with T-shirts for my two boys and their documents. I put food out for the cat, fed the fish, and we left."

Nadia and I sat on a terrace of a coffee shop in Slavutych, the city built to replace Pripyat. Nadia paused and looked out at the empty space before us on the large square where there should have been a statue of Vladimir Lenin, the vaguely lost sleepwalker at the center of every Soviet city. In October 1986, architects from seven republics designed Slavutych in a surge of patriotism that followed the Chernobyl accident.

They laid out grand plans for a comfortable city, but during the three years of construction, the Soviet economy fell flat. By the time Slavutych was ready to be inhabited in 1989, the scant supply of money and patriotism prevented the installation of the usual statue of Lenin pointing to a communist future. And so, as she talked, Nadia stared at empty space. "We boarded a bus. It took us to a village not very far away. I figured there had to be radiation there too, but no one knew anything."

Nadia's hunch was correct. Nuclear bomb target maps suggest that distance is the main factor in nuclear accidents, but radioactive fallout from the smoking reactor swept along in a patchwork fashion. Rain laced with fallout fell on April 27 on several regions of northern Ukraine and southern Belarus. The day before, these regions had normal background levels of radiation. After the rain, they became covered with hot spots of contamination.[1] Of the 44,000 people who left Pripyat the day after the accident, most were resettled to two regions that registered levels of contamination greater than Pripyat by the time they got there.[2]

An employee at the Chernobyl Nuclear Power Plant, Nadia knew that she had been dropped in a place extremely ill-equipped to protect against radioactivity. In the village, there were no showers to wash up, no monitors to measure radiation, and no extra beds. They slept on the dusty floor of a drafty hut. The villagers were going about their business as usual, gardening, feeding livestock, and milking their cows. Nadia understood that she and her kids were likely in the path of radioactive nuclides coming from the burning reactor and that the emergency evacuation had likely increased their exposure.

Nadia shot me a look. "I decided that we'd better get out of there."

A bus passed through the village heading for Kyiv. Nadia got on it with a few other mothers and their children, but when they arrived at the train station in Kyiv they had no place to go. Nadia stood on the platform with her backpack and her two boys. They had no coats, no cash, and no plans. A conductor at the train station pointed her to an office door. She told the clerk at the desk they had come from Pripyat. The clerk appeared to know all about it, though the accident had not yet been reported in the news.

"She asked where we wanted to go," Nadia said. "I had a sister in Moscow, the only place where I knew someone who could take us in. The clerk wrote me tickets."

Nadia was part of a gathering flood of people streaming to Moscow to seek help. Moscow, at the center of the Soviet Union, had a magnetic attraction. It was known to have the best hospitals, the best specialists, and the best supplies. In early May, scores of people were showing up daily in Moscow train stations asking for medical help. Medical officials ordered a halt to what they dubbed a "self-evacuation" from exposed regions. Despite their assurances that the Chernobyl accident was under control, Politburo bosses were morbidly anxious about radioactive contamination spreading to their ruby-jeweled capital. They decreed a ban on radioactive food products going to Moscow and ordered an inspection of passengers for radioactivity on trains arriving from the accident zone.[3] These measures backfired. Within a month, radioactive veal and milk were reported in Moscow, and when monitors in jumpsuits waving beeping wands showed up on trains from the Belarusian city of Gomel, the passengers "panicked," exactly what Moscow officials sought to avoid.[4] So they called off the monitors, and the refugees kept coming. By the end of May, three hundred people were arriving daily.[5] In June, after the end of the school year, even more people poured into the capital.[6]

Of 120,000 people resettled in the first weeks after the accident, 93,000 were from Ukraine. Emergency measures largely focused on Ukraine, the site of the accident. Moscow leaders were surprised to find that nearly 20,000 disaster migrants came from Belarus.[7] Doctors in Moscow measured people arriving from Gomel, 200 kilometers from the disaster site. Why were their thyroids so frightfully radioactive? Some had measurements of radioactivity in their thyroids ranging from 30 to 50 Sv, a dose large enough to destroy the organ.[8] Somehow, with no general alarm, no real news, and no experts with sensitive equipment, Belarusian villagers figured out they were in danger and they boarded trains out of town.

Perhaps their children showed symptoms as alarming as those of Nadia's son. Arriving in Moscow, Nadia stayed with her sister until her younger son came down with a fever and started to vomit. Nadia's sister called a friend who was a nurse. The nurse phoned someone she knew in radiation medicine, and in no time there was a knock at the door. A medical team rushed in.

"They were all bound up," Nadia wryly noted, "in Saran wrap." A radiation monitor held a wand that started beeping the moment he crossed the threshold. The medical team told Nadia and her two sons to pack: "Are you sure you have everything?" they kept asking. As they were leaving, a medic ordered Nadia's sister to mop up three times, wash down every surface, and then do it all over again. They sped Nadia and her sons to Hospital No. 7, one of two facilities designated to treat Chernobyl refugees rounded up from train stations.[9] In the daily reports of the operative group managing the disaster four days after the accident, 468 evacuees had been hospitalized, 38 with radiation sickness.[10] By the time Nadia arrived in Moscow, 911 people had checked into Moscow hospitals No. 7 and No. 15 for radiation injury. By May 5, that number had grown to 1,346, including 330 children, 64 of whom showed signs of radiation sickness.[11] The following day the number of Chernobyl-exposed in hospitals doubled to 2,592.[12] And the numbers kept growing. By summer's end, Moscow hospitals had treated 15,000 people exposed to Chernobyl radioactivity. In Kyiv, Gomel, Zhytomyr, and Minsk, a total of 40,000 people had checked into hospitals for the same reason. Half of the 11,600 people treated in Belarus were children.[13] Most of the hospitalizations were for screening, but thousands stayed for long visits because doctors noticed troubling symptoms from radiation exposures.[14]

In the press coverage, no one mentioned hospitalized evacuees, among them children. Orders were to report only on patients—firemen and operators—admitted to Hospital No. 6 and nowhere else.[15] Soviet officials spoke publicly only about the firemen with the worst cases of acute radiation poisoning. This fiction became part of the Politburo's alternative universe. Soviet minister of health Evgeny Chazov reported to the Central Committee of the Communist Party the mistruth that 299 people were hospitalized, just after he received reports from the Ukrainian and Belarusian ministers tabulating over 40,000 Chernobyl patients. The Soviet government, in other words, lied not just to the world but also to itself.[16] Why?

Firefighters willingly rush toward a fire. They accept the risky nature of their profession. The Chernobyl first responders were heroes because they gave their lives to save the world from the greater disaster of more explosions at the plant. But children with radiation injury—that's trickier. Kids don't know what radiation is. They are generally scaredy cats who shrink from life-threatening risks. If children are exposed, it is as victims of some careless party, in this case, the Communist Party, which had long promoted itself as the protector of all children.[17] The fact that thousands of minors were sidelined in hospital beds because the party had failed them was too incongruous for party leaders to acknowledge, even to themselves.

I came across a letter in a Minsk archive from a despairing mother, Valentina Satsura, who fled the Khoiniki region of Belarus in early May with her infant and toddler. She went to the Gomel Hospital, which was unprepared, understaffed, and overrun with mothers and children in a "desperate condition." There her baby suffered a brain inflammation. Gomel doctors counted the gamma rays coming from their bodies. Satsura learned that at her baby's thyroid the radiation reading was 37 µSv/hr. At that level of radioactivity, nestling her child to her neck was a perilous act of love.[18]

In the first few weeks after the accident, the Chernobyl fires emitted radioactive iodine, which the body mistakes for stable iodine, an element that human thyroids need to function. Although local soils were poor in natural iodine, it was not added to commercial salt. As a result, people's bodies craved iodine and their thyroids took up its radioactive mimic readily.[19] Radioactive iodine comes in various forms, some with whisperingly short half-lives of just a few hours. Iodine-134 has a half-life of just fifty-two minutes, but there was enough of it in the atmosphere that it was detected in Sweden five days after the reactor exploded.[20] Because these man-made elements are so fleeting, scientists know little about them and tend to discount them because they dissipate so quickly, but the shorter the half-life of a nuclide, the greater its rate of radioactive decay. Monitors in Ukraine were still detecting radioactive iodine in air, drinking water, and milk until the end of June 1986.[21]

The evacuations of the thirty-kilometer Zone of Alienation around the power plant took several days to organize and two weeks to carry out. During that delay, evacuees breathed in radioactive iodine, drank it in milk, and wiped it from dusty sweat. Children, whose bodies are smaller and who absorb minerals more efficiently than adults, took in three to five times more radioactive iodine than adults.[22] Prophylactic (stable) iodine tablets block the uptake of radioactive iodine to the thyroid. Scientists in Ukraine and Belarus requested that iodine be distributed. Moscow leaders waffled, not wanting to cause panic. They finally agreed, but often too late—several days, sometimes weeks after the accident.[23] Doctors later noticed that the children who had taken iodine immediately did much better than those who had not.[24]

Maria Kuziakina, a doctor in the southern Belarusian town of Ulasy, described to me how she and her neighbors watched the reactor burn from their village a few kilometers from the plant. Helicopters flew overhead. One landed in a field. Radiation monitors in suits and respirators jumped down and turned on their devices. "We wanted to talk to them," Maria remembered, "but they looked at the needle on their box, rushed back to the helicopter, and flew away." A few hours later, the collective farm chairman called to say they would be leaving their homes in four days. All fieldwork had to stop. Everyone was told to stay indoors. Maria went door-to-door and gave villagers iodine. They left their homes a week after the accident. During that week, radiation levels ranged from 400 to 1,900 $\mu$Sv/hr.[25] Each day for a week, Kuziakina had the equivalent of two computed tomography (CT) scans or fifty times the annual amount of background gamma radiation. Kuziakina did not know the radiation levels, but she did notice her neighbors didn't look right. "On our bus ride out of the village we were all sunburned, a strange purple suntan." Maria sighed and added, "They are all gone. I can think of only ten people from Ulasy who are still living."[26]

As evacuees arrived in resettlement locations and kids landed in summer camps, medical brigades sorted people into categories A through G, according to how much radiation issued from their thyroids. Children who had more than two sieverts were hospitalized for twenty-one days and given a battery of tests.[27] Doctors checked people who had over seven sieverts into special antiseptic tents in separate wards.[28]

Nadia remembered her time in the hospital. "They took our blood and urine, measured and prodded us, and then they did it all again the next week." The medical staff did not tell Nadia her estimated doses. Her medical files were classified "for office use only." Hospital staff had to get permission from KGB officers to see them.[29] Later, most of the dose records of Pripyat residents disappeared under what the hospital director called "mysterious circumstances."[30]

Nadia and her sons' doses must have been high because they stayed in the hospital for two months. Pripyat residents averaged a 500 mSv dose, a serious dose by Soviet standards.[31] According to a manual issued in June to doctors for the Chernobyl emergency, patients were to be released from hospital care when their blood indicators returned to normal and their symptoms of radiation poisoning passed.[32]

The secretive Third Department of the Soviet Ministry of Health wrote the doctors' medical manual for the Chernobyl disaster. It differed from the vast body of literature on radiation medicine in the West, which derived largely from the large, long-lasting Life Span Study of Japanese bomb survivors in Hiroshima and Nagasaki. The U.S. Atomic Energy Commission began the Life Span Study in 1950. Investigators eventually included in the study 120,000 survivors plus about 75,000 offspring. They tracked vital statistics, cancers, and causes of death and coordinated that information against elaborate estimates of doses of radioactivity based on subjects' reported location at the time of the bombing, as survivors remembered it in subsequent years.

The atomic bomb delivered radioactivity in two ways. With the blast, Japanese survivors received a single, large dose of exposure, like a very big X-ray, lasting less than a second. After the mushroom cloud dissipated, radioactive fallout filtered down around Hiroshima and Nagasaki and spread on air currents farther afield to deliver a second exposure. Japanese medics puzzled over the fact that people who arrived in the bombed cities after the attack grew ill from radiation sickness. American doctors noticed that U.S. soldiers working on reconstruction in Hiroshima and Nagasaki suffered mysterious burns, and their white and red blood cell counts dropped by half.[33] The Japanese press attributed these post-bomb symptoms to a mysterious "atomic poison."

General Leslie Groves, head of the Manhattan Project, grew extremely agitated on hearing this charge. If the new atomic bomb was placed in the same category of banned ordnance as chemical and biological weapons, then the $20 billion investment (in 2016 dollars) would be wasted and Americans would look morally as bad as the Germans who introduced mustard gas during World War I.[34] Fearing this outcome, Groves issued orders to confiscate Japanese medical records, notes, slides, and films on bomb victims. He orchestrated a media campaign to refute claims of lingering radioactive toxins. He sought to frame the atomic bomb as just a very powerful conventional explosive with the massive number of Japanese deaths caused by nothing more mysterious than thermal burns. Censorship on the atomic bombs' radioactivity remained in place for decades. To this day the medical section of the U.S. Army report on physical damage in Hiroshima is missing in the U.S. National Archives. Strangely, the medical section appears to have been reclassified in the 1990s when the controversy over Chernobyl health effects was at a peak. "For years," historian Janet Farrell Brodie writes, "radiation remained the least publicized and least understood of the atomic bomb effects."[35]

American investigators later admitted to the atomic bombs' fallout. They called it "residual radiation." Even so, American investigators did not include fallout in their estimates of doses survivors received because they believed reconstructing that dose would be "impossible," and they dismissed it as "on average small," something "to be disregarded."[36] Yet estimated doses from fallout in just the first two months after the bombing could be substantial—from 100 to 1,000 mSv, enough to induce radiation sickness and later cancers and thyroid disease. Hematologist William Maloney worked in Japan in the first stages of the Life Span Study. He puzzled over the physicists' low reported doses against the high incidence of leukemia he found among people who were several miles from the blast. He suspected that repeated low doses from fallout would eventually cause more cancers than the bomb's single big dose.[37]

The Life Span Study, then, was useful at estimating damage to a population directly subject to a nuclear attack, the question that most interested its American backers during the Cold War. Because of the study's failure to account for "residual radiation," results were less certain in determining health damage at chronic, low levels, such as those caused by nuclear accidents, residual fallout from bomb explosions, or emissions from bomb factories. The Life Span Study had one other major drawback. The study started in 1950, which was five years after the bombs had fallen. Deaths, miscarriages, birth defects, and disease among survivors before 1950 were not recorded. That left a blank spot about the health effects of chronic low doses of radioactivity in the first months and years after exposure.

Despite these problems, Western specialists in radiology considered the Life Span Study to be the gold standard for radiation epidemiology because of its size and long period of followup. The study was also invaluable in producing probable doses and risk estimates for general populations, expressed as excess relative risk (ERR) relative to the pre-existing risk. Western epidemiologists used the ERR to create a dose-response model, which was basically a prediction of health effects at various doses, a useful tool for regulation of nuclear enterprises and for talking to the public about exposures from medicine and fallout from bomb tests. By 1986, the Life Span Study showed that bomb survivors and children exposed in utero had an increased risk for some cancers but had no extra risk for birth defects or other diseases. The researchers determined that people experienced no radiation sickness at doses below 1 Sv and only a small increase in expected cancers at that dose.[38] They found that only those survivors who had received high doses over 1 Sv had statistically significant health impacts.

Soviet researchers could read published literature on the Life Span Study but had little access to data from the study because of Cold War secrecy. In 1986 when Russian scientists asked UN officials for "precise information" about how the survivor study was carried out, they were given instead data about a chemical explosion in Italy.[39]

In the decades before Chernobyl, Soviet scientists and physicians like Angelina Gus'kova developed radiation medicine in isolation in clinics classified as top secret. Soviet scientists were engaged in a forty-year study of three generations of villagers who lived along the radioactive Techa River in the southern Urals near the massively polluting Mayak plutonium plant. That secret research followed 1.5 million people (including a large control group). By 1986, Soviet researchers had found that people living on the Techa River exposed to chronic low doses of radioactivity had significantly increased death rates and cancers that occurred two to three times more frequently than among Japanese bomb survivors.[40]

The hard-won wisdom of the Techa tragedy made its way into the classified medical manuals issued for the Chernobyl emergency.[41] Unlike the Life Span Study, Soviet manuals did not include charts with predictions of cancers for each organ, ERR risk estimates, and dose-response computations. Instead, they described the immediate symptoms from exposure to radioactivity and the accompanying changes in the body at doses ranging from high to low. Instead of a threshold at 1 Sv, Soviet doctors envisioned a continuum of health problems along a scale from severe to mild. Children and fetuses, they found, were especially sensitive to medium doses of radioactivity (100–400 mSv).[42] They called the lower level of long-term exposure chronic radiation syndrome (CRS), defined as a complex of unspecific symptoms that could include malaise, headaches, lower work capacity, loss of appetite, sleepiness, insomnia, bleeding gums, and disorders of the liver, kidneys, thyroid, reproductive system, and respiratory and digestive tracts. American scientists had neither this nuanced understanding nor a category for chronic exposure, such as CRS. When American military scientists learned of it in classified diplomatic channels, they saw collaboration with Russian investigators as a "unique opportunity" to find out more.[43]

As Nadia and I talked on a hot summer day, I could see the scars ringing her neck from thyroid surgery. I learned over the years visiting Slavutych that Nadia, decades after the accident, had good days and bad days. On bad days, migraines, heart problems, and fatigue kept her in bed.

In 1986, doctors could do little to treat their patients. Radioactive isotopes do not readily leave the body. The best medical staff could do was give their patients vitamins, antibiotics, and diets of nonradioactive food. In this respect, Soviet citizens were well taken care of at the hospital. Nadia was served red wine while the boys enjoyed equally exotic orange juice. They ate red meat, bananas, dumplings, white bread, chocolate, and candies. Stuffing themselves and moving very little—never leaving the hospital compound—was the cure. Nadia and her sons were treated like Soviet cosmonauts or Olympic athletes preparing for a first flight or a big match. Nearly every material wish was granted. Their sacrificed bodies were treated to a cure that simulated life in an idealized Soviet society, one that did not experience chronic shortages, limited consumer choices, and nuclear accidents.

Nadia laughed thinking about it. "We got sick of all those luxury foods! They really took care of us."

One day her ten-year-old, Slavik, was crying in the ward when the chief doctor walked by. He asked Slavik why he was upset. Slavik spoke up boldly. He told the doctor that his mother had taken a trip abroad to Bulgaria and bought him a pair of jeans that had metal rivets and were almost as good as American jeans.

"So what is the problem?" the doctor asked impatiently.

"They took my good jeans and gave me this shit to wear," the exasperated boy explained gesturing to his Soviet-issue polyester slacks and shirt.

Nadia held her breath, waiting for the chief's reprimand about sacrifice, nation, duty, and profanity, but instead the doctor ordered an attendant to go out and get the boy a new pair of jeans and a sweater to match. The attendant returned in a few hours with clothes for Slavik and his older brother.

For Nadia, this story was about the state and its relations to its citizens. Hospital No. 7 and its cornucopia of delicacies embodied the patronage-style economic system of Soviet socialism. Citizens gave everything they had to the state, sacrificing their health if need be and, in exchange, the state took care of them.

On release, Nadia was given her passport and a document attesting to her stay as a ward of Hospital No. 7. The doctor asked her where she was going. "Well," she answered, "I have to get back to Pripyat and to my job. I need to work toward my pension."

The doctor laughed. "You've given enough to your country for ten pensions."

It was August in Moscow. Nadia set off to the Ukrainian Republic consulate to retrieve a promised 200-ruble subsidy to help pay for the family's lost possessions. She was passing through Pushkin Square where a crowd of Muscovites had gathered to read Pushkin's poems. She asked a police officer where the consulate was. He curtly gave her directions, but she could not find the consulate and returned in a half hour to ask him again. At that point, the policeman turned on her and demanded her identity card, a sign that she was under suspicion.

Slowly it dawned on her why he was so brusque. She looked like a woman just released from prison. Her shaved hair was growing back in patches. She had on government-issued clothing and shoes several sizes too small that bloodied her feet. Realizing that the policeman thought she was a bum made her furious. She had lost her apartment, furniture, money, job, and community, and this man thought she was a parasite. Her rage metastasized into an uncontrollable and dangerous urge to shout in the face of a Moscow police officer. She screamed at such a pitch that the Pushkin-lovers turned to watch. They formed a circle around her, while she bawled through tears that she came from Pripyat and had lost her home, that they had cut off her hair and taken her clothes, and that she had nothing left and just wanted to find the Ukrainian consulate.

As she yelled and kicked, her too-small shoes flew from her blistered feet, she dropped her purse, and in her agitation she tore open her shirt. Someone picked up her documents, and the crowd passed them around. The bystanders turned into witnesses at seeing the hospital certification, and they quietly began to curse the policeman, who grabbed his walkie-talkie to call for reinforcements. A woman helped Nadia to her feet, straightened her clothes, and buttoned her shirt. The policeman suddenly ushered her into a black sedan that appeared on the street to drive her to the consulate. Someone in the crowd pressed a wad of cash into her hands, a collection the bystanders had taken up.

"None of that would have happened," Nadia told me in 2015 with uncharacteristic bitterness, "had the accident occurred now. We would have been left," she spat, "to die in Pripyat."

As she pushed around melted ice cream in a bowl, she wiped tears from her cheek. Nadia wasn't happy about the Maidan Revolution in

2014 when civil war had simmered in eastern Ukraine. She identified as a Russian in Ukraine, where she no longer felt at home.

Gesturing around the square, she continued, "When this was the Soviet Union people had respect for one another. They didn't call each other *Khokhly* or *Moskali* (Ukrainian hayseeds or Moscow snobs). We were just citizens of the same Union, and we spoke a common language. Only that one policeman ever treated me poorly."

After Nadia was checked out of the hospital, a relocation official offered her a plum job and a much desired apartment in Moscow. She turned both down. She just wanted to go home to cozy, peaceful Pripyat. In 1986, as today, that desire was as fanciful as wishing to live on the moon. She tried to get hired at the Chernobyl plant when two reactors went back online in the fall of 1986, but at the time they were not taking women, especially women with children. She found a job instead at the Rivne nuclear power station and transferred to the Chernobyl plant when the new city for nuclear workers, Slavutych, opened to residents in 1989. Many other former residents of Pripyat also settled in Slavutych. It was as close as Nadia could get to home.

We paid at the cafe and took a walk in Slavutych. We strolled by a preschool. Each quarter in Slavutych has its own preschool, designed by architects from different Soviet republics. The school in the Yerevan Quarter was painted a sunburnt orange, punctuated by red stone cut from Armenian cliffs. Octagonal patterns in the wrought-iron fence repeated in the school's doors and windows. The building was surrounded by a large, parklike garden with spacious outdoor classrooms and covered verandas. It was inviting, even festive.

I admit I was envious. I had handed over a good portion of my monthly paycheck for my child to attend a preschool in a dark church basement in Washington, DC. My son had nothing of the luxury and thoughtfulness exhibited in these light-filled, spacious chateaus for children. Soviet planners and educators had long focused on children, yet there was more to these buildings than the usual socialist fixation on behavioral adjustment. Slavutych's elaborate, ornate preschools signaled how children's well-being and adult fears for their future stood at the center of post-Chernobyl society, one that was dedicated to overcoming the accident.

And that is what Nadia did. Rather than fleeing the catastrophe, she returned to it. She worked at the Chernobyl Nuclear Power Plant until she qualified for a pension. She raised her sons in Slavutych. She started a folk choir and knew just about everyone in the brand-new town, the last one built in the USSR. She sang whenever an occasion called for it.

That evening we made a trip to the countryside near Slavutych and wandered into a field with her friend Olga from choir. The sinking sun slanted crosswise in bright ochre beams that gilded skin and hair and flashed Nadia's lipstick a brilliant crimson. We watched the shadows lengthen and the light drain from the red pines edging the meadow. The day as it ended became one of those sadly sweet terminations that signpost life as it passes. In that shimmering, sunset radiance, I felt the majesty of the summer evening. I watched Nadia also breathe it in.

And then slowly, she picked up her arms, and rotated east and west. On cue, Olga followed. Her long wrists levitated overhead, punctuating her willowy figure. The two friends, one short, the other tall, danced in dresses of floral prints that wildly mimicked the meadow flowers surrounding us. Their voices joined, keyed in the atonal harmonies of Slavic folk song. Just as it was meant to, the song rang out in the wide open space of the evening pasture, echoing against the walls of the enclosing twilight. In that meadow, in a mostly emptied village, in a depopulated land, we experienced a joy written just for the moment.

# Rainmakers

Yuri Izrael had a regrettable decision to make. He ran the powerful Soviet State Committee of Hydrometeorology. It was his job to track radioactivity blowing from the smoking Chernobyl reactor and deal with it. Forty-eight hours after the accident, an assistant handed him a roughly drawn map. On it, an arrow shot northeast from the nuclear power plant and broadened to become a river of air ten miles wide that was surging across Belarus toward Russia.[1] If the slow-moving mass of radioactive clouds reached Moscow, where a spring storm front was piling up, millions could be harmed.

Izrael's decision was easy. Make it rain.

That day, in a Moscow airport, technicians loaded artillery shells with silver iodide. Soviet air force pilots climbed into the cockpits of TU-16 bombers and made the easy one-hour flight to Chernobyl, where the reactor burned. The pilots circled the ten-kilometer zone. Following the weather, they flew farther—thirty, seventy, a hundred kilometers—chasing the inky black billows of radioactive waste. When they caught up to a cloud, they shot jets of silver iodide into it to emancipate the rain.

Seeding clouds was not out of the ordinary for Red Army pilots. Soviet scientists began work on weather manipulation in 1941. In the early fifties, they set up an institute dedicated to "cloud physics." Like their American rivals, Red Army generals dreamed of torqueing the weather to both win battles and feed the masses.[2] Weather manipulation had lots of other uses as well. The TU-16 bombers first went into active duty to clear the skies over the 1980 Moscow summer Olympics. (The boycotting U.S. athletes missed that fair weather.) The big Soviet holidays on May 1 and November 7 occurred during the rainy Russian spring and the rainy Russian fall, so pilots chased off storms over tribunals of leaders reviewing the marching columns of socialist workers and soldiers. The Moscow city government massaged the clouds to lower the cost of snow removal. Soviet airports shooed away fog for smoother landings.[3]

In the sleepy Belarusian town of Narovlia, forty-eight kilometers north of Chernobyl, villagers remember looking up to see planes with strange yellow and gray contrails snaking across a pewter sky.[4] On April 27, at 4 p.m., powerful winds kicked up and raked the dusky pine forests around the town. Across a level plain of plowed fields, cumulus clouds mushroomed upward. The silver iodide attracted moisture and bonded with other waterlogged pellets. The seeded clouds stacked higher and higher, forming a craggy mountain of gas. At 8 p.m., thunder rolled out and rain poured down in a deluge. Precipitation fell all night until 6 a.m. Similar weather hit the Belarusian towns of Khoiniki and Bragin but largely spared the city of Gomel with half a million people. The raindrops scavenged radioactive dust floating 200 meters in the air and sent it to the ground.[5] The pilots trailed the slow-moving gaseous bulk of nuclear waste beyond Gomel, into neighboring Mogilev Province where they again made it rain.[6] Where pilots shot silver iodide, drops usually fell along with a toxic brew of a dozen radioactive elements mixed with heavy metals used to try to smother the fire in the reactor.

Tampering with the weather is tricky. Lots can go wrong, but this mission worked. No rain fell on the large Russian cities of Moscow, Voronezh, and Yaroslavl.

In 2006, President Vladimir Putin awarded Alexander Grushin, commander of the air force Cyclone-N Brigade, a medal for his role in saving Russian lives after the Chernobyl accident. With his medal, Grushin was finally free to speak about his secret mission. "We were young," Grushin told a reporter, "and we didn't think about radiation. We didn't know how to understand the dose we were getting." Grushin started to worry only at the end of a flight when he landed at an airport in Ukraine.

"Radiation monitors in protective suits approached our plane," Grushin recalled. "They had Geiger counters. They looked at their needles and all at once, as if on command, they turned and started running from the plane! I've never seen anyone run so fast." On their missions, the pilots received acute doses of radioactivity, which ransacked their digestive tracts and caused tissue damage in organs and exposed extremities. In 2006, Grushin, in his forties, walked with crutches. Two men in his command had amputated legs. Five pilots had portions of their stomachs cut out. Many of the rest of his crew were invalids.[7]

In Ukraine, pilots also made flights to manipulate the weather. Instead of precipitation, they sought sunshine. Civilian pilots from the Institute of Hydrometeorology ran missions to dispel rain clouds on the approach to the thirty-kilometer Zone of Alienation. They were worried that heavy summer thunderstorms would swamp the Pripyat River, which flows into the Dnipro River, the main artery pumping fresh water through Ukraine.[8] On weekly operations, from May to June and September to December, Ukrainian pilots let loose nine tons of cement 600, a mix of cement and reagents that dried up moisture from clouds in an eighty-kilometer loop around the plant.[9] The operations in Belarus and Ukraine accomplished opposite ends. During the summer and fall of 1986, air force pilots continued to make it rain over portions of Belarus, while civilian pilots navigating over portions of Ukraine induced a drought that lasted five months.[10]

No radioactive rain on Moscow and little radioactive rain in Ukraine. If Operation Cyclone had not been top secret, the headline would have been spectacular: "Scientists using advanced technology save Russia and Ukraine from technological disaster!" Yet, as the old saying goes, what goes up must come down. The rain that did not fall on Ukraine migrated with the prevailing winds north and east to Belarus, where summers are normally cooler and wetter.

No one told the Belarusians that the southern half of the republic had been sacrificed to protect Russian and Ukrainian cities. In the path of the artificially induced rain there lived several hundred thousand Belarusians. In deciding what areas to evacuate after the accident,

Moscow commissioners followed the simple logic of a bomb target map. They drew a thirty-kilometer-radius circle around the smoking plant and ordered it cleared of residents.[11] They relocated 90,000 people from Ukraine but only 20,000 from much more heavily contaminated Belarus. It is a myth that Soviet leaders resettled people from "the most contaminated territories."[12] Communities in Mogilev Province, 400 kilometers from Chernobyl, topped the chart of exposures. No one informed Mogilev leaders that the rainmaking pilots were spreading a blanket of radioactive fallout around them.[13] No soldiers showed up for evacuation. People remained home. Izrael kept secret the maps showing the carefully planned spread of radiation beneath the pilots' flight paths.

But some secrets are hard to keep. On April 28, Belarus's leading physicist, Vasily Nesterenko, appeared in the office of Nikolai Sliun'kov, the boss of the Belarusian Communist Party. Nesterenko asked why the radiation counters at his nuclear research institute were going crazy. Sliun'kov had no idea. He phoned the Ukrainian party leader, Volodymyr Shcherbytsky, in Kyiv, who told him that the Chernobyl Nuclear Power Plant, three kilometers from the Belarusian border, was on fire. Moscow leaders, he said, were holding a meeting about it that day. Hanging up, Sliun'kov barked, "They call themselves neighbors and they didn't even bother to tell us! Everyone just saves himself."[14]

Sliun'kov ordered Nesterenko to form an "operative brigade" and head south to see what was going on. Nesterenko's team started to piece together a map of radioactive contamination. In the three regions bordering the plant, he estimated that the air a meter from the ground from April 27 to May 5 measured 350–4,000 $\mu$Sv/hr.[15] At the time, the permissible dose for an adult nuclear worker was 28 $\mu$Sv/hr. Before the Chernobyl accident, the maximum exposure for adults and children was even lower.[16] The numbers Nesterenko recorded were a thousand times too high.

In Moscow, a commission of Russian and Ukrainian leaders met daily to manage the disaster. No Belarusian authorities were included in the commission.[17] A group of radiation monitors went to southern Belarus to take measurements in the areas bordering the Chernobyl site. It took them a week to notice the contamination of distant Mogilev Province where air force bombers were wringing the clouds dry.[18]

Nesterenko estimated that in Belarus those who remained at home in radioactive fallout for ten days after the accident had doses of 500–1,500 mSv, which explains why villagers felt dizzy and weak and experienced signs of radiation sickness.[19] Those doses were enough to produce other symptoms too, such as nausea, vomiting, changes in the blood, and increased susceptibility to infection. These were the people besieging local clinics and boarding trains for hospitals in Moscow. "Many residents require medical attention," Nesterenko wrote Sliun'kov. "And we need to widen the Zone of Alienation to fifty or seventy kilometers."[20] Three weeks after the accident, blasts of radioactive iodine and cesium-137, the most prevalent radionuclides emitted in the accident, continued to pulse powerfully. In a few weeks residents received more than today's permissible annual limit.[21] Several hundred thousand people lived in this highly contaminated territory.[22]

Villagers in Gomel Province sent group letters to their leaders. In these sorrowful missives, the petitioners, most of them with an eighth-grade education, informed Belarusian leaders of news that Moscow officials neglected to pass on. Several hundred women wrote that their husbands and teenagers were sent to fields in evacuated areas against their will. "They [the bosses] downplay how high the radiation levels are. They worked in fields that measured 3500 µSv/hr." That level exposed the farmers every hour to three times more than today's annual limit. The women continued, "The military tells us that the zone of evacuation should be eighty kilometers wide. Why were we not given the chance to move? We, residents of Bragin, Khoiniki and Narovlia regions, are doomed to extinction. Save US!! Don't leave us in this zone for the sake of a medical study."[23]

The voices are haunting. They represent an early whisper of a problem that would reach outsized proportions in the coming years. Understanding the possible harm radiation could cause, Nesterenko directed an urgent letter to Belarusian leader Sliun'kov with recommendations to issue iodine pills, monitor food, cease farming, and inform the public about safety measures.[24] Sliun'kov, who clearly disagreed with Nesterenko that they were dealing with an emergency, did nothing of the kind. Instead, he issued an order to confiscate radiation-counting devices from Belarusian research institutes as a prophylactic against panic.[25] He told his scientists to sit tight and wait for orders from Moscow.[26]

By that time, Sliun'kov should have known that waiting on Moscow was dangerous.

# Operators

n December 1991, the Soviet Union was collapsing, and Alla Yaroshinskaya was standing in a sharp wind on a Moscow street holding a heavy package stuffed with top-secret government documents. Yaroshinskaya had been a member of a parliamentary commission to investigate official misconduct associated with the Chernobyl accident. In 1987, six plant operators had been found guilty of causing the accident, but Yaroshinskaya suspected there was more to the disaster than a few men pushing the wrong buttons. For two years, she and her colleagues had special access to padlocked filing cabinets belonging to the Communist Party and they had compiled a record revealing a long trail of criminal negligence at the very top of the Soviet hierarchy. In the chaos of the USSR unraveling, Yaroshinskaya showed up at the commission offices just before Christmas in 1991 to find haulers removing boxes of files. She asked where they were going. The movers shrugged their shoulders. Fearful that her work would disappear, she planned a rescue mission.

Yaroshinskaya was no stranger to espionage. She started out as a beat reporter at a newspaper in the decidedly provincial Ukrainian city of Zhytomyr, 150 kilometers from the blown Chernobyl Nuclear Power Plant. In the months after the accident, she believed officials' accounts that courageous firemen had contained the disaster, radiation levels had dropped, and the emergency was over.[1] But then she began to hear rumors from people in nearby villages. She asked her editor if she could do some reporting. The editor, a classic party hack, emitted only hostile noises in reply, which made her even more dubious. On her own time and in secret, Yaroshinskaya visited the countryside. She describes these trips in her books and, like most Soviet-era memoirs, they include long passages about the tangled contortions necessary to do simple things: get a car, get gas for the car, get permission for a day off work, get past the scowling fat-armed secretary at the front desk, and find chocolates to smooth out all those transactions. Once she reached villages zoned as contaminated, Yaroshinskaya talked to people who told her stories that her editor would deem typical of ignorant, unreliable peasants. The villagers spoke of milk too radioactive to drink, dizzy children passing out in school, and slack-jawed animals. Yaroshinskaya wrote up these interviews in articles that no paper would publish. Even at newspapers promoting glasnost, the mandate to shake up Soviet society with more transparent reporting, editors sat on her story or passed it on to the KGB.[2]

Only three years later, when Premier Mikhail Gorbachev lifted the official ban on the Chernobyl accident in 1989, did her articles see the light of day. When that happened, Yaroshinskaya became an instant hero in Zhytomyr Province. Petitioners flooded her office and her fame spread widely. When, a few months later, Soviet citizens had the first chance to vote in relatively open elections to the national parliament, Yaroshinskaya was elected on a campaign promise to bring Chernobyl problems to national attention and solve them. She was doing just that when five military leaders appeared on national television during the sleepy August holidays in 1991. The men, one of them clearly drunk, announced that they were taking over the government to save the Communist Party and the USSR. The putsch lasted just a few days, but the officials' actions totaled the Communist Party, which had ruled the Soviet Union for seventy-five years. When Boris Yeltsin took power over the Russian state a few days later, he declared Communism defunct. With the Soviet government slated to dissolve at the end of the year, Yaroshinskaya knew political winds could shift, and she wanted to be sure the Chernobyl files didn't disappear.

Waiting till after hours, she flashed the guards her deputy's pass and waltzed into the parliamentary commission offices on Novyi Arbat, a boulevard of glass-and-steel high-rises in the midst of medieval Moscow. She opened the commission's safe, took out the papers marked top secret, and tied them up with twine. As a commission member, she had the right to reproduce the documents, but the in-house KGB officer blocked her access to the copy machine. No matter. She had a friend at the newspaper *Izvestia* that had a rare Xerox. The two women stayed up late passing documents through the grinding copier. Yaroshinskaya returned the originals to the safe and took the duplicates home. She kept one original, the first document, stamped and signed, in case she had to prove that the files were not fakes.[3] Among the records was a forty-five-page transcript of a Politburo meeting convened to discuss guilt for the Chernobyl catastrophe. I had never seen anything like it. On top of the piece of paper in uppercase letters were the words: "TOP SECRET, ONE COPY ONLY."

Only one copy in the world. I paid Yaroshinskaya 600 euros to make me a copy. I had never before paid anyone for a document. I had reason to believe this one was not a fake. I checked other documents she reproduced in her books with those I found in archives and they were exact copies. Yaroshinskaya had access to high-level records as a parliamentarian. In 1992 she was awarded the Right Livelihood Prize, known as the alternative Nobel Peace Prize, for her journalism.

The transcript reads like a drama. It includes a cast of men who were members of the top body of the Soviet Communist Party, the Politburo. The men were leaders of politics, industry, and science. On July 3, 1986, the Politburo men gathered in a Kremlin situation room around a vast table of polished mahogany that reflected the crystal chandelier above. They were there to hear a report of the Chernobyl investigative group, appointed to determine the cause of the accident. The men, in effect, were serving as a grand jury to determine who should be put on trial for the world's largest nuclear disaster. The meeting was at the "conspiracy" level. The "conspiracy" principle in the Soviet tradition meant that the trusted party leaders could speak in confidence that their secrets would not go beyond the gilded plaster appliqué walls surrounding them. The men were burly and aging with an air of well-fed calm, despite the emergency. Some had shaved heads, a sign they had recently returned from the accident site and were relieved of their radioactive hair.

The last to arrive in the conference room was First Party Secretary Mikhail Gorbachev. Short, stocky, with the well-known birthmark, Gorbachev was the only man in the room who did not visit the Chernobyl plant after the accident. His baldness, in other words, was genetic. Like most of the men at the table, Gorbachev had been born in a lowly village and had climbed, one rung at a time, to the peak of power. He was younger than the others and spoke with a village accent, yet he was something of a know-it-all. Gorbachev was the kind of person who assumed the moral high ground as if it was his alone but then reversed himself to choose political expediency when necessary. For his colleagues, this was an annoying trait. In a few years, many would openly refer to him as "that blabbermouth," and after the fall of the USSR, his comrades called him far worse, "a shifter and traitor." But that was still to come. At this meeting, Gorbachev was at the top of his game. The others were deferential and cautious.[4]

As Gorbachev took his place at the head of the table, I imagine the Politburo men squared their shoulders and pointed their doubling chins down, readying for the upbraiding that they expected after this major screwup. (An exploded reactor and melting core were bad enough, but worse still it was Swedish, not Soviet, news sources that told the world about it.) The Politburo men knew to expect harsh words and to admit their mistakes. The trick was to show contrition, limit one's guilt, and, if need be, implicate someone else. That was how Kremlin politics worked. These men had advanced to the ranks of general or minister not only through hard work but also by nimble maneuvers, making sure others paid the price for mistakes. By the end of this meeting, it stood to reason that several jobs would be lost.

Gorbachev opened the meeting and handed the floor to Boris Shcherbina, head of the special commission to manage the Chernobyl accident. Shcherbina had made a career in pipeline and oil rig construction. He had a reputation as a man who met his deadlines, even impossible ones. Wearing a well-cut suit, he looked like the oil man he was. He stood to read from a prepared text:

The accident was the result of gross violations by the operating staff, which were exacerbated by serious flaws in the reactor design. But these two causes are not on the same scale.
The commission believes that operator error triggered the
accident. Plant employees focused their attention solely on
producing electricity [for salary bonuses] at the expense of
safety.[5]

I can see in my mind's eye the burly Shcherbina looking around at
the men's faces, making sure they were listening, while beads of sweat
pooled on his large forehead, his skin a gray hue. By this meeting,
Shcherbina had received a high dose of radioactivity in the weeks he
spent at the accident site as the reactor burned and volatized millions
of curies of fission products that floated from the fire.[6] Shcherbina died
four years later. His death was not recorded in the official roster of fifty-
four Chernobyl fatalities.

In the transcripts, he described how the accident took place because
of a "routine safety test." Because the test put the reactor in a highly
unstable state, Anatoly Diatlov, the second chief engineer at the plant,
told two operators (both would die a few weeks later in Hospital No. 6)
to turn off the reactor's alarm system. They did and then proceeded to
manually slow down the reactor to see if the turbines would generate
electricity with the reactor coasting to a stop. Once they finished the
test, Diatlov gave the order to activate the SCRAM button for a complete
shutdown. Five seconds later the reactor blew.

The Politburo men must have breathed a sigh of relief on hearing
Shcherbina blame the catastrophe on six employees who had the bad
luck of clocking in that night for their shift. Many of the men in the
room—the engineers who designed the reactor, the scientists who drew
up the experiment, the officials in charge of energy output and nuclear
safety, the businessmen who managed construction, the ideology
experts who directed public relations—had played a role in the systemic
failures that led to the accident. But the Soviet state had long functioned
on the premise that problems were caused by "hooligans." To solve prob-

lems, they simply unmasked the "deviants," sentenced them, and order was restored.[7] Shcherbina too was guilty of mistakes. He appeared at the accident site while reactor No. 4 was still burning and demanded that soldiers shovel ragingly radioactive graphite blocks by hand from the roof of reactor No. 3.[8] His eagerness to get electricity production back on track rewarded hundreds of eighteen-year-old conscripts with health problems that would last throughout their now shortened lifetimes.[9]

It is clear that when you don't own up to your mistakes, you repeat them. Reading the Politburo transcripts, I was astounded to see Shcherbina acknowledge that fact:

> In addition to operator error, there are other reasons for this, the largest accident in nuclear history, reasons related to construction and design of the RBMK [Chernobyl-style] reactor.
>
> In the last five years, there were 1,042 accidents at our nuclear power plants, yet, since 1983, the Ministry of Energy and Electrification has not had one meeting to discuss reactor safety. The Chernobyl Power Plant alone had 104 accidents. The RBMK reactor is potentially dangerous. The reactor does not pass modern safety standards. No one abroad would ever use that reactor. We need to make some tough decisions on whether to halt new construction of all RBMK reactors.

Minister of Energy and Electrification Anatoly Mayorets agreed with Shcherbina:

> The reactor model is no good. There was a similar accident at the Leningrad nuclear power plant in 1975. No one ever dealt with it. And the same thing already happened in Chernobyl in 1982, except that there was no release of radioactive material at that time. We didn't learn anything from that accident either. Foreign sources show that the West has already simulated the Chernobyl accident. What do we do—do we continue to lie to the International Atomic Energy Agency [IAEA]?[10]

I knew from archival sources the answer to that last question was positive. Soviet leaders would continue to tell half-truths to the IAEA.[11] Generally Soviet officials fibbed and covered up mistakes if they could get away with it. Physicists at the secretive Medium Machine Building Ministry, in charge of nuclear weapons, had known that the RBMK reactor had design problems and was hard to control because of a positive void coefficient; the graphite tips of the boron control rods, inserted to slow down the chain reaction, caused the reactor to speed up momentarily.

Secrecy was second nature. When the accident occurred on April 26, 1986, Soviet leaders sat on the news for two days. Gorbachev asked his leadership why the world found out from Sweden about the accident. Yuri Izrael, in charge of reporting environmental pollution, shifted the blame to his comrade Nikolai Ryzhkov, chairman of the Council of Ministers. "The report was written right away," Izrael stated. "We sent it to the Council of Ministers at 11:30 a.m." Piling on, Belarusian party leader Nikolai Sliun'kov chimed in: "We in Belarus received no news."[12]

Midway through the Politburo meeting, Gorbachev called in Viktor Briukhanov, the disgraced director of the Chernobyl Nuclear Power Plant. Testing his honesty, Gorbachev asked Briukhanov: "How many mishaps did you have at the Chernobyl plant?"

"One, maybe two a year," answered Briukhanov.

Shcherbina pounced on him. "Of 104 accidents that occurred at your plant in the last five years, 34 were caused by operator error."

"One hundred four accidents in five years!" Gorbachev repeated. "Why have you been so complacent?"

Briukhanov, whose training was in hydropowered plants, not nuclear reactors, fell into the expected role of the contrite communist: "We had no idea from the earlier accidents that something like this could happen. We would make repairs, but we should have analyzed every accident as it occurred."

Briukhanov was the fall guy. A month later, the Soviet government officially explained to the IAEA that the primary reason for the Chernobyl accident was operator error. Briukhanov and his two deputies were arrested and held in isolation.[13] A year later, Briukhanov and five of his subordinates were tried and found guilty of gross negligence. Briukhanov was sentenced to ten years.

Case closed; the world could breathe easily again. The irresponsible adventurers who sent clouds of radioactive gas careening over Europe were behind bars.

// 

IN AN ARCHIVE in Moscow, I found an anonymous letter to the Soviet Central Committee of the Communist Party. The writer, identifying himself as a nuclear engineer, mocked the Chernobyl verdict: "Yeah, right, the plant employees blew up the reactor. Who else? No one but us worked there." He blamed the accident on the "grim defects of the RBMK safety system." Like a person recovering from trauma, the engineer returned again to the moments before the reactor exploded.

> On April 26, 1986, at 01:23:04, the operator started a routine experiment. The power block was more or less stable. After 36 seconds, the operator, seeing that the experiment had nearly finished, calmly, without panic—because there was no reason for panic—pressed the SCRAM button to stop the reactor. And then, against all logic and rationality, the reactor's power at first decreased just a little, and suddenly and with growing speed sharply increased until the reactor underwent a surge that ended in a massive explosion.

Even the designers at the Moscow Kurchatov Institute of Atomic Energy, the engineer wrote, admitted the "paradox" of a reactor speeding up on shutdown.[14] "That's like having a car where the brake becomes a second gas pedal. That scenario should only occur in an operator's nightmare."[15]

Soviet physicists knew the RBMK reactor was hard to control and had a fatal flaw, the positive void coefficient. They had other reactor designs. Why did they use the unreliable RBMK, plugging in dozens of them across the USSR? They built RBMK reactors only in the USSR, not in Eastern Europe where they constructed safer, light water reactors. As the historian Sonja Schmid points out, the RBMK was cheap to build on-site, easy to scale up, and "uniquely Soviet," a point of pride.[16] The top-secret Politburo transcript reveals one more reason to deploy the RBMK.

At the Politburo meeting, Valery Legasov, a nuclear chemist, agreed with his colleagues' poor assessment of the Chernobyl-style reactor: "The RBMK reactor does not meet international and domestic standards on several levels: There is no safety system, no structure for radiation monitoring, and no containment. We are guilty for not following up on a new design for this reactor. The poor rating of the RBMK reactor has been known for fifteen years." So why use the faulty reactor? Legasov explained: "We are carrying out the *Zaslon* mission, and the RBMK is more reliable for that project."[17]

The code word *Zaslon* means "screen" in Russian. At the time the American president, Ronald Reagan, was spending billions of dollars to create what eventually was abandoned as a mythical "Star Wars" missile defense system that was hoped would protect the United States from a nuclear attack. Star Wars consisted of deploying yet more land-based missiles to try to shoot the adversaries' approaching rockets from the sky. Historians argue that Gorbachev refused to be lured into Reagan's expensive revival of the nuclear arms race. The Soviet chief negotiated drastic arms reduction instead. Yet the Politburo transcript shows that before the Chernobyl accident Soviet leaders secretly were building their own missile defense system. In 1984, Reagan ordered that the long-mothballed Hanford Plutonium Plant in eastern Washington be dusted off and returned to production for the Star Wars project. In his cryptic comment, Legasov was alluding to the major advantage of the RBMK reactor over other designs. In addition to generating energy, the reactor could produce plutonium, the fissile material at the core of nuclear bombs. The more Soviet engineers ran "peaceful" RBMK reactors, the more plutonium they could potentially stockpile for the "screen." There is no evidence that they used Chernobyl fuel to produce plutonium, but for a paranoid state in an age of fear, running the flawed RBMK reactor made all the sense in the world.

In deciding to prosecute a few operators rather than designers and industry bosses as scapegoats for the accident and in resolving to continue to operate accident-prone RBMK reactors, Politburo leaders took a vote for secrecy because there was no other way to justify these decisions other than by covering up these basic facts. That was a risky decision for the new Premier Gorbachev, who had in 1986 introduced a reform policy of more transparent and accountable governance. When the truths of Chernobyl came to light three years later, the load of public skepticism and doubt would speed downhill like a runaway truck, knocking out Gorbachev and his administration with it.

I read the forty-five-page Politburo transcript again and then again. There was something in addition to the fabrications that bothered me. Where were the people fighting the fire? What about the farmers preparing to plow contaminated fields held in the family for more generations than anyone remembered? Why did the Politburo men fail to mention the radioactivity that was still spreading in tongues licking at the thatch roofs of cottages in Belarus and Ukraine? The ashy fallout was spreading with the winds and rain like the gray, smudgy Angel of Judgment that flies through Marc Chagall's paintings. Rather than the Hebrew characters in Chagall's clouds, there are figures from the table of elements: a toxic brew of radioactive iodine, cesium, ruthenium, plutonium, and strontium. The angel flapped her wings and her shadow darkened a larger and larger portion of the map of Belarus, heading south and east, then north toward Chagall's hometown of Vitebsk. The Politburo men worried a lot in the months following the accident. They were apprehensive about the mounting costs, about damage to the Soviet Union's reputation abroad, and the loss of electric power, but in the records they rarely dwelled on the contaminated territories or the people exposed in them.

To my surprise, that concern was left to others. The leaders of the Ukrainian Communist Party, later discredited as hard-line, old-school Stalinists, were the first authorities to realize just how dangerous the disaster was and take action.

# Ukrainians

U ntil I met Natalia Baranovska, I thought historians had pretty safe jobs. She proved me wrong. In the 1990s, Baranovska became obsessed with the Chernobyl accident. At the Ukrainian Academy of Sciences where she worked, no one seemed to care about what she saw as a major event in Ukrainian history. She worried that evidence documenting the accident would be lost if someone didn't do something.

In her spare time, Baranovska looked through government archives and then searched in institutes and offices. Her hoarder's reflex kicked in. Year after year she hunted for Chernobyl records in any repository she could get to—all over Kyiv and in the Ukrainian provinces and Belarus. In Moscow, she flipped her long blond braid over her shoulder and played the role of a Ukrainian hayseed to charm her way into restricted archives. At the time she was a single mother. Her work hardly paid and won her no prestige. She kept going. She lived in one room with her son. While he slept, she read late into the night at a tiny desk surrounded by mounting stacks of paper. When her son left home, she rented a room in the nuclear city, Slavutych. From there she commuted to the still functioning Chernobyl plant to coax papers out of the hands of engineers. (The Chernobyl plant was finally closed in December 2000.[1])

In 2014, I first looked her up. Baranovska gave me directions to her dacha on the outskirts of the city. We sat down for tea and fruit from her orchard in the garden. Natalia wasn't looking so good. She'd had surgery for thyroid cancer a few weeks before. She was pale with a bandage on her neck and spoke in a raspy whisper. I thought of the records I found in a Paris archive. In them, Soviet archivists had begged the UN agency UNESCO for help in cleaning up 123 archives contaminated with radio-active dust from Chernobyl. Some important medical files were too hot to touch.[2] They asked for personal dosimeters for archivists and funds to decontaminate archival paper. A medical exam showed that the major-ity of archive employees had substantial medical problems. They asked for help in July 1991, just a month before the August putsch that ended the USSR. Help did not come. The UNESCO reviewer determined that decontaminating paper was an extravagance, "considering the fact that a hundred thousand people more urgently need to be evacuated for rea-sons of health."[3]

Baranovska told me about one of her first trips in the mid-1990s inside the Chernobyl Zone of Alienation, where three reactors were still operating. From Slavutych, she took the commuter train loaded with employees traveling across lonely forest and swamp to the nuclear com-plex. At the station, workers changed from street clothes to work clothes at a bank of lockers. Baranovska fumbled, figuring out what to do. By the time she had pulled on a jumpsuit, everyone had left on a depart-ing shuttle. She asked the conductor when the next bus was coming. "That was it," he told her, "but you can walk there. It is only a kilometer." Baranovska set out. A mix of snow and rain fell from a gunmetal gray sky. The road ran right past the blown reactor decaying in its massive, concrete grave. Baranovska knew that in front of the reactor the level of radioactivity was especially menacing. When passing that spot, driv-ers would gun their engines. A car appeared up the road. Drivers were required in the Zone to pick up pedestrians, who were vulnerable in the open air. Baranovska waved to the motorist, who kept right on going. "I was so angry," she remembered, "and I was wet and cold. I walked by that monstrous object and cried my eyes out."[4]

At the dacha, Baranovska had a friend visiting. Irina had worked as a doctor in Kyiv. "You are researching a history of Chernobyl?" she asked. "I have a story to tell you." I settled back on the bench under an apricot tree and told her I was all ears.

In 1986, Irina said, she was called up for duty in a special medical brigade commissioned to go into the disaster zone. She piled into a small car with a senior doctor, a few nurses, and their luggage. They arrived in a village where a makeshift clinic had been set up in a grammar school. In assembly-line fashion, the brigade examined a hundred people a day.[5] I asked her what they looked for. "Oh, this and that," she said, waving her hand vaguely.

"What did you know about radiation medicine at the time?"

"Nothing!" she replied in exasperation. "My father, who was also a doctor, told me to drink vodka to cleanse radioactivity from my body."

"But when I had been there for a while," she confided, "an army doctor gave me some good advice. He told me not to drink vodka all day, starting in the morning. That it is better to just have a drink at night."[6]

I surmised Irina had not read the manual issued in August 1986 for doctors treating Chernobyl patients. The booklet debunks the popular Soviet notion that alcohol purges radioactive substances from bodies. "Alcohol," the authors soberly state, "is always a stress on the organism."[7]

I was starting to understand why Irina was fuzzy on details of the team's medical practice that summer. Day-drinking was only part of the problem. The lack of training and the demanding volume of work also contributed. After examining hundreds of anxious villagers, the doctors would pile into a Soviet Lada, the faint smell of benzene and mold pushing up from the sedan's vinyl seats, and drive to the next village to sleep in another country school, eat more meals from cans, and hope for a bath.

Mostly what Irina wanted to talk about was her own medical problems that dogged her after she returned. She pointed to her left eye, which was clouded over and tilting wildly to the right. "Four years after Chernobyl, I started to have trouble with this eye," she said. "No specialist can explain it." Irina described her thyroid disease, pains in her legs, and joint problems that made it hard for her to walk. She attributed her health problems to her time in contaminated villages treating people whose bodies had become radioactive sources. I knew she had grounds for making that claim. I had seen the estimations of doses for doctors in the summer of 1986. Aware of the risks, Moscow commissioners sent orders to rotate medical brigades frequently.[8]

You assume after a nuclear emergency that people move away from dangerous radioactive clouds. The best safety against radiation is time and distance. Strangely, the Chernobyl disaster drew people toward it with a centripetal force. As 120,000 evacuees fled the Chernobyl Zone of Alienation, Irina and 600,000 other emergency workers were sucked into its vortex to serve as a workforce for the relief efforts. Steelworkers arrived from eastern Ukraine to manufacture a sarcophagus to cap the accident site.[9] Thousands of miners from Donetsk burrowed under reactor No. 4, where the miners dodged neutron streams and choked on radioactive dust.[10] Crews rolled into villages to pack up villagers' livestock, tools, and households. KGB officers fretted that much of the work "is carried out with poorly mechanized methods, which results in considerable doses."[11] In the first month, 40,000 Red Army conscripts were called up.[12] Most were adolescents doing two years' mandatory service after high school. Thirty-one thousand young soldiers camped out near the burning reactor, lived in tents, ate outside and waited, exposed to the ambient radiation, sometimes for weeks, for orders.[13] They helped civil defense brigades hose down roads, schools, and houses in the slim hope of decontaminating them.[14] Construction crews built hundreds of nuclear waste depositories, dams, and dikes to try to stop the flow of radioactive water into drinking reservoirs.[15] Police stood guard around a newly established perimeter fence. Even a priest served as a cleanup worker. He led a procession in southern Belarus to pray for deliverance.[16] Irina was one of 9,000 medics called up in Ukraine starting in early May to go into the hot zones fanning outward from the blown reactor.[17]

Such was the power of the Soviet state. It had the capacity to expose millions of people to radioactive contaminants and the authority to resettle and examine them. The effort was unimaginable for any but a state highly evolved in the art of mass mobilization. In Ukraine alone, doctors examined 70,000 children and more than 100,000 adults in the summer following the accident.[18] In the next few years, they performed half a million medical exams.[19]

Other states had exposed populations to radioactivity. No state had ever mustered a mass examination of exposed bodies on such a scale. After the core melted at the Three Mile Island Nuclear Generating

Station in Pennsylvania in 1979, scientists ran computations to assess potential health damage. They ran some residents through whole-body counters to detect radioactivity and predicted there might be one, maybe two, extra cancer deaths caused by the accident. When Pennsylvania State Health Commissioner Gordon McLeod announced nine months later that child mortality in a ten-mile radius around the plant had doubled, the governor did not order an investigation of the problem but instead fired McLeod.[20] The Soviet mobilization of doctors was the largest of its kind in the world. Doctors in Ukraine led the effort.

Ukraine was the Soviet Union's bread basket, a major producer of grain, dairy, and fruit, and it was an important Soviet arsenal, a source of rockets and military hardware with intercontinental ballistic missiles lining its periphery. While most contaminated areas had to wait weeks for radiation monitors to show up from Moscow, Kyiv could take care of itself.[21] Oleksandr Popovych, the scientific director for the Kyiv Communist Party, organized radiation monitoring control points around the city. Popovych told me that his team's measurements came in more quickly and accurately than those of the Moscow State Committee of Hydrometeorology managed by Yuri Izrael.[22]

The Kyiv scientists watched in the first days of May as radiation levels spiked and stayed high.[23] Within a week of the accident, they reported detecting radioactivity in drinking water and grass that exceeded by a thousand times the norm for ambient radioactivity. At these levels, the scientists estimated residents of Kyiv, a hundred kilometers from Chernobyl, would get in three months the annual permissible dose for nuclear workers.[24] Popovych quickly realized it wasn't just a matter of turning on a radiation meter and writing down the number on the needle. His team needed to know what was going into the food chain and so into bodies. The cow, they grasped, could serve as an important dosimeter. Cows, grazing in fields, picked up radioactive cesium and iodine on grass and released it in their milk. The young physicists sent to dairies to measure radiation in milk were motivated by the thought of their own children at home. They reported high levels in milk.[25] Dairy products made up a major portion of the Soviet diet. Doctors in mobile medical brigades learned that a quick and crude way to measure exposures was to hold a radiation counter to people's thyroids.[26] They found that nearly every child examined had some kind of dose to the thyroid.[27]

Popovych's group recommended ending the school year immediately and shipping all children to safe locales, a drastic and expensive measure that would shuffle several hundred thousand bodies.[28] Ukraine's party leader, Volodymyr Shcherbytsky, ordered the evacuation be taken "immediately," but two visiting Moscow experts, Leonid Ilyin and Yuri Izrael, who were in charge of disaster control, disagreed. Calm should prevail, they advised.[29] But calm did not reign in Kyiv. People were racing to get their children on trains and buses to relatives out of town. Tens of thousands had already left, among them the family members of the party elite. The parks were empty. Shoppers were stockpiling canned goods and shying away from fresh vegetables at the markets.[30]

Consistently, experts in distant Moscow doubted the information that Ukrainian scientists supplied. Russians tend to view Ukrainians as slightly inferior "little brothers," a bit as Americans see their Mexican and Canadian neighbors. It's no surprise that Ukrainians greatly resented this attitude. When the two Moscow scientists arrived in Kyiv, these dynamics were at play. Izrael, the minister of hydrometeorology, had started his career studying the atmospheric impact of nuclear bomb tests in Central Asia.[31] For him, a blown reactor was nothing compared to the blast of a hydrogen bomb on the Kazakh steppe. Izrael, the architect of the Belarusian cloud-seeding program, appeared to be in no rush to figure out where fallout had landed in order to protect people from more exposures. Nor was he inclined to share the information his agents gathered with other ministries.[32] Before Chernobyl, Ilyin, a leading Soviet biophysicist, criticized American scientists for underestimating the health problems of Japanese bomb survivors and Marshall Islanders who were exposed in American nuclear blasts. He became a different man after the Chernobyl accident and consistently downplayed the disaster.[33]

Ilyin and Izrael refuted the Ukrainian scientists' map of contamination. Their numbers, they said, were too high.[34] They spurned Kyiv doctors' diagnoses of hundreds of cases of radiation illness, including dozens of children.[35] Moscow scientists consistently argued that damage was limited to the accident site and the two hundred liquidators treated in Hospital No. 6. They admitted publicly to that number of patients only because Robert Gale and his colleagues had worked in Hospital No. 6.[36] This limited picture of damage they created for world media.[37] Moscow leaders' strategy was to admit only what could not be denied.

Izrael and Ilyin directed the Ukrainian leadership to give the podium to a well-known expert to assure the public that they were safe.[38] The trusted official was the pliant, eager-to-please Anatoly Romanenko, the Ukrainian minister of health. When the accident occurred, he was enjoying his first trip abroad to the United States. He returned to Kyiv several days after the accident. He knew next to nothing about radiation, but he found himself pushed in front of microphones to tell a dubious public that the danger had passed. "They put him there as the sacrificial lamb," one of his colleagues remembered.[39] Romanenko made a lot of soothing statements about Chernobyl over the next few years. At the same time, he worked behind the scenes to protect exposed people as best he could. Romanenko was a player, but he had a conscience. Meanwhile, over the next decade, Ilyin consistently minimized the estimation of health damage. The Ukrainian Communist Party condemned him, and a Greenpeace lawyer labeled him an "apologist." Izrael would later make a name as a climate change denier in the Putin administration.[40]

Ukrainian officials grew ever more nervous about the cavalier attitude of Moscow scientists who lived hundreds of miles away from the contaminated streets of Kyiv.[41] More than half of the first children arriving from evacuated territory showed signs of "radiation trauma."[42] In Kyiv Province alone, 30,000 children were placed under medical observation because their exposures were high.[43] Kyiv city leaders and scientists had children and grandchildren who went off to school each day in the sunny Ukrainian city, the air charged with ionizing radiation. Hundreds of letters were pouring in from worried citizens, and hundreds showed up each day in the reception halls of their leaders. The Ukrainian party leadership was under pressure. Waiting on orders from Moscow, they had delayed evacuating Pripyat. They had held the

May Day parade because Moscow commanded it so. They stalled on giving out iodine pills because Ilyin told them to wait.[44] All of that was bad advice.

By mid-May, Ukrainian leaders no longer submitted. Without waiting for consent, they cut the school year short and arranged for schoolchildren in Kyiv and contaminated provinces to go to summer camp early and for the whole summer.[45] They cleared sanatoria and resorts and ordered up a phalanx of buses heading to Crimea.[46] Kyiv children went first.

Moscow leaders were angry about Ukrainian leaders' "unsanctioned" and "emotional" evacuation. They upbraided Ukrainians for "spreading panic and slanderous rumors." Look at Belarusians, they said, who remained calm.[47] Gorbachev personally chewed out Shcherbytsky. The veteran party boss did not back down.[48]

Ukrainian leaders next insisted that, against the recommendations of Moscow scientists, pregnant and nursing mothers, infants, and toddlers be sent to resorts in noncontaminated regions.[49] That measure involved another titanic shuffle of vehicles, bodies, and emotions. Mothers could go with their infants, but toddlers went to camps alone. Many mothers told me of the trains heading out of the city filled with crying children, one long rolling wail.

As levels of radiation spiked yet higher in mid-May, hundreds of anxious parents besieged officials in Belarus asking questions no official could answer publicly. "When will radiation levels go down? Children are leaving Kyiv. Why are mothers and children staying in Gomel? Are Ukrainians better than us?"[50] Belarusian doctors reported that 25 percent of children they examined had enlarged thyroids. Every fourth child had a compromised immune system.[51] The Politburo Chernobyl commission partially relented and decreed the summertime removal of 2,700 children and pregnant women from the regions of Belarus with the highest levels of radiation, a small percentage of people in danger zones.[52] No one suggested that children and pregnant women leave Minsk and Gomel as they had Kyiv, though the doses children had received were considerable.[53]

Not all the camps were a pleasure. While urban children went to resorts in sunny, seaside Crimea, village kids went to village camps, which were often just repurposed collective farms. One camp had no real counselors and no nurse or doctor. The farm chairman, short on workers because so many were conscripted for the cleanup, ordered the children, who were supposed to be resting and staying indoors, out to the fields. Like a class of indentured servants, they weeded and blocked until they finished their daily quota. At the farm, the food was terrible: last year's cucumbers, cream that had soured, and mouse droppings in the porridge. Four children ran away. Three were recovered, but one, Stanislav Lisitsky, was still at large at the end of the interminable summer of 1986.[54]

What were Ukrainian leaders thinking? Did they indeed panic as Moscow scientists charged? Health Minister Romanenko received reports from doctors working in contaminated villages. By May, he was starting to get a sense of the medical landscape. The doctors wrote that most children were in good condition, but a quarter of the examined kids showed strange symptoms—nervous tremors, flushed faces and throats, the slowing of motor skills, and weight gain. Physicians held counters to children's thyroids to measure gamma rays emitting from them. Most children (89 percent) in the first weeks after the accident had absorbed doses to the thyroid of 0.3 to 2.0 Sv. Several hundred had absorbed doses between 2 to 5 Sv and others up to an alarming 50 Sv.[55] (There should be no radiation coming from a child's thyroid.) Two-thirds of the children had enlarged thyroids and 60 percent had thyroids functioning on overdrive.[56] As the summer wore on, doctors noticed that 20 percent of the exposed children had anemia, chronic tonsillitis, or gastritis. Increasingly the children suffered from respiratory illnesses and severe infections.[57] From 10 to 25 percent had "cataract conditions" developing in their eyes.[58] The head of the science sector of the Communist Party of Ukraine, Mr. V. Sokol, made a visit to Crimea in June 1986.[59] Sokol reported that children who had received the highest doses of radioactivity slept continually, ate practically nothing, and were limp and listless. A month after the accident, doctors in Leningrad with the most serious pediatric cases wrote, "Severe infections and respiratory diseases among the children under observation are significantly increasing."[60] The doctors found that children who had resettled from near the plant showed signs of damage from radiation: toxic activity in their blood, lung complications, and damage to their small intestines.[61]

Lungs and intestines are gateways to the body. People breathed radioactive aerosols and radioactive dust. Nearly everyone in the disaster areas in the first weeks after the accident coughed and had runny noses and diarrhea. Doctors in Ukraine observed reddening and slight ulcers in oral cavities and on gums and tonsils.[62] Investigating, they found radioactive aerosol particles seated in lungs. As people coughed, particles came up with mucus, which was swallowed in saliva and flowed to the intestines. In one clinic, doctors held sensitive measuring devices to their patients' spleens and large intestines. More than a third of the patients' torsos emitted gamma rays.[63] The more radioactive isotopes that were found in the intestines, the greater was the amount of radioactivity absorbed in the bloodstream. Intestines are radiosensitive and easily damaged. Ukrainian researchers found a close connection between the high levels of radioactivity in the intestines and problems with anemia and other disorders of the blood-forming system.[64]

When doctors in Ukraine examined pregnant women, they also did not look well. Women from contaminated areas were anemic and half had enlarged thyroids. Doctors recommended that women with signs of exposure have abortions.[65] Despite the increase in pregnancy terminations, the number of miscarriages, hemorrhages, complications at birth, and premature babies in contaminated regions rose alarmingly during the summer of 1986. The newborns were sicker, smaller, and weighed less than the average.[66]

Most of the doctors in the emergency clinics were, like Irina, not specialists in radiation medicine. They received a crash course after the Chernobyl disaster.[67] I wondered whether pediatricians reporting about sick children in Kyiv were projecting the diagnoses of chronic radiation syndrome (CRS) onto their patients after reading the Soviet manual issued for the emergency that described CRS. The manual categorized what they would find—nebulous symptoms at low doses—and so they found it. Certainly, no Western radiologists would have expected to see the wide range of symptoms doctors listed in their reports at such low doses. There was no category in Western medical literature for noncancerous symptoms at doses below a whole-body count of 10 mSv. As far as Euro–North American literature was concerned, either a person had acute radiation syndrome, or they had nothing at all.[68]

The problem puzzled me. The fact that children and pregnant women got sick and the fact that they were exposed to Chernobyl radiation is a correlation. A correlation does not prove a connection. Radiation could be the cause or it easily could not. Maybe the children, pregnant women, and cleanup workers were sick, as Romanenko asserted, not because of Chernobyl radiation but because of the stress of evacuation or because they were far from home and living in miserable camps on Spartan rations with barbarian hygiene.[69] Or maybe something else entirely was going on. The only way to know for sure if radioactivity was the cause was to measure it.

# Physicists and Physicians

During the summer of 1986, doctors, radiologists, and scientists met daily as part of a Politburo-organized Chernobyl medical commission. The commissioners grappled with big decisions that would affect millions of people. Soviet leaders had decided to evacuate a thirty-kilometer ring around the plant as a temporary measure. Were there other people in danger who should be moved?[1] Could residents return to towns and villages inside the circle after the fires were put out? Repopulating abandoned communities, including the large city of Pripyat a few miles from the burning reactor, was a major bullet point on the commission's agenda. Moscow leaders pushed hard for it because returning people home showed that the disaster had an end, and life could go back to normal.[2]

A commission member, Dr. Andrei Vorobiev, recounted in his memoir that Soviet scientists in the nuclear weapons industry had worked out before the accident that people could safely live on ground contaminated with five curies per square kilometer.[3] They understood this to be an optimistic number for people who purchased food from elsewhere, but the contaminated territory around Chernobyl was rural. Everyone ate fresh, local produce. Just one curie per square kilometer could deliver unacceptably high doses of radioactivity to a person who consumed local produce. The commissioners looked at the classified radiation contour maps and realized that if they chose even five curies per square kilometer as a threshold, soldiers would have to eject over a million more people from their homes. Where would they all go? Who would pay for it? During those meetings in distant Moscow, after the fires were extinguished and crews were building a sarcophagus over the smoldering reactor, the mammoth scale of the accident suddenly hit home.

Arbitrarily the commissioners selected a new number. They decided that fifteen curies of cesium-137 per square kilometer would be safe or, at least, safe-ish. Uneasy with this decision, the Soviet minister of health insisted that if people were going to live with fifteen curies, they had to ship in clean food. And what about the 300,000 children and pregnant women, he asked, who were sequestered in summer camps? Should they be allowed to return to areas with fifteen curies? A physicist piped up that radiation had the same effect on adults as on children. There was no difference. The medical doctors on the commission strongly objected. Everyone knew, they argued, that children were far more vulnerable to radioactive exposure than adults.[4]

At the meeting of the Politburo medical commission, Yuri Izrael asked Vorobiev how curies of radiation recorded in soils translated into doses measured in sieverts (or rems) in bodies. That was a tough question. Ionizing radiation emanates from radioactive decays measured in disintegrations per second called a becquerel or a curie. A measurement of ten becquerels means ten bursts of radiation every second.[5] Becquerels and curies are just counts of radioactivity that come from the environment, for example, from the ground, a tree, or a dusty old truck. What matters to humans is how the radiation acts in bodies.

Radiation manifests in different forms—in gamma rays and in alpha and beta particles. Gamma rays are like X-rays but have higher energies. Gamma rays penetrate skin and damage cells. Beta and alpha particles are knocked off an unstable atom's nucleus.[6] Alpha and beta particles cannot penetrate skin, but if ingested, they bash into cells, crushing and blasting them. Alpha radiation is rated twenty times more damaging to humans than gamma rays and beta particles, and so scientists multiply its potency by twenty when estimating doses to a whole body or part of a body. The units for measuring doses are rems and rads in the United States and the Soviet Union and sieverts and grays in the rest of the world. Radiologists realized that some parts of the body are more radiosensitive than others, so they tried to differentiate between them. When it all became too confusing, they decided to divide the amount of energy deposited in tissue by the weight of the tissue in kilograms to get rems or sieverts.

To estimate an internal dose, scientists often ask people to remember how much time they spent outdoors breathing contaminated air, what they ate, how much they ate, and where they got their food. This is a dicey science because of the extreme patchiness of both human memory and radioactive contamination. Take, for example, two villagers. Each has a cow. One farmer pastures his cow on top of the riverbank. The other stakes his along the river bottom. The two farmers live a few meters apart, but their doses differ radically because the lowland cow eats flood-washed grass laced with high levels of radioactivity and produces extremely contaminated milk. The milk from the cow on the bank is within permissible limits. The first farmer remembers precisely where and when he staked his cow. The second man points vaguely across the field. The investigator then writes that both farmers pasture their cow on the bank.[7]

You can see that the numbers that emerge from this science are rough. Ukrainian health minister Anatoly Romanenko noted that dose estimates from environmental data differed by two orders of magnitude depending on who calculated them.[8] Curious about how dose estimates were made, I contacted Lynn Anspaugh, a U.S. Department of Energy scientist, who in 1987 helped draw up the Chernobyl damage estimates for Europe. Over the phone, Anspaugh told me he had been assigned the dose estimate for Romania. "Romania at the time was a closed society," Anspaugh explained. "No one wanted to go there. Even the IAEA inspectors wouldn't visit because the hotels were cold and the people were cranky."

Anspaugh called someone in Romania who gave him two numbers, one describing the concentration of cesium-137 deposited on the ground from fallout and one for cesium concentration in milk. "I took those two numbers," Anspaugh told me proudly, "and derived an estimate for the entire country."[9] Anspaugh's two numbers show the guesswork and speculation involved in this science.[10] From these untidy calculations scientists estimate a number, which in that miraculous way of modern policy takes on a numerical existence as a kind of fact, although it is shot through with uncertainties.[11] Once a dose estimate is born, it becomes a little numerical prodigy emerging in the world as an important historical actor. People quote the number and use it to establish regulations, to depopulate territories, or to do nothing at all. This has little to do with science and everything to do with expediency and politics.

In calculating doses, scientists naturally bicker. Picking one set of data or parameters can move the slide ruler extravagantly to the right or left of risk. You may ask what sense it makes to measure radioactivity in environments in order to translate it to bodies. Why not just measure bodies directly? Soviet scientists tried that after Chernobyl. They had several ways to estimate doses in bodies. Because Soviet KGB agents barred doctors from information about radiation exposures, the doctors got very good over the years at reading bodies for radioactive exposure.[12] Their most tested method involved looking for chromosomal changes in cells of bone marrow a few days after radiation exposure. Chromosomes do not alter with temperature, diet, age, or mood, so they are stable markers to record damage from decaying radiation. The trick is that these indicators only last a few weeks. Then the scent goes stale. The enamel of teeth complements blood work because it preserves the count of radiation deposits for many years. Researchers use teeth enamel to cross-check data derived from chromosomal analysis. Used together, Soviet doctors were confident they had a reliable method to establish an individual's dose.[13] A decade later scientists in the West started to adopt this method.[14]

A second way to estimate dose in bodies is by using a machine invented in the 1950s called a whole-body counter. The counter uses sensitive detectors to count gamma rays coming from the body. Whole-body counters are faster and cheaper than recording biomarkers in chromosomes or teeth enamel. The downside of whole-body counters is they measure gamma rays only, not the more harmful alpha and beta particles. Another problem is that the body is a pretty good biological shield so whole-body counters designed for standardized humans don't measure all radiation hiding in the crevasses of bodies that differ greatly in size and weight. The machines can miss a lot, especially in children.[15] Soviet doctors complained that whole-body counters gave results that were ten times lower than chromosomal analysis.[16] The machines are useful though. People can emerge from a counter feeling relieved that their dose is not so bad. "Whole-body counters," nuclear researcher Lucas Hixson explained to me, "are not medical, but political machines."[17]

The two ways of measuring radiation in the body reveal a divide in disciplines. Usually doctors perform chromosomal analysis. Physicists build and operate whole-body counters. The hematologist Vorobiev advocated that they send teams to Chernobyl regions to take blood samples and study chromosomes. He stressed urgency. Time, he repeated, was running short.[18] Physicists Leonid Ilyin and Yuri Izrael threw their weight behind whole-body counters and claimed there was no rush.[19]

It mattered a lot which method researchers used. Estimates of doses slid up or down depending on it. At their daily meetings in the summer of 1986, the Politburo medical commissioners argued about doses and how high they could go before they caused unacceptable levels of harm. Today the internationally recognized annual limit is 1 mSv. Representatives from the Ministry of Health drew on the limited data Soviets had of the American-directed Life Span Study of Japanese bomb survivors. With that reference point, they suggested that women and children could safely get a hundred times more—100 mSv as an emergency one-year dose.[20] The physicists said a temporary emergency threshold could be set much higher—at 500 mSv for the first year, at least for adults.[21]

As the summer wore on, these emergency limits appeared to be more and more optimistic. Researchers in Ukraine calculated that children in contaminated zones had already accumulated 100 mSv in just the summer months.[22] Vorobiev's teams took blood samples and found that half the residents examined in the most contaminated areas had doses of cesium-137 ranging from 200 to 400 mSv. Another third of residents had doses yet higher, from 800 to 1,000 mSv—doses high enough to cause radiation injury, including nausea, fatigue, and damage to organs and immune systems. Chromosomal analyses also showed some residents had ingested hot particles of plutonium and strontium, elements that are excreted very slowly from the body and can zap single cells with very high jolts of radioactivity.[23] Tooth enamel examinations in the same regions confirmed the dose estimates. Doctors from the secret Third Department of the Ministry of Health calculated dose estimates for thousands of the patients who checked into hospitals in Moscow, Gomel, Kyiv, Minsk, and Bryansk. Unfortunately, their data were, and apparently still are, classified as secret.[24]

The high estimated doses were bad news. Soviet researchers determined that people begin to have symptoms of chronic radiation syndrome at doses from 200 to 500 mSv.[25] Biophysicists fought back against the high-dose estimates. They sent out whole-body counters, used them en masse, and recorded lower estimations. Of thousands of people examined, they reported that 90 percent did not exceed 10 mSv. Only in the most heavily contaminated communities did residents get doses of 300 to 400 mSv.[26]

Vorobiev remembers the scientists on the medical commission fighting about dose limits all night long. Izrael, Ilyin, and other physicists at the Institute of Radiobiology asserted that, after decontamination, evacuees could move back to their homes by fall. The doctors were doubtful. One scientist cautioned the group that the experience in the radioactively contaminated territories of Siberia showed the "extremely insignificant effects of decontamination measures."[27] The physicians asserted that no one should return to homes inside the thirty-kilometer Zone of Alienation. They wanted several hundred more contaminated communities evacuated. They sought strict federal guidelines and a definite threshold of fifteen curies per square kilometer for further evacuations.[28] The physicists preferred vague formulations and a wait-and-see approach.

The physicists largely won the argument. Ilyin drew up the document that determined what to do next. He used guidelines that Izrael had sketched out a month before. The new regulations sorted territories into three zones according to estimates of how many sieverts they guessed residents would get in the first year after the accident and then in a lifetime of seventy years. In the least contaminated regions (with doses up to 100 mSv/year), Ilyin wrote, children and pregnant mothers could return home and lead an "unrestricted life." In areas of more contamination, of 100 to 200 mSv per year, residents could stay, children could return, but they should not eat locally grown food and the communities would need to move in a few years. Forty-five villages, where scientists estimated people would get more than 200 mSv per year, would need to be evacuated immediately.[29] They determined that places inside the Zone of Alienation that had levels where people would receive not more than 100 mSv could be reopened for habitation. The instructions were vague on how the doses would be determined, whether by direct measurements in bodies or extrapolating from environmental data.[30] Vague formulations were useful because the count of curies in soils was high. After Ilyin's decree, the Politburo shut down the Chernobyl medical commission. They apparently no longer needed the counsel of doctors.

The instructions went out via classified pouch. Ukrainian leaders read them in late July and discussed them in their own daily Chernobyl operative group meeting. They refused to have children and pregnant women return to regions with more than fifteen curies per square kilometer. Children in those areas had already gotten a year's dose, they declared. Oleksandr Liashko, the chair of the Ukrainian Council of Ministers, ruled to send schoolchildren to boarding schools and to resettle the mothers and toddlers elsewhere. Izrael, sitting in on the meeting, objected to splitting up families. Liashko countered, "We know it is a hardship for children to go to boarding schools and to break up families, but we have no other option. If the [Ukrainian] Ministry of Health determines people are getting a dangerous dose," insisted Liashko, "then we need to evacuate."[31]

Izrael repeatedly questioned the Ukrainians' dose estimates and asked for more time to take more measurements. Even in areas with high counts of curies in soils, he argued, his ministry's estimated doses in Belarus and Russia did not add up to 100 mSv, as they did when Ukrainians calculated them. "If that is so, then we can regulate life there through organized measures."[32] Organized measures meant shipping in clean food and restricting villagers' time outdoors.

"Our measurements show that if you are going to resettle children in those areas," a Ukrainian scientist replied, "then they have to have clean food. If we can't guarantee clean food, we can't send them back." The Ukrainians did not see how farmers could live in areas that they could not farm. "What would they do?" Liashko asked Izrael. "You could have them convert to specialized, livestock farming with imported, clean feed, but who would pay for that?"[33]

Moscow wanted to return residents to eleven Ukrainian villages inside the Zone of Alienation. Party leaders in Kyiv halted this measure too. The chief sanitation doctor of Kyiv Province, V. V. Malashevsky, had been tracking radioactivity inside the Zone since May.[34] He reported the counts were too high for habitation. Izrael didn't believe him. He sent his own team. Army and civil defense units also took measurements. A battle of dosimeters broke out. Five teams sent from Moscow agreed that the levels were safe. Malashevsky's unit recorded rates ten times higher and declared that to return people was dangerous. The lowly province sanitation doctor did not stand down before the powerful minister. Moscow authorities were furious. They told Romanenko to punish the parties responsible for presenting "subjective information." "Take special note," they added, "of the incompetency of the chief doctor Malashevsky."[35]

Romanenko did no such thing. The Ukrainian leadership refused to repatriate villagers to homes inside the evacuated Zone of Alienation. Nor did they terminate Malashevsky. A couple years later, I found him still at his job and still in a battle with Izrael's ministry over its persistent secrecy and slow-moving pace.[36] In the end, Malashevsky was right. In 1989, those villages were indeed judged once and for all to be too radioactive for habitation.[37]

Across the border in Belarus, party leaders faced the same decisions, but they did not show the same backbone. At the end of the summer, the Belarusian leadership obediently complied with Moscow's orders to return children and mothers to their radioactive homes and reopen a dozen villages inside the Zone of Alienation. With that 1,400 miserable farmers returned to their villages seven months after leaving them.[38] Their animals, seed, preserves, furniture, and tools were all gone. They asked for help. Province officials gave them a few sacks of potatoes and told them to stop complaining.[39] Belarusian leaders commanded factory workers and prisoners into the Zone to harvest produce.[40] The Belarusian physicist Vasily Nesterenko tried his best to stop the reimportation of several thousand villagers, but alone he got little traction.[41]

The differences between Ukraine and Belarus could not have been starker. While Ukrainian leaders banned farming in territory with more than forty curies per square kilometer, Belarusian officials condoned people living and farming with counts ranging from an eye-popping forty to a hundred curies of cesium-137 per square kilometer.[42] Ukrainian villagers received subsidies to purchase food in territories with more than fifteen curies per square kilometer. Until 1988, Belarusians only saw a payout at levels higher than thirty curies per square kilometer.[43]

I leafed through a stack of transcripts of the Politburo medical commission meetings. No one mentioned Mogilev, the Belarusian province where pilots worked to bring down radioactive rain in the days following the accident. Izrael, Ilyin, and the other big decision makers learned that villages in Mogilev were really hot, yet they gave no orders to resettle.[44] The summer of the accident, the newspaper *Izvestia* sent a trusted correspondent to Mogilev Province with the mission of reporting how life had returned to normal after the accident. Nikolai Matukovsky wrote a rosy article about happy farmers in fertile fields. Then, bothered by a troubled conscience, he sent an alarming telex to the paper's chief editor:

> Secret teletype. Do not show this to anyone but the Editor
> in Chief. Destroy copies.
> The radiological circumstances in Belarus have become
> greatly complicated. In many regions of Mogilev Province
> they have discovered radioactive contamination which is
> considerably higher than that which we wrote about [in
> our articles]. By all medical standards human residence
> in these regions constitutes a great risk to one's life. I
> have the impression that our comrades here are at a loss.

> They do not know what to do, especially since the relevant
> Moscow authorities do not want to believe what they report.
> I am writing this to you via telex because all telephone
> conversation on this topic is categorically forbidden. July 8,
> 1986.[45]

If the editor did send Matukovsky's letter up the chain of authority, there is no evidence that anyone acted on it.

Vasily Nesterenko sent more letters urging Belarusian leaders to do something about the tens of thousands of people who were living in communities counting over forty curies per square kilometer. He provided tables showing levels of radioactivity to be three to five times higher than the emergency permissible limit.[46] The celebrated Belarusian writer Ales Adamovich sent a personal missive to Mikhail Gorbachev alerting him to the fact that food was contaminated and that women and children needed to be resettled. "This Republic," he lamented, "is too small for so great a catastrophe."[47] Adding to the chorus of voices was a Red Army general. "No one," General A. Kuntsevich pleaded, "should be living in areas over 80 curies per square kilometer. These people must be moved."[48] Full stop.

General Kuntsevich requested a commission of experts.[49] The commissioners, sixty scientists from Moscow and Minsk, went to Mogilev Province. They looked around, took samples, ran their calculations, and decided residents could stay.[50] That decision, among dozens of other resolutions made in secrecy, left several hundred thousand people to live and farm in levels of radioactivity similar to those at a nuclear power plant.[51] To live there safely, residents would need shipments of clean food, the well-fed policy makers resolved. How they would find in the hungry Soviet Union reserves for people who normally fed themselves did not figure in their calculations.

WHAT DID IT mean to live, work, and subsist on territory with such high levels of radioactivity? Trying to figure out what the choice of leaving people in those places meant, I hunted down records from regional health departments, agricultural bureaus, and labor unions. Those documents described a new, quotidian danger spreading outward from the red splotches on the top-secret radiation maps. The new radioactive threat expanded outward because humans are social and productive creatures, engaged in trade, barter, and travel. They took the radioactive fallout that rained from the sky and carried it with them to the most unexpected places.

# PART II

//

# HOT SURVIVAL

# Woolly Truths

I n Kyiv near a sprawling market, I picked my way around the "beware falling bricks" sign and went into the Central State Archive of Supreme Bodies of Government. Not a lot had changed in the archive since I had last worked there twenty years ago—the same worn parquet floors, sickly green walls, and oriental runner. I recognized the woman at the reception desk mounted on her stool as impassively as an Egyptian sphinx. I asked her for public health records on Chernobyl, and she laughed. "Chernobyl was a banned topic in the Soviet period. You won't find anything." I asked for the finding aids anyway. "You never know," I said with a deferential smile.

Flipping through the large catalog, I quickly identified whole collections labeled in plain Ukrainian, "On the Medical Consequences of the Chernobyl Disaster." I looked up in surprise. The receptionist shrugged and handed me an order form. It's not that she was trying to deceive me. She did not know about these records because no one had ever asked for them before. I could see on the library card that I was the first to sign them out.

I sat down among the rubber trees in the cozy, sunlit reading room and waited for the bound volumes. In the 1990s, through the open windows I used to hear trumpets playing a funeral dirge from the nearby cemetery. Now the rumble and beep of construction vehicles filled the room. The files arrived. I turned to the first tall stack. The papers, hundreds of them, contained medical and farm records, statistical reports, transcripts of meetings, official correspondence, petitions, and letters narrating how officials in Ukraine came to understand the effects of the Chernobyl catastrophe. I took notes, as the archivist stacked more papers on my desk. I understood that first day I would be at this job for years.

I soon came across a document that left me bewildered. It was a petition requesting "liquidator status" for 298 people who worked in a wool factory in the northern Ukrainian city of Chernihiv.[1] "Liquidators" was a term reserved for people who received significant doses of radioactivity while employed to clean up the Chernobyl accident. I was confounded. How could wool workers, most of them women, have been liquidators?

And Chernihiv? Maps show the city lies outside the main path of Chernobyl fallout. In my imagination, liquidators were men, wearing suits of lead, rushing bravely at invisible gamma waves. They were not female laborers in the textile industry in a quiet "clean" town eighty kilometers from the accident. What were they doing to get such doses? I puzzled over these wool worker liquidators and searched for more information about them. I found more records, but I was still confounded. In July 2016, I rented a car and with my colleague, Olha Martynyuk, headed north to Chernihiv.

Chernihiv does not usually appear in Chernobyl films and media. There is a good reason for that. The city is too beautiful to be a stage for disaster. The medieval city of golden cupolas sits high on a precipice overlooking the Desna River. It is a local tourist attraction. Go there if you ever have the chance.

The wool factory was another matter. Olha and I found it with some difficulty on the margins of the city. The plant consisted of a dozen large brick buildings shot through with railroad tracks. In the Stalinist period workers built factories in manic, mass construction campaigns in an effort to industrialize fast. The wool factory was a product of such a building spree. It looked like it hadn't been touched up much since 1937. The factory is a "primary" works where employees sort and wash wool. That is all they do. The cleaned wool next goes to another factory to be made into yarn and cloth.

We snooped around the plant, which was laid out like a walled citadel. Inside, the complex had its own streets, shops, museum, public bath, health clinic, and a gate that swung open each morning to let in low-paid, low-skilled laborers from wool town, a settlement of squat, Bauhaus-style apartment buildings exclusively set aside for wool workers. In 1986, over a thousand employees, most of them women, worked at the plant. They shopped in corner stores and sent their children to the wool town elementary school.[2] Housing was meager. One woman complained at a union meeting that she had been on a waiting list for an apartment for twenty-six years. In the meantime, she had lived with five generations of her family in a single room so cramped that, she was embarrassed to say, she slept each night on the kitchen table.[3] A person could pass a lifetime within the wool factory's brick fortress and never escape.

We started asking questions at the manager's office. The current factory owner was the nephew of a local oligarch. He had long sideburns and wore cowboy boots. Despite his dude ranch appearance, he didn't know much about the wool business. He called up a factory old-timer, Tamara Haiduk, a retired shift supervisor, to supply answers. We sat on a bench outside in the sunshine. Haiduk looked every bit the sweet old Ukrainian grandma. She remembered dates and production volumes from thirty years before. She could list regulations by code and statute. Haiduk said that in June 1986, a month after the accident, the factory was operating at full capacity. Every spring after the annual shearing, 21,000 tons of wool arrived at the loading dock from all over Ukraine. To manage the deluge, she had three shifts going around the clock, everyone working twelve hours, seven days a week. "Very stressful work," Haiduk remembered. "Two hundred trucks and train cars would be backed up, waiting to unload." Soldiers and students were conscripted to help. I asked Haiduk about radioactive wool.

"Some wool came in from contaminated areas," Haiduk said in her matter-of-fact way, "and two dock workers had nosebleeds. Another man, a supplier, felt nauseous. So we called up Moscow. Moscow sent a commission. They measured, and we changed our process. After that, any wool that measured over 1,000 microroentgen (10 µSv) an hour, we pulled off the line and stored it."

Sometime in the fall of 1987, Haiduk recalled accompanying a dozen drivers to bury the most radioactive wool in an evacuated village inside the Chernobyl Zone of Alienation. "That was it," Haiduk concluded. "After 1987, hardly any wool was radioactive."[4]

So, that was the story: a bit of exposure to a few workers that lasted a short time.

I was baffled. What about the two hundred women given liquidator status? Haiduk waved a hand. "That was just a formality," she said. I asked if we could see the factory's production line. A woman named Tamara Kot showed up to give us a tour. She was tall, robust, and warmly expressive. As soon as we were out of earshot, Kot started talking, and her story differed disturbingly from Haiduk's account.

"I came to work here in 1986. My girlfriend and I both had just graduated from the Institute of Light Industry. A year later, my friend got diagnosed with leukemia. She died soon after that. I think she got sick from the radioactive dust we breathed in. All the dock workers and the drivers, those guys are all dead. They got dosed."[5]

I should not have been surprised by the difference between the stories managers tell and the stories workers tell. They rarely match up. While managers wield a clipboard to count, measure, and direct, workers come to grasp the products they work with viscerally and sometimes painfully as they pick up bales of hay and hug them to their chests, pull handfuls of wool close for a good look, or sweep up small dust storms at the end of a shift.[6] Fine particulates, like wool fibers, coal soot, asbestos, and radioactive particles, clog lungs and throats. Chemical powders and radioactively charged wastewater burrow into the cracks of skin and under fingernails. Skin, lungs, and oral cavities serve as gateways into the body. Physical labor entangles bodies with raw materials so that the borders between bodies, produce, and environments dissolve. Dirty work is left to people with the fewest choices; when I asked Kot about the kind of people the factory employed, she replied, "They were not of the best quality. I mean," she added, nodding at some passing workers, "who wants to spend their life washing shit?"[7]

At the warehouse, porters in coveralls milled around a loading dock that opened onto a barn-sized receiving shop. The men heaved bales of dirty wool onto a conveyor belt. The large bales rode up to the sorting

loft through a shaft of dust glistening in the sunlight. Inspectors measured that dust in 1987. It was six times thicker than allowed by Soviet
labor safety laws.[8] We followed the wool, climbing the stairs into a cav-
ernous room where women, in blue smocks and masks, stood over tables
of wire mesh pulling apart great wads of dirty wool. Air ventilators and
lights hung over the tables. They were turned off. In the dark and dusty
loft, the women worked quickly, pulling and grading the wool, then tossing it into bins that would roll on to the washing and packing shops.

The women stopped on seeing us enter and gathered around the list I
held out. "That's my name, and this is Svetlana here. That's Maria." The
women picked themselves out of the past from a category created for the
disaster. They were the "liquidators" I had found in the Kyiv archives.
Of two hundred women on the list, these ten were the only ones left. For
the past thirty years, they had stood at their stations day after day sorting dirty wool with sweat beading behind their masks, eyes dry, backs
aching, and their hands calloused and rheumatoid. They showed no surprise at our appearance. They acted as if they had been waiting these
three decades for someone to turn up to record their story.

In a chorus of voices, the women described the slowly dawning realization in the summer of 1986 that the distant nuclear accident had
entered their lives. They first noticed a trickle of blood slipping from
the mouth of a young female colleague. The women described how
they started to feel dizzy and nauseous at their tables. They had to take
breaks outside in the fresh air, away from the dusty loft. Their boss, the
grandmotherly Haiduk, who gave fines to workers for being two minutes
late, would goad them back to work. Haiduk was aiming, as she always
had, as she managed to do even in the disastrous year of 1986, to not only
make the annual production quota but also to surpass it.[9] By the end of
May, many workers suffered mysterious nosebleeds. They complained
of scratchy throats, nausea, and fatigue.[10] Union records show that a couple of drivers, after helping out in the fields, sought medical treatment.[11]
In the sorting shop, the hay bales measured up to 30 μSv/hr. The wool
workers did not know that picking up the most radioactive bales was like
embracing an X-ray machine while it was turned on.[12]

Every large Soviet institution during the Cold War had an employee in charge of civil defense whose mission was to prepare for a nuclear attack. Hearing the complaints at the loading dock, the retired colonel, V. I. Goroditsky, dusted off his DP-5A, a counter that measures gamma rays at high, nuclear-war-sized doses.[13] As soon as he turned it on, the device began to tick and then screeched in alarm. Thick columns of dust in the warehouse were two times more radioactive than the elevated, emergency permissible levels.[14] Goroditsky reported the news to Mikhail Shesha, the factory manager. Shesha, like most Soviet managers, lived in fear that problems reported in his enterprise would be blamed on him. To stay out of trouble, his first managerial impulse was to cover up the problem. He told Goroditsky to stop measuring.

As the days passed, workers' complaints mounted. Shesha relented and called Kyiv.[15] Inspectors in Kyiv didn't believe his wool could be so radioactive. They ordered Shesha to bring the DP-5A to Kyiv to check its calibration. The machine worked fine. Naturally alarmed, the Kyiv inspector called Moscow. The Moscow officials ordered a team of specialists in veterinary science to take a trip to Chernihiv. The monitors arrived with boxes of tampons, which were impossible to buy in Soviet stores. They swiped walls, equipment, and clothing with the tampons. The monitors found radiation everywhere—at the entrance to the plant, in managers' offices, in the lunchroom, in the drums of the washing machines, on the sorting tables, and hottest of all, at the loading dock. They measured the arriving bales. The most ragingly radioactive wool came from northern Ukraine.[16]

That is when the Ministry of Health learned that officials in the State Committee of Industrial Agriculture had given an order to slaughter 50,000 sickly animals rounded up during evacuation from farms inside the Zone of Alienation.[17] The animals' highly contaminated body parts were sent to a string of factories for dismemberment and processing as consumer goods: wool to Chernihiv, hides to Berdychiv, and meat to Zhytomyr.

The Zone of Alienation was just a circle drawn on a map. It didn't stop radiation from transgressing its borders. Farmers outside the Zone also had radioactive sheep. As they sheared that summer, monitors measured the wool and found some to be very radioactive at 32 µSv/hr. State

Committee of Agriculture officials kept that news to themselves and sent the hot wool on to the factory.[18]

Once they recognized that the plant was contaminated, factory director Shesha put shift supervisor Haiduk in charge of managing three young women as radiation monitors to take daily readings.[19] They recorded that the gamma rays in the warehouse measured from 1.0 to 180.0 $\mu$Sv/hr.[20] This number—180 $\mu$Sv/hr—requires translation. It helps to break it down and relate it to the body, as Soviet managers of the disaster struggled to do. At 180 $\mu$Sv/hr workers at the loading dock received in just one week seven times the annual dose recommended today for civilians. After Chernobyl blew, Soviet leaders set a much higher emergency dose of 100 mSv for citizens during the first year after the accident. Wool workers received the annual emergency dose in just the four busy summer months after the spring shearing.[21] And 180 $\mu$Sv/hr was only a partial measurement. Monitors started counting gamma rays after the most powerful isotopes of radioactive iodine had already decayed, and they counted only external exposures. They did not include the dust workers breathed in or ingested in the food and water they took in during their breaks, nor the radioactive isotopes they brought home to wool town in their hair and clothing. Radioactive isotopes once ingested do not readily dislodge from the body. The most prevalent radioactive substance in the dust at the wool factory was ruthenium-106, an element that migrates to bone marrow and remains until it decays.[22]

In other words, the factory management had an emergency on their hands. Commission after commission from the Ministries of Light Industry, Health, and Justice arrived in Chernihiv from Moscow and Kyiv. They confirmed and reconfirmed the measurements and drew up safety measures. In late July, Haiduk ordered new ventilation fans to repair the broken ones above the sorting tables. She requested respirators and jumpsuits.[23] Plumbers flushed out the factory's pipes and fixed the drainage canals that flowed into ponds where kids played in hot Ukrainian summers.[24] They set up a radiological lab.[25] The visiting experts wrote new regulations to sort the wool, not by quality as before but by the amount of ionizing radiation emanating from the bales. The bales radiating 1-10 $\mu$Sv/hr were to be washed and measured again. After

cleaning, the bales measuring below 1.0 μSv/hr were to be processed
as normal wool. The most radioactive bales were to be fenced off with
barbed wire and stored until further instructions.[26]

From August 1986, time at the wool factory came to be measured
differently—in the height and width of the mound of wool that grew
by the day and in the power of the gamma rays pulsing from the heap.
The half-life of ruthenium-106 is 373 days, meaning approximately
half of the ruthenium-106 lost its radioactivity each year. I imagine the inspectors ordered the wool to be stored, hoping to wait out
ruthenium-106's half-life. Once it had decayed, the wool would be fit
for soldiers' uniforms, carpet, socks, or warm coats. But, honestly, it
is hard to imagine what they were thinking. By November 1987, 2,400
tons of wool had piled up in the open, uncovered, just off the loading
dock, where workers warmed themselves in the sun on their cigarette
breaks. Every day, the ruthenium and cesium particles decayed, emitting gamma rays that passed into and out of workers' bodies. Even
when the factory was processing clean wool that came from abroad,
the levels of radioactivity remained high because of that pile of wool
increasing in volume each day.[27]

The point of the elaborate Chernobyl Zone of Alienation—the evacuations, fencing, and guards—was to put a stop to the migration of Chernobyl radioactive isotopes to places where people were living. But the
mission of businessmen at the State Committee of Light Industry was to
maximize production, meet their quotas, and feed and clothe the country. In the USSR, as in many places around the globe, safety gave way to
production and profits. After the accident, regulations stipulated that
radioactive wool, hides, meat, fat, and bones were to be fed into production lines as usual with some new precautionary measures to process
them.[28]

Once they discovered contaminated wool at the factory, business continued as usual. As the radioactive wool came in, porters unloaded it.
Sorters sorted it. Radiation monitors measured it. In the greater countryside, the sheep kept walking through the muddy spring, dragging
their bellies across puddles swimming with radioactive particles. The
wool in 1987 was only slightly less radioactive than the wool in 1986.[29]

For over a year, no one resolved the problem of the heap of wool near the loading dock. Nor did they know what to do with the washing machine drums, trucks, and other equipment that had grown radioactive, or how to treat contaminated wastewater gathering in a two-acre pond near the factory. And then there were the risks involved in reporting each day for work. "The question remains," the inspectors wrote in June 1987, "as to the danger of this work for workers who haul, sort, and wash the wool."[30]

"Oh, we were full of radiation. Ping, ping, ping," the sorters remembered. "We took off our smocks, and they balled them up and threw them away. We asked what kind of dose we got. They said, 'You don't need to know.'"

The women in the sorting shop recalled that six months after the accident a young colleague went to the hospital for severe anemia. She needed a blood transfusion. The women remembered how the factory's chief engineer, the elegant, educated Maria Nogina, visited the ward and warned her dying employee, "Don't tell anyone where you work!"[31]

If news got around of Chernobyl health problems, the leadership, people like Nogina, would pay for it. In a hotel lobby in Chernihiv, I met with Nogina, now retired. She did not remember going to the hospital or threatening her subordinate. She did recall the nosebleeds and dizziness as well as her own stress and anxiety about them. "All those who directly worked with wool started to feel sick," Nogina recalled. "We were not ready for that."

Hoping to get a better grasp of the wool factory's problems, Olha and I stopped at the local archive, which is located in Chernihiv's monastery, built in the ninth century. The monastery, surrounded by a leafy park overlooking the river, had thick white walls and tiny windows and was crowned by a riot of small golden cupolas, sparkling in the sun. Inside the church, over the centuries, monks had painted colorful frescoes from floor to ceiling, as if they could never have their fill of God, saints, and miracles. A newer, eighteenth-century building held the archives. We sat down in the crowded room to leaf through files we had ordered. The dissonance of the soul-jarring chimes of the church bells clashed against the electronic beeping sounds I imagined as I scanned long lists of gamma readings taken around the city in the summer of 1986. The

Chernihiv Department of Health recorded radioactivity in the schools of wool town, but had nothing to say on the contamination of the wool factory.[32] The labor union did, and that is strange. Labor unions in the Soviet workers' state were wholly captured in the 1930s by management.[33] They largely served state enterprises to speed up employees and goad them to work more for less pay, or to volunteer their days off for no wage at all. A few documents contained transcripts of meetings where workers spoke about their problems. The transcripts record these workers' voices in what I would call historical whispers. The workers' complaints did not rebound upward to division chiefs or local leaders, much less to Kyiv or Moscow. They remained in lowercase letters in union documents, buried in that old monastery.

In that way, the factory management kept health problems under wraps. They certified in a report to Kyiv, which Nogina signed, that the factory staff had no illnesses in connection with radioactivity. Nor, they wrote, had any workers suffered from any occupational health problems in the previous four years.[34] Union records flatly contradict this statement. Those records describe a 200-pound bale of wool falling on a worker's limbs and another bale clocking a woman on the head.[35] More troubling, from 1987 to 1989, union representatives grew anxious about the "worrisome" increase of illness at the factory.[36] In regional farms, doctors reported a peculiar "guttural swelling" among shepherds.[37] Thirty years later, the women sorters and Nogina fingered their throats and described to me problems with thyroid disease, adult-onset diabetes, and cancers. They mentioned other symptoms that sounded strange—aching joints, "legs that don't go," migraines, fainting spells, and painfully twinging nerves. They attributed these unspecific health problems to Chernobyl.

I was at a loss with what to do with these associations. Many of the symptoms the workers reported could be attributed to plain old senescence—the gradual slowing of cell reproduction in the body that leads to symptoms commonly known as old age. Maybe, as critics would charge in the debates over Chernobyl health effects, the wool workers attributed normal symptoms of aging to radiation. On the other hand, the sorters knew more about radiation than the average person. They grasped that some radioactive isotopes, such as ruthenium-106, settle in bone marrow, whereas others, like cesium-137, target muscle tissues, and radioactive iodine-131 gathers in thyroids. They visualized the isotopes in their bones, joints, and crumbling teeth and radionuclides sending off radioactivity that was blasting cells, causing damage and pain.

I compared the workers' testimony against published literature on radiation medicine. Researchers know a lot about the health effects of large, single blasts of radiation at levels over 1 Sv because symptoms at those exposures are hard to miss and also hard to deny. Over the decades, researchers who created the field of radiation medicine rarely considered low, chronic doses of radiation.[38] The first doctors who studied radiation noticed that people with high doses suffer from nosebleeds, nausea, dizziness, and headaches, much like the wool workers experienced hauling and sorting radioactive wool in the first days after the accident. At acute levels of exposure, decaying isotopes attack the sensitive blood cells, causing anemia. Marie Curie, the Polish French scientist who first isolated radium, had frequent bouts of anemia.[39] She learned that when she and her lab assistants grew ill they needed to take a break from lab work and wait while blood cells regenerated and they regained their health. People can recover from high exposures to chronic radiation and lead a normal life. But if the exposures continue, then illness becomes complex and persistent and can be fatal. That was the case for Marie Curie, who died of aplastic anemia.

It was also the case in the 1920s for radium dial workers who licked the tips of their brushes as they painted glow-in-the-dark numbers on instruments and watches. The dial painters felt no high-dose symptoms as they worked, but gradually radium from the paint accumulated in the workers' oral cavities, bone marrow, joints, and blood. The "radium girls" first started feeling aches and pains in their jaws, hips, and knees. Their teeth wobbled painfully and crumbled away. When dentists removed the rotting teeth, the lesions did not heal. The women developed anemia, lost weight, and felt chronically fatigued. As the ingested radium decayed, it broke down bones into honeycomb configurations and ate into joints. Whole sections of the young women's jawbones gave way. Femurs snapped. Hip joints froze in place. Bewildered doctors treated the women with surgical knives, casts, and metal braces, measures that only increased their pain.

When a few radium dial workers died and their relatives filed lawsuits, managers at the Radium Dial Company and U.S. Radium claimed the women's doses were too low to cause health problems. They had the backing of university researchers and local public health officials, both of whom in the 1920s generally bowed before the power of corporations. After several more women died and others became invalids, company officials hired their own medical doctors to investigate. When those physicians ruled that radium could indeed be a factor, the company managers hid the reports or found other, less competent "experts" to vouch for worker safety. When the lawsuits picked up speed, company businessmen courted public health officials, lobbied for restricting workmen's compensation laws, produced their own misleading health statements, and hired teams of lawyers who did their best to sow confusion and stall legal rulings.

Finally, Dr. Harrison Martland, the new chief medical examiner in Orange, New Jersey, devised a way to ash the bones of a radium dial worker recently deceased and test the remains with an electrometer. These were the first measurements of radioactivity in the human body. Martland surmised that radium inside patients' mouths promoted the growth of bacteria that led to chronic infections and loss of teeth. He guessed radium in bones and joints made them brittle and painful. He exhumed the body of a painter to dramatically make his point. The bone fragments in the grave glowed in the dark. It took fourteen years for the women to win the first lawsuit.[40] A medical researcher, Robley Evans, studied the radium dial workers in the 1930s and determined that trace amounts, as little as one to two micrograms of radium, caused death.[41]

There was no such study of wool workers in Chernihiv, at least not one I could find, but the women, as they spoke to us, listed health complaints that were similar to the radium girls'—aching bones and joints, chronic dental problems, anemia, fatigue, and failing mobility.[42] The problem with extrapolating from one radiation incident to another is that radiation exposure is not one universal event. Radioactive nuclides come in great variety, and each nuclear emergency releases a unique cocktail of radioactivity. And within one emergency, each place delivers up a different spectrum of exposure. Each body absorbs and reacts individually to radioactivity. The radium workers were exposed mainly to radium, a bone-seeking element. The wool workers walked through powerful rays of gamma radioactivity and ingested dust laced with radioactive iodine-131, cesium-137, and ruthenium-106. After the Chernobyl accident, workers in various professions were exposed in particular ways. Laundry women repaired and washed linens for legions of contaminated households.[43] Farmers plowing in Ukraine in the exceptionally dry summer of 1986 churned up clouds of dust laced with radioactive cesium and strontium and sometimes hot particles of plutonium.[44] Loggers in forests surrounding the Zone were exposed to gamma rays coming from fallout that settled in leaves and trees.[45] Researchers eventually discovered twelve different radionuclides in the bodies of people living in southern Belarus and northern Ukraine.[46] Each ingested combination of radioactivity was unique: like snowflakes, no two were alike. Figuring out patterns of disease in this complicated postaccident terrain was terrifically complex. One thing was certain. Work, after April 1986, suddenly became a lot riskier.

The West had a wealth of data on Japanese bomb survivors and a bit less on nuclear industry workers, who tend to be male, but there were and are very few studies focused on people who received chronic low doses of radioactivity day after day and reported subtle, "unspecific" symptoms. Yet I was impressed by the depth of the women's grasp of the spreading radioactivity at their plant. These women with no more than high school degrees described to me how radioactive wastewater draining from the plant's giant washing machines poured into a pond, and from there it recycled into the municipal water treatment facility to either return to the factory or float down the soft, brown currents of the Desna River, where Olha and I swam in the warm summer evenings after work. I corroborated this account of contaminated wastewater in the records. No manager I talked to remembered that detail.[47]

It would have been easy, the sorters reasoned, to have avoided the contamination of the plant altogether. "They should have never unloaded that wool here," one worker pointed out. "They could have measured it on the trucks first." The women knew the places around the factory that were the most radioactive—the loading docks and their own sorting tables—and they understood the significance of the mountain of radioactive wool. "Why," the sorters asked, "did that wool sit there for so long?"

That's a good question. The wool piled up for eighteen months, six months more than the half-life of ruthenium-106. The plant engineer, Maria Nogina, recounted that the officials in Moscow and Kyiv stalled and stonewalled, refusing to give them permission to throw out the wool as radioactive waste. "They made us document it all, every bit— radioactivity, weight, value," she said. "They even sent in a prosecutor to investigate possible corruption for attempting to dump good wool."[48]

Finally, just before Christmas 1987, Haiduk supervised ten drivers who worked round the clock, loading up the wool and stashing it in pits inside the thirty-kilometer Zone of Alienation. After a year and a half, the radioactive bales were finally buried.[49] For their trouble, the drivers got a bonus of fifty rubles. The sorters received three extra rubles in their monthly paycheck. Later they were awarded status as liquidators, which enabled them to retire early, have extra medical checkups and longer vacations, and ride the city bus for free.

Before we left town, Olha and I made a last visit to the wool factory. In their stained smocks, the sorters looked over the liquidator list again.[50] "After Chernobyl, a lot of us are gone," one woman sighed. "They didn't all die in one day," she continued. "They took sick and passed away gradually, from heart problems, from cancers." Another woman added, a finger on the list, "Look, none of these drivers are alive. They died when they were just forty or fifty. Volodia is gone. Victor too. And Kolia." They went on like that, sounding out the name of each dead comrade.

# Clean Hides, Dirty Water

Not everyone went along with the dismemberment of radioactive animals and the lavish redistribution of body parts that had become nuclear waste. I noticed in the correspondence of the State Committee of Industrial Agriculture the letters of a certain Dr. Pavel Chekrenev, the chief doctor of the Department of Health for Zhytomyr Province. Northern sections of the province were heavily contaminated with Chernobyl fallout. Chekrenev and his staff discovered in July 1986 that wastewater streaming from the Berdychiv Tannery into the Hnylopiat River was up to six times more radioactive than the already elevated emergency thresholds.[1] The Hnylopiat River, Chekrenev pointed out, flowed into the Zhytomyr drinking reservoir. "There is no way," the doctor fumed, "that long-lasting radioactive isotopes can be drained into a drinking water reservoir."[2] Citing a 1980 pollution law, Chekrenev did something no one had done at the wool factory in Chernihiv. He issued a stop work order, shutting down the tannery and holding up the processing of 19,000 hides.[3] He refused to let the cleaned skins leave the factory.[4]

The officials at the State Committee of Industrial Agriculture were fit to be tied. They had given the order to kill the contaminated animals. And now those skins languished in the open, rotting and losing value. The powerful businessmen counted on the unexpected profits from 19,000 extra animal carcasses brought to them free of charge by the Chernobyl disaster. Experts in Moscow had drawn up a plan that called for warehousing the hides for several months so that the most powerful, short-lived isotopes decayed and then processing the leather in extra baths of water and chromium.[5] The USSR Surgeon General himself had signed off on the special protocol. Everything was set until the provincial doctor with a funny-sounding name came along and defied their orders.[6] Chekrenev disrupted their calculations by showing the flaw in plans for "decontamination," which is a word to be placed in the same aspirational vocabulary as "liquidation" and "permissible dose." Radioactive isotopes can only be moved from one place to another, while they decay on their own schedule. The cleaner the leather emerged from the factory, the dirtier the water flowed from the factory's drainpipes.

Curious about the leather factory, Olha and I drove from Chernihiv to Berdychiv. The factory languished on an emptied plain, behind a large, weedy Jewish cemetery. If you have heard of Berdychiv, it is usually because someone's grandparents came from there. Berdychiv was, in its day, a major center of Jewish culture and one of the birthplaces of Hasidism. Not much of that legacy remains. Jewish Berdychiv was first dismantled by Communists in the 1930s. In the fall of 1941, Nazis intent on a *Judenrein* Europe trucked Berdychiv's Jews to an airstrip and shot them into pits. Finally in the 1990s, Zionist groups finished off Berdychiv's living Jewish legacy by setting up programs to sponsor Ukrainian Jews' migration to Israel, which was hungry for bodies to inhabit disputed territories.[7]

Pulling up to the Berdychiv Tannery, I noticed a worker in coveralls dumping a large vat of foul-smelling brown liquid into a manhole cover. That didn't look right. I asked the man, slouching over his cigarette, what he was doing. He scowled in reply.

We moved inside to make our appointment with Volodymyr Tsymbaliuk, the factory's general manager. I soon realized our timing was lousy. Tsymbaliuk, a short man with a stout face and a hedgehog brush of white hair, was shouting into his cell phone about wastewater and sanitation norms. Olha had read in the newspaper that the swimming beaches in Berdychiv were closed because of chemical toxins in the water. It appeared that the man outside dumping wastewater into the sewer had something to do with Tsymbaliuk's angry conversation on the phone. Clearly, he was in no mood to talk about the factory's history of disgorging radioactive refuse into the Hnylopiat River.

In truth, he didn't know much. He did volunteer that Chernobyl was no accident. "CIA agents," he said, "sabotaged the plant. I am sure of it." Tsymbaliuk pointed out the window at the mammoth leather factory, most of it in ruins. "Three thousand people used to work here until Americans brought capitalism and wrecked the USSR."

There was no arguing with that. We left our hostile witness, got back in the car, and drove down the road to Zhytomyr where on Sunday morning I located Chekrenev's widow, Nina Aleksandrovna Chekreneva, also a doctor. She fed me cake and tea, and showed me photos. She called her late husband "Chekrenev." "Everyone did," she said. "No one used his first name." Nina Aleksandrovna described her late husband as "strict and exacting," and also "kind and scrupulously honest."[8]

"He worked a lot," she reflected. "The phone would ring day and night. Others didn't allow calls at home, but he did." She grew quiet, then added, "He lived for his work. He wasn't much of a family man."

The day after the accident, Chekrenev volunteered with five other men from the Zhytomyr Department of Health to help out at the Chernobyl disaster site. The physicians spent a month on the cleanup, where they did everything but work as public health officials. They bagged sand to load onto helicopters to dump on the fire. They scrubbed down buses. They hauled furniture from houses in towns that were being evacuated.[9] Helping to clear one village, Chekrenev carried on his shoulders an old woman who refused to leave her home.

Surprisingly, Nina Aleksandrovna had never heard about the conflict over the Berdychiv leather factory, nor did she know that Chekrenev had been censored for his defiance. She suddenly understood the reason for his demotion in 1986 from head of the Department of Health to a lowly inspector. "He didn't talk much about his work," she said.

Years after the Soviet Union dissolved, leading Soviet scientists reported that they kept silent about Chernobyl because if they spoke up they would have been thrown in jail. Chekrenev's experience shows the stakes were much lower. Pressured by the Ministry of Light Industry to reopen the factory, Chekrenev refused. He took his case to his superiors in Kyiv.[10] They sent it on to Moscow to review. Finally, Anatoly Romanenko, the Ukrainian minister of health, ruled in Chekrenev's favor.[11] The rest of the radioactive hides would not be processed but dumped as radioactive waste. The drinking water was saved. The already radiologically burdened bodies of Zhytomyr residents were spared this one additional source of exposure.

Unfortunately, Chekrenev did not live to tell me his story. He expired in his late forties, a few years after the Chernobyl accident, of organ failure that Nina Aleksandrovna believed was caused by the dose he received working at the Chernobyl disaster site.[12] His death is not included in the official count of fifty-four Chernobyl fatalities.

# Making Sausage of Disaster

A t first it was simple. "If the meat is no good," commanded Oleksandr Liashko, in charge of Chernobyl relief in Ukraine, "then kill the animals and bury them."[1] In a secret operation, Soviet cowboys went back out and herded an additional 50,000 head of livestock from a sixty-kilometer radius around the Chernobyl plant.[2] But they didn't bury the carcasses.[3] Teamsters drove the bleating animals to slaughterhouses in what became a new distribution pattern in Ukraine and Belarus. Animals from radioactive regions went to meat factories in contaminated provinces.[4] Uncontaminated animals were driven to stockyards in clean territories.[5] Moscow agronomists sent out a special manual for meatpackers with instructions on how to process radioactive meat.

The instructions ordered butchers to grade the meat by radioactivity. Packers were to grind up radioactive flesh and mix it with appropriate proportions of clean meat for sausage.[6] The experts in accident logistics were thinking along the commonly understood belief that diffusion was the solution. Spread the contaminated meat broadly so each person across the vast USSR unknowingly ingested their own small part of the tragedy. Preparing the goods for sale, the packers were told to "label the sausage as you normally would."[7]

The emergency instructions directed the sausage makers to soak the organs for several minutes in water, add sodium and nitrates, and measure again for radioactivity. They were to use all soft flesh except for udders, spleens, lungs, and lips of cattle. Bones, hooves, and horns required extra radiological control. Liver sausage was especially tricky and the new instructions went on for pages. Carcasses measuring above the permissible level were to be disposed of as radioactive waste.[8]

Somehow the careful calibrations went wrong. The busy slaughtering season normally occurs in late fall. The packinghouses had not planned for a surge of raw material in May and June. The directors sped up the lines of overhead trollies as the animal carcasses moldered in the summer heat. The butchers hurried to cut and disassemble the dead animals swinging on meat hooks. Women in head scarves rushed to gather innards into rolling bathtubs and push them toward the sausage shop. A few strangers drifted around the workers. The radiation monitors, press-ganged schoolteachers, were new at their jobs, having just passed an express course on how to take a bead on radioactive flesh. They aimed the wands of radiation counters at the carcasses spinning around them.[9] The monitors had only a handful of shared devices as they were in demand all over the region.[10] They sloshed through a mash of blood, fat, and bone, carefully staying out of the way of the flashing knives. The heightened field of radioactivity added to the already miserable work conditions of a packinghouse.[11] Now even safe jobs were not safe. Drivers hauling the animals on their second and third runs got spooked. The frightened animals left behind a radioactive trace inside the cabs measuring 3 µSv/hr. Seeing the needles on the counters, some drivers quit right there. They turned their trucks around and sped off.[12]

The monitors missed some radioactive meat. Meat locker superintendents mislabeled it, and polluted goods moved on to markets.[13] When highly radioactive sausage and veal turned up in Moscow stores, the Soviet minister of health sent a top-secret emergency-gram to Kyiv: "We ask you to take all possible measures to stop the shipment of such [radioactive] produce to Moscow, and possibly limit its distribution in Leningrad."[14] Ukrainian officials had already secured their own dining tables by banning the sale of radioactive meat in Kyiv for six months.[15] The KGB also had in place its own supply of clean food sources for "special distribution." KGB buyers went to stockyards to personally select animals for slaughter. These goods went to the top leadership, military installations, and KGB employees.[16]

It's true. Everyone who could save themselves did.

Fortunately, it was Anatoly Romanenko's job as the Ukrainian minister of health to worry about the general public. He wrote the Politburo's Chernobyl medical commission for advice about the radioactive meat. The commissioners considered Romanenko's telegram in their daily meeting. "The meat can be eaten," one commissioner speculated. Another chimed in just as optimistically, "It's getting less radioactive every day."[17]

Once again, the Ukrainians took their own counsel. A few days later, Romanenko's assistant banned the sale of all meat from packinghouses in Zhytomyr Province.[18] With that, he shut down the nonsense of trying to salvage highly radioactive beef and pork for human consumption.

Yet meat, especially beef, was hard to buy in the USSR and it was greatly valued. If you were lucky, you might purchase a couple of pounds a month. Officials in the State Committee of Industrial Agriculture, loath to toss out the valuable produce, issued this order: "In order to avoid a buildup of raw materials and so that the meat factories of Zhytomyr Province do not have a stop in work, we ask that you redistribute the meat with higher than permissible levels around the provinces of the [Ukrainian] Republic to the sausage divisions of other factories. Meat higher than permissible and labeled 'undesirable' also send to the sausage division of other factories."[19]

With that, animals continued to flow to the slaughterhouses.[20] In a great circle around the blown plant, livestock stood for weeks in pastures saturated with radioactive gases and contaminated dust. At that time, people got radiation burns just from weeding their gardens.[21] Cows pulled the same grasses with their lips and drew them into their intestines. Packers placed highly contaminated carcasses in meat lockers. They stuffed in more, as more bodies of contaminated animals cascaded from the stockyards. By early June, 3,500 tons of meat had accumulated in Zhytomyr; by September, 6,300 tons.[22] Officials in Kyiv shuffled the frozen product to freezers around the republic. In Belarus, the Gomel packing plant quickly stockpiled 16,000 tons. The director begged for more freezers.[23]

Maybe the plan would have made sense if the freezers were made of lead to block the gamma rays coming from them or if the only contamination in the meat was short-lived isotopes that would decay in a few months. But no. The meat lockers were normal freezers, and the radioactive isotopes that settled into flesh have long half-lives.[24] The raw, radiating meat in Zhytomyr and Gomel was much like the mound of radioactive wool in Chernihiv. As it piled up, meatpackers toiled in ambient radiation a least a hundred times higher than background radiation. And when they processed it, consumers ate meat that was six to ten times more radioactive than international norms considered safe.[25] Those exposures were gratuitous.

The frozen flesh sat in freezers in southern Belarus and northern Ukraine for a long time. The next year, more meat over the limit amassed, and Belarusian officials repeatedly made requests to ship the product from the Gomel meat factory to other packinghouses in Moscow or Leningrad. Because the requests were repeated, I take it they were not fulfilled.[26]

I spent some time in Gomel with my research assistant Katia Kryvichanina. Gomel is a nice city—pretty, prosperous, comfortable. Katia and I were working in the regional archives and every day we passed the gleaming gates of the Gomel Meat Factory. I told Katia I wanted to make a visit to the factory. Her eyes grew wide at my stupidity. "No one there is going to talk to you."

I have to admit, I got a little irritated. I already had three speeding tickets that week in Belarus and the rental car towed. "I feel like Belarusians," I said to Katia, "are constantly telling me what isn't possible,"

"Sure," Katia replied reasonably, "there is a reason they call us the Germans of the Slavic world."

I drew some insight from that generalization. Belarus is clean and orderly in a Germanic way. In Minsk, the streets are swept. Buildings are in good repair and glisten with fresh paint. Minsk looks more like a film set than a place where two million people live. I noticed some people in street clothes painting fences and sweeping sidewalks. They didn't look like municipal employees. I asked a cab driver about them. He said they were jobless "parasites" who were put to work to pay off their debt to society. He thought forced labor was a good idea.

At least Belarus's dictatorship is a tasteful one. There are no big billboards of the great leader, no outward show of a police presence (unless people openly protest, as they did in 2010). Aleksandr Lukashenko has ruled Belarus since 1994. His wife cloistered in a village, he appears at public functions with his young son Nikolai. Like a ventriloquist and his doll, the two dress alike. Most Belarusians I talked to endorsed their dictatorship. They pointed to their southern neighbor, Ukraine, in a war with itself and with Russia and rife with corruption, as the example of what unharnessed democracy brings.

With Katia looking on doubtfully, I called up the meat factory, which is a blue-ribbon industry in Gomel. I asked the receptionist if I could have an appointment with someone in public relations to ask questions about Chernobyl-contaminated sausage. Silence. The voice asked me to repeat the question. I did. Once she understood my query, the woman could not get off the phone fast enough.

Disappointed, I turned to Katia, who tried to hide a justifiably smug look. I suggested we hang out around the factory gates at the end of the shift to try to talk to workers. Katia quickly put the kibosh on that idea. I got it. I could leave Belarus or get deported, and my life would go on. Katia, a Belarusian, had little choice but to live in Belarus.

There was nothing for us to do but go back to the archives.

In March 1990, the head of the Southwestern Railroad sent a coded telegram to Moscow. The manager reported that three years before someone had come up with a solution to the mounting volume of radioactive meat at the Gomel Meat Factory. In 1987, packers loaded 317 tons of frozen radioactive meat onto a train and sent the dubious gift to the Georgian Republic. There monitors detected the radiating mass inside the wagons and rejected the shipment. Managers at the North Caucasus Railroad passed the four train cars back to the Southwestern Railroad, which also refused the contaminated shipment at station after station.[27]

For three years the cars floated here and there in limbo, a hot potato no one wished to touch. Finally, without permission, the cars ended up in the train station of Ovruch, a community in northern Ukraine already radiologically overburdened. Leaders in Ovruch asked the director of the Chernobyl cleanup site to take the meat as radioactive waste. The director replied he was up to his neck in such waste and had no more storage space. The train cars moved on to the town of Poliske, also highly contaminated. Railroad workers staked a fence around the train, posted warning signs, and the train sat there, in a public transit hub, for months, radiating.

The KGB stepped in when the cooling equipment of the refrigerated cars failed, and electricians refused to service them. Railroad workers on the Southwest Line walked off the job.[28] A news photographer snapped a few shots, drawing public attention to the glut of contaminated meat.

Finally, four years after the overexposed animals were slaughtered KGB officers supervised their disposal in southern Belarus. In a deep, cement-lined trench they created the final resting spot for the meat that was toxic waste.[29]

I wish I could say that other branches of the food processing and farming industries of the western half of the Soviet Union better managed the catastrophe. Unfortunately, the dairy, grain, vegetable, and fruit industries struggled with similar outsized problems as public health officials tried to both feed people and meet their production quotas.

Sanitation inspectors quickly learned that almost everything was contaminated—milk, berries, eggs, grain, spinach, mushrooms.[30] Even tea imported all the way from Georgia was over the raised emergency-level permissible limits.[31] As happened with contaminated meat and wool, Soviet officials were unwilling to toss out contaminated agricultural goods, so they issued more manuals, also very detailed, about how to process radioactive provisions. Contaminated milk was to be dried or turned into butter or caramels. Irradiated sugar beets would be repurposed into animal feed, contaminated potatoes into starch, dirty berries became preserves, and over-the-limit vegetables transformed into paté.[32] The processed food was to be stored for months or years until the most pernicious isotopes decayed.

The insistence on selling radioactive food was not uniquely Soviet.

Chernobyl fallout spreading across Europe in patchwork fashion hit Greece particularly hard. Farmers in Greece harvested grain saturated with Chernobyl radioactivity and exported 300,000 tons to Italy. The Italians didn't want the wheat. The Greeks refused to take it back because they were "afraid of the reaction from Greek wholesalers." The two Mediterranean neighbors started fighting. Finally, the European Economic Community agreed to buy the contaminated wheat. They mixed it with clean grain and shipped it to Africa and East Germany as "aid."[33]

In the USSR, milk was especially important because children drank it and because it is so easily contaminated. Grasses concentrate radioactive isotopes.[34] Grazing cows streamline radionuclides in their milk, and a relatively low level of radiation in soils biomagnifies into harmful quantities in milk. Within twenty-four hours of swallowing radioactive grass, cows produce radioactive milk. Planners had a fix for this problem too. The Soviet State Committee of Industrial Agriculture issued orders to confine free-ranging animals in barns, cages, and feedlots. They instructed that meat would be cleaner if steer were kept inside and fed on concentrated feed for several months prior to slaughter.[35] Dairy cows eating clean hay while lactating would produce cleaner milk. Half of the agricultural fields in the Polesia regions of southern Belarus and northern Ukraine were pasture land, and large swaths of them were contaminated. Suddenly Belarusian farmers required 850,000 tons of feed every year to nourish the newly confined animals.[36] Farm bosses sent letters to the State Committee of Industrial Agriculture requesting feed. Receiving little, they asked again in urgent telegrams to the Politburo.[37]

To house the formerly grass-fed animals, bosses wrote directions for thousands of square feet of new barn stock.[38] They planned on building acres of greenhouses to produce hothouse vegetables kept out of dusty wind and contaminated soils.[39] These orders came at the same time that state construction firms were racing to build new communities for the 120,000 displaced farmers from inside the thirty-kilometer Zone of Alienation. It took many years to build the barns. In the meantime, agronomists elaborated temporary instructions for decontaminating milk. The instructional manuals became complicated with

formulas, charts, and drawings. "Milk had to pass through cellulite filters, pumped from pressure tanks, in a steady stream measuring 1/10th
of a second, or 30 liters a minute."[40] Even when it was possible to farm
with these expensive new procedures, the filters trapped only a portion
of the radioactivity.

In 1986, a third of the milk in thirteen regions in Ukraine measured
over the permissible emergency level. Oleksandr Tkachenko, at the
Ukrainian State Committee of Industrial Agriculture, issued an order
to buy up the contaminated milk to turn into butter.[41] Following the
meat, 18,900 tons of radioactive butter filled the state reserves. That
left a shortage of milk for children to drink. The minister of health in
Ukraine, Anatoly Romanenko, directed dairies to replace fresh with
powdered milk to supply the population.[42]

At that, trucks revved up. Drivers raced over dusty roads to deliver
dried milk to farm children who normally drank milk fresh from the
family cow, rich, warm, and after April 1986, highly radioactive. The
taste of powdered milk paled in comparison, but at least it wasn't contaminated. Well, that was true only until someone tested the powdered
milk and found it also measured above permissible levels.[43] It turns
out that the country's dried milk plants were in northern Ukraine and
southern Belarus, the USSR's main dairy region.[44] KGB officials quietly
withdrew the powdered milk from shops. As shelves emptied, managers
complained that their food stores "lost the look of a store."[45]

Inspectors asked about honey, herbs, currants, gooseberries, mushrooms, blueberries, and cranberries and even about bed linen and
underwear. Romanenko kept asking for norms for each category of food
and household goods, and he wanted yet more safety manuals.[46] No one
in Moscow had ready answers. Biophysicists got on the job. The calculations took years.[47] But Romanenko didn't have that kind of time.

At least he could save his city, or pretend to. A week after the accident,
state troopers set up three control points on roads leading to Kyiv. Police
checked every arriving car and truck. If too radioactive, they washed the
vehicles down and measured again. The Ministry of Health ordered that
100 percent of all food arriving in the city be screened before sale.[48] City
officials created labs in the dairies that supplied Kyiv.[49] Kyiv officials

placed orders for vegetables from southern territories of the Ukrainian Republic that were comparatively clean. They mined warehouses for foodstocks grown before the accident.[50] They carefully monitored the city's drinking water.[51] They shut down open air markets and set up radiation labs in shops.[52] They devised sealed packaging for goods and airtight trucks to ship them.[53] Street cleaners sprayed streets daily to rinse away radioactive dust.[54] Conscripted high schoolers raked up contaminated falling leaves. Dump trucks sped them away.[55]

In the tenth century, Kyivans created their city high on a bluff surrounded by palisades to withstand attacks. In this 1986 strike by a new-age invader, city officials turned again to urban infrastructure to build what they called a "radioactive shield" around Kyiv. A network of roads and police forces, an army of scientists equipped with labs, centralized plumbing and heating, a food distribution network, and a phalanx of new regulations helped to reduce the number of radioactive isotopes entering the bodies of city dwellers.

The measures also had a soothing function. The technicians in lab coats and police officers with clipboards transmitted the message that the state was doing everything possible to "liquidate" the accident. A TV documentary showed a physicist testing a chicken. The voice-over explained, "Radioactive food that exceeds the norms, as much as we regret it, must be destroyed."[56]

Plenty of food that exceeded the permissible thresholds slipped into Kyiv and circulated from markets to kitchens.[57] Even so, it must have been a relief to believe the reporter's claim that the state was taking care of the problem. Party leaders showed concern for the people who counted most in Soviet society—urbanites.

Villagers, on the other hand, were like medieval serfs left outside the city walls. They had to fend for themselves.

# Farms into Factories

M odern disasters require a modern state to clean them up. The problem was that rural Ukraine and Belarus, where the greatest volume of fallout from the burning reactor landed, had few of the resources necessary for overcoming a high-tech disaster. Most villages within eighty kilometers of the plant lacked plumbing and central heating. The electrical wires that pulsed from the Chernobyl reactors missed many hamlets. Villagers carried water from open wells in open buckets. Taking a bath and doing laundry involved a lot of work and so occurred only occasionally.[1] Heating came from burning contaminated peat and wood. Dirt roads generated dust carrying radioactive particles. The region was swampy. During spring floods, some villages were cut off for weeks or months. Rural shops carried salt, kerosene, and matches but little else. Farmers ate what they produced on their private plots. Everyone worked the fields, including pregnant women and children. The few hospitals and clinics were understaffed. Much of the industrial age had passed this part of the world by, until it rained down all at once in the spring of 1986.

As regulations from Moscow stipulated that collective farmers become modern consumers of food, fuel, and medicine, while following safety regulations designed for workers at nuclear power plants, it became clear this would be a losing battle.

Half a million people resided after April 1986 in areas of Belarus and Ukraine with high and very high levels of radioactivity.[2] Radiation levels in Belarusian villages surpassed those at nuclear reactors No. 1 and No. 2, restarted in the fall of 1986 at the sullied Chernobyl Nuclear Power Plant.[3] Measures to combat radioactivity telescoped onto the village—usually a cluster of two dozen to several hundred houses surrounded by vegetable patches, sheds, and small barns with an outer ring of fields. A village household doubled as a site of food production. Large kitchen gardens generated fruits and vegetables. Sheds held chickens, cows, goats, sheep, and pigs. Farmers hauled hay to store in their barns. They herded animals from pastures to stalls in the village. Residents gathered mushrooms, berries, and herbs, carrying them home. They cut wood from the forest and brought it to the village. They used ashes and manure, exceptional radioactive concentrates, for fertilizer. Trucks, horse carts, and buses tracked mud into the village. They washed the trucks and tractors, and radioactive puddles formed near garages.[4] Flies that fed on manure swarmed into kitchens and onto food. A farm's battle against flies took on new import after Chernobyl; so did dirt tracked home on boots. Day by day, radioactive isotopes were drawn into the village, the center of the rural economic vortex. Residents' households became the spleen where radioactive isotopes moored. To deal with this exceptional circumstance, agronomists churned out survival manuals to teach farmers how to live and work in a postnuclear world.[5]

Farm workers in areas with more than two curies per square kilometer were advised to dress like nuclear operators in jumpsuits, respirators, hats, gloves, and personal dosimeters. Farm bosses, who had neither dosimeters nor the training to use them, were told to somehow set up radiological control posts in every village.[6] Farmers were to shower after work, could not lie down on the grass, eat outdoors, or ride in horse carts or in trucks with the windows rolled down. They could not burn branches or leaves, graze livestock in June and July, or use local wood or peat in stoves. They could not fertilize their gardens with manure and ashes or gather herbs, mushrooms, and wild berries in the forests. Better not to even enter the forests, which were often the most heavily contaminated areas.[7] To follow the survival manuals, in short, farmers were expected to give up the means of survival they had relied on for generations.

Soldiers and prisoners conscripted for dirty work dug up tainted topsoil and dumped it on the outskirts of villages in "temporary waste treatment areas," which were just pits that were not lined, fenced, or posted as radioactive.[8] To make up for lost nutrients in the topsoil that had been removed, agronomists told farmers to add more nitrogen fertilizers to the fields. Farmers poured on hundreds of thousands of tons of chemical nitrates.[9] Agronomists noticed that in abandoned fields inside the evacuated zone pests that fed on unharvested produce proliferated "massively" after the accident. Collective farm managers sent workers into the Zone of Alienation to gather crops to stop the parasite explosion, while pilots from the farm-chemical service buzzed the fields, sending down a shower of pesticides, including DDT that had been formally banned.[10] The volume of fertilizers and pesticides spread on radioactive land grew impressively in the postaccident years.[11]

The Chernobyl Nuclear Power Plant was built on the southwestern edge of the great Pripyat Marshes, Europe's largest swamp, in an area called Polesia. The surrounding terrain was marshy and boggy. Village wells in the region were shallow and open to the sky. Radioactive fallout landed in the wells from above and seeped in from the groundwater. Agronomists wrote more regulations, stipulating the need to dig deeper, artesian wells, cover them, and construct sewer systems for indoor plumbing. One of the first rules of nuclear emergencies involves simple hygiene. A contaminated person should wash up in the first hour or minutes after exposure. The longer radioactivity lingers on hair and skin, the more damage it does. Farm administrators wrote up plans to build village bathhouses and laundries for a new regime of "radiological hygiene." It took years to build those bathhouses.[12]

In the spring, ice dams clogged up the hundreds of rivers and streams of Polesia. In the floods, water spread horizontally across the marshy plains. In a healthy environment, seasonal floods are good. Inundating water spreads silt from river bottoms and refreshes soils with nutrients. Farmers staked their animals on the fertile floodplains where the grasses grew best. After 1986, spring floods became a problem because they liberated and spread radioactive isotopes that had settled in river bottoms. Hydrologists decided it would be better not to have any more

seasonal floods, so they ordered crews to build dams lined with filters
to channel water into dozens of holding ponds.[13] In the summer of 1986,
bulldozers rolled into contaminated territory to build new dams. Moving masses of earth for dam construction kicked up a lot of dust, which
was picked up in the breeze, rode along on airstreams, and returned
radioactive isotopes to communities that soldiers had just stripped of
contaminated topsoil.[14]

The conscripts fought the dust using a green, foamy chemical sprayed
on houses, fences, tractors, barns, and schools. They pulled down thatch
roofs and installed corrugated steel that insulated poorly but was easier to scrub with long iron brushes. As rainwater rolled off the roofs,
radioactivity concentrated in the puddles pooling under downspouts.
Engineers sketched plans to lay gas lines so villagers could heat and
cook without radioactive wood and peat.[15] Instruction manuals directed
mayors to pave dirt roads, streets, and playgrounds, entombing the
earth under asphalt.[16]

The new plans called for farms to specialize in either meat or dairy
production. They drew up schemes to monitor village food. They
announced that villagers must give up their private livestock because
villagers had no funds to purchase commercial feed. Leaders outlined
budgets for village shops to receive "clean" food supplies. (Officials in
Ukraine referred to "clean" food in quotation marks, expressing their
understanding that all food had become at least a bit dirty.) This was
a fast-tracked modernity. Polesians were told they should no longer
grow their own food as they had for centuries through plague and war,
drought and famine. In the postnuclear age, rural producers would
become consumers.[17] The only problem was that consumers are not
made overnight.

Half of village stores had no refrigeration. Some villages had no shops
at all. Nor did the trucks that delivered the stocks have refrigeration.[18]
Food arrived in towns in open truck beds. Teamsters carted produce
from towns to villages in horse-drawn wagons. In hot weather, the milk
soured on the road.[19] The food arrived spoiled, dusty, and radioactive.[20]
Logistics experts ordered closed trucks, freezers, and hermetic packaging.[21] This new infrastructure took years to materialize, if it did at all.

Schools had to recalibrate too. Planners ordered educators to provide students four meals a day. To do so, school administrators had to outfit new cafeterias. To keep children from wandering outdoors into field or forest, they extended the school day from breakfast past the dinner hour.[22] Instead of playing in radioactive fields, lakes, and forests, they drew up plans for paved playgrounds, gyms, and indoor swimming pools.[23]

The sleepy farm communities of northern Ukraine and southern Belarus had not fully kept pace with the postwar trend, led by American farmers, in industrializing agriculture. By the 1980s, large American landholders had simplified, specialized, and maximized agricultural production so that farms functioned like factories.[24] Employees in agribusiness concerns no longer toiled in the dirt. They ran machines and applied chemicals. American farmers arrested chickens in battery cages stacked in chicken houses that grew to the size of airplane hangars. They led sows and steer into airless, dark barns for speedy fattening. Free-range animals lingered as a nostalgic remnant for hobby farms and petting zoos.

I witnessed this transformation as a child on my grandparents' farm. My grandfather, John Brown, had a quintessential name and a textbook historical trajectory for an American farmer. In the early 1970s, he converted his dairy farm in northern Illinois into a steer-fattening yard. He bought Texas longhorns specially bred to survive the harsh conditions of the feedlot. The steer stood all day on a cement slab, packed flank to flank, ankle-deep in manure. Chewing corn they could not fully digest gave them ulcers. I would climb the fence and watch as the animals would eat and moan. My grandfather's pigs never left the twilight of the barn. Little, underworld creatures, they scampered between the hooves of the steer and fed on undigested matter in the cow pies. The animals fell ill in these conditions. My grandfather administered antibiotics to keep them going until they reached the slaughter yard.

Farmer Brown didn't last long in the steer-fattening business. He began to suffer from asthma, a common occupational illness caused by swirling clouds of bacteria and dust from animal hair and manure. As the investment costs of industrial farming grew, my grandfather's profits dwindled. Refusing a large bank loan, he retired just before Chernobyl blew. This was the "progress" Soviet agronomists at the State Committee of Industrial Agriculture ordered up for farms in the Chernobyl territories in 1986.

I can't tell when I read the instructions planning for an expensive new infrastructure to combat radiation in the Polesian countryside if the authors knew their schemes and budgets were utopian. I admire the effort, the intellectual brainpower and dedication of billions of rubles to repair the mess. No state had ever before attempted a nuclear cleanup effort on that scale. Yet the volume, the complexity, and persistence of radioactivity, I fear, doomed the project from the start.[25] It didn't take many curies of radioactivity in the soil for it to show up in the food chain. Scientists had known that for decades. "We could give 'nature' an apparently innocuous amount of radioactivity," the American ecologist Eugene Odum exclaimed in 1959, "and have her give it back to us in a lethal package!"[26]

Six months after soldiers scrubbed down villages and scraped away topsoil, the villages were again just as polluted. Dust and radioactive nuclides migrated from surrounding forests and fields into villages, drawn like everything else into the humming centers of regional economies.[27] Soldiers returned to clean again, three and four times. The conscripts moved the radioactivity around, shoving it from one place to another. Mounds of contaminated dirt and roofing material piled up on the outskirts of towns. Eight hundred unmarked radioactive graves leaked radioactivity into the water table.[28] In 1987 and 1988, radiologists added more communities to the list of contaminated population points to be "strictly controlled."[29] Republic leaders started to suggest resettling more people for safety's sake.[30]

With decontamination bringing no relief and most fields still contaminated, Belarusian province bosses calculated that they could produce only a fifth of their usual quota of meat and dairy. They begged to have their quotas lowered and for more food to replace contaminated produce for villagers to buy.[31] The requests went unanswered. The same provincial bosses pleaded for more commercial feed.[32] Agricultural officials allocated a quarter of what they needed. Farm directors were especially hamstrung in Mogilev Province, which though 400 kilometers from the disaster site, had been in the path of intensely radioactive rain. A year after the accident, the province still had no radiation monitoring posts, which were supposed to exist at every large farm, in every town and market.[33] Farm directors had no way of knowing just how polluted the fields were. And so, they kept plowing and harvesting.

The man in charge of food production in Belarus, K. Ukrainets, wrote in 1987 a formal complaint about this problem. Despite a lot of regulations and work to clean up contaminated territory, he noted, a third of the milk and a fifth of the meat in Gomel and Mogilev provinces were too contaminated to use. In areas of "strict control," defined as areas with over fifteen curies per square kilometer, those percentages doubled.[34] Even cows pastured on decontaminated fields produced polluted milk. One reason for this, he wrote angrily, is that local officials ignored the federal order banning farms with over forty curies a square kilometer from engaging in agriculture. They kept farming, and "they feed that radioactive hay to all sorts of animals, and those animal products are contaminated. They are planting fields right now in the Zone of Alienation," Ukrainets wrote in exasperation. "And, as a result, they will again have contaminated feed and again they will have to figure out what to do with spoiled animal products."[35] They weren't even trying, Ukrainets complained.

Ukrainets proposed that people living in territory with contamination measuring over forty curies per square kilometer be resettled. That would have meant moving 12,000 people in Belarus and twice as many animals. Ukrainets's recommendation was a practical compromise. Even in areas with ten curies per square kilometer, he pointed out, you can't generate clean food.[36]

In the 1990s, experts in nuclear medicine argued that moving people from radioactive territory causes stress, which in turn leads to health problems.[37] Better, opinion makers argued, to leave people in place. Yet in 1986 and in the years that followed residents of Chernobyl-contaminated communities understood clearly the danger they were in and they begged to be relocated. Many officials like Ukrainets, who witnessed firsthand the consequences of the accident, backed them.[38]

Forced to meet their production quotas, farm bosses had no choice but to send their workers to plow contaminated fields, cut radioactive hay, and milk exposed animals.[39] The results were predictable. In Gomel and Mogilev provinces, meat and milk in 1987 were more radioactive than in 1986, hotter again in 1988 than in 1987. It took about fifteen curies per square kilometer for nearly all the milk to be over permissible levels.[40] Mushrooms, meat, and wild boar also became more, not less, contaminated in subsequent years.[41] The same dynamic played out in Ukraine.[42] Even in "clean" areas, 30–90 percent of milk came in hotter than permissible. "Despite restrictions," a regulator wrote, "most of this produce is consumed by the population. The rest is given over to the state for sale."[43] The too-radioactive food slipped into the food chain. It traveled widely to large cities and towns.[44] Sometimes farmers in contaminated areas purchased clean food that was delivered to them but then sold their homemade radioactive produce in neighboring communities where no one was checking for radiation.[45] The orders to monitor and modernize repeated about every six months and continued into the 1990s.[46] All those instructions to clean, replace, and rebuild, the reprimands again and again about the persistence of contaminated food, sounded out a sad, dull echo year after year.[47]

When I read those repetitive orders, I doubt the sincerity of the survival manuals and the people who wrote them. If Belarusian meat was too contaminated to eat, why didn't M. S. Mukharsky, the Soviet minister of agriculture, lower production quotas for contaminated provinces? Why didn't he supply more feed? Why in 1990 were people still living in areas measuring over a hundred curies of cesium-137 per square kilometer?[48] Why write up orders if lower-level managers had no choice but to disregard them?

There are several answers to these questions, none of them very satisfying. First, Soviet leaders had a habit of sending sunny reports to the top leadership about what was happening in the localities.[49] Party leaders knew they were being lied to, but they happily read reports affirming that the grand plans flowing from their pens were in motion. Second, if refrigerators and extra food supplies did make it to the at-risk regions, the district committee had a habit of divvying up the lot, handing out extra rations according to favors owed, not based on the needle ticking on the counter. Third, the USSR suffered from an agricultural crisis. In the 1970s, Brezhnev-era financiers took out loans to import wheat, feed, fertilizers, and other agricultural goods. In the 1980s, debt financing became painful, just as the cost of oil, the USSR's most lucrative export, fell. Feeding the nation ranked as a top priority, higher evidently than safety. After Chernobyl, Soviet markets only got hungrier.

Not that the postaccident measures made economic sense. It cost a million rubles to decontaminate a square kilometer.[50] Planners eventually spent millions more to build bathhouses, new clinics, schools, and kindergartens and lay down pavement and pipelines.[51] Spending the money instead on moving people out of contaminated territory would have been a saving. And not just in rubles but in exposures, anxiety, and human dignity.

Residents wrote thousands of collective letters to their leaders.[52] A group of villagers penned a sad appeal in the margins of an instruction pamphlet containing a long list of restrictions for their new life on contaminated ground. "There is no food in the stores, and we are not supposed to eat our own homegrown produce, yet we are banned from leaving. How are we supposed to live?" Steeped in the Soviet ethos of sacrifice, the petitioners were willing to give their lives for their country, but this kind of pointless offering mystified them: "In honor of what cause will we perish?"[53] One hundred thirty-two people pleaded: "We don't want to stand with a string bag in front of the store and wait for a shipment of clean food. We are farmers, yet we cannot feed ourselves. We want to be where we can live like human beings and see the fruits of our labor."[54]

As the people in the Chernobyl lands went about their lives, cautiously or not, they slowly shifted location without ever moving. What I mean is, as they ate and breathed and slept in linens that were ten times more radioactive than permissible, their biochemical composition changed. Picocurie by picocurie, they were becoming a part of reactor No. 4, the reactor that no longer was.

# PART III

// 

# MAN-MADE NATURE

# The Swamp Dweller

The stacks of documents I was gathering in the archives piled up. Many reports referred to the special qualities of the swamp and the swamp-dwelling Polesians. I found the landscape perplexing. I had trouble making out what was "nature" and what were hurried man-made transformations to it. Whether reading in archives or in scientific and engineering literature, I ran into roadblocks caused by a wall of secrecy, the descriptive shortcomings of people who created the documents, or the paucity of my imagination. I realized I needed a different kind of education, a very specific one in the rural culture of Polesia and the particular ecosystems of the Pripyat Marshes. I typed keywords into search engines in English, Russian, and Ukrainian and came up with little. Finally, in 2014, I boarded a bus heading north from Kyiv. The ticket was cheap, a few dollars for a ride on a faded, circa 1960 bus.

The bus was packed. We rode under a simmering July sun, airless and hot inside. I opened the window and counted to ten, waiting until the older riders touched their necks and sternly lectured me about the draft, as if it were a force of pure evil. I closed the window. Pushing aside the sooty tweed drapes, I watched the unregulated sprawl of Kyiv dissolve into suburbs and exurbs. The new houses of the newly rich and the middle-rich yielded to fields of lemon-yellow sunflowers intersected with winter rye of a laser green. Overhead the sky was a watery blue. The love of bright colors comes naturally in this part of the world. The farther north the bus heaved the number of houses, roadside taverns, and crates with berries for sale dwindled. I saw fewer kids skipping around the dusty roadside. As we neared the Chernobyl Zone, I made out the remnants of abandoned dairy and poultry farms in the distance. I watched a hawk circle an untended field, the forest invading its perimeter.

On that first trip, I ended up in Nedanchychi, a village that was not evacuated after the Chernobyl accident because it had low levels of contamination. In the 1980s, the village tottered on a fulcrum between existence and disappearance until residents, given the choice, voted against resettlement. In the years that followed, young families left on their own, part of a more general depopulation of the post-Soviet countryside. When I arrived, four elderly villagers remained. The oldest was Halia, who, when I first met her, was ninety-seven years old.

I returned to find Halia in the next two summers. As she had no phone, I could not warn her I was coming. Sometimes it was hard to locate Halia. I'd never met a near centenarian who kept so busy. I'd look for her in her cottage but find it empty. I'd head to the nearby town where she sold herbs at the market and search among the stalls of berries, dried mushrooms, and pungent-smelling pickles. I'd ask for Halia. The vendors said she'd already left. I'd head back to her cottage. Still no Halia. Her neighbor said she might be out in the forest.

Tired of searching, one day I sat down to wait on a log outside her back door. Halia's vegetable patch looked good. Beans, corn, yellow squash. Her cottage was less well tended. The door hung off-kilter from its hinges. The windows had no glass panes, just two sheets of plastic stapled to the frames. I tried to imagine the winter frost inside those makeshift windows. I plucked a few raspberries and watched the meadow. After a good wait, Halia's figure emerged from the forest, a bent frame supported by a walking stick, a bag on her shoulder, her long skirt blowing in the breeze. She was barefoot, her scarf pushed low on her forehead. As she got closer, Halia smiled at me broadly, toothlessly. I was never sure if she remembered me from year to year or how many details she made out through her bottle-thick glasses. She greeted me with a hug, as she did everyone. Halia was always happy to see a visitor.

She dragged out a chair for me and sat herself on a low plank, her knees tucked under. Halia heard my questions lucidly, then answered with a wave of her hand at the impossibility of expressing what she had experienced in a century of living.

I was fascinated by Halia's persistence. Except for a short stay in Kyiv, she had remained in her village for nearly a hundred years. Holding on in one place doesn't sound remarkable until you consider the violent history of the territory surrounding the Pripyat Marshes. When Halia was born in 1918, the marshes were a vast, watery landscape where seventeen rivers and streams met and entangled across a bowl-shaped swamp that extended for 167,000 square kilometers.[1] Floods made the territory impassable for months out of the year, especially so for strangers who did not know how to navigate the bogs and quicksand. The giant swamp disoriented invaders, slowing cavalry and tanks alike so that the major twentieth-century conflicts came to a standstill on the edges of the Pripyat Marshes and festered—World War I, the Russian Civil War, the peasant wars over collectivization, and World War II.

Three times I asked Halia to tell me her life story. Three times she narrated nearly the same account in her Polesian dialect. One day, as she repeated her biography, I had the feeling of levitating, as if I were floating up and over her cottage to survey the flat expanse of the marshes. As my imagination scaled higher, it mixed with what I had read in archives and libraries. I could see the village of Nedanchychi below, huddled in an elbow of the Dnipro River where the waterway sweeps along powerfully, dividing and bending, so that from the perspective of wartime reconnaissance photos, the river appeared less as a single artery and more as a hydraulic network of pumping capillaries.[2]

In stopgap motion, Halia's life spread before me as she talked. I saw her as a child in the 1920s going about her daily chores among the remnants of trenches dug by the tsarist army still visible a decade after World War I ended, strategic hillocks entangled with barbed wire, and bleached bones exposed on a sandy embankment.[3] Young Halia was up in the morning to tend to the animals in the shed. Out in the garden while the morning was cool, she hoed, sowed, and pulled weeds. As children do on the farm, she matured rapidly. She disappeared into and out of the forest, ranging surprisingly long distances. I can see her leading the calf to the riverbank, staking it, and searching the grassy shallows for greens to make a sour fermented soup. Sometimes children were with her. As time passed, she was often alone. Halia's peregrinations changed seasonally but over the century remained steadfast: to the pasture and back, to the forest and back, to the river, to town, to the village square and return. Over the seasons, across the decades, her perambulations remained surprisingly resistant. What changed was everything else around her.

In the Russian Civil War, the Whites fought a ragtag Bolshevik army surging up from the south. As the armies swept in, they quartered and requisitioned food, animals, and bodies. Children like Halia were kept close to home. In time, the soldiers withdrew. Ten years later, during the great famine of 1932–33, an emaciated Halia slowly trudged a long way, arriving finally at a bakery where a "Jewish" woman slipped her three loaves of bread for her family. As she walked home, Halia furtively devoured one loaf herself. Eight decades later, she still felt guilty about that loaf of bread.

Later as a teenager on the village square, she lifted her arms to dance, drawing her hands to her hips. A young man materialized at her side, on a sandy embankment, their limbs entangled. After a year of marriage, her husband was drafted into the Red Army and left for the World War II front. Halia remained with his parents. One day the roar of tanks interrupted the birdsong peace. A German officer in a black uniform commanded the women and children to the village square. They were searched, manhandled. Gunfire barked from the forest.

Later, the Germans burned the village. Halia stood in the smoking remnants, pulling her toddler toward her. With liberation, Red Army investigators arrived. They asked questions and took notes. As during the war, now after the war, people hung between poles in the square. In a letter, Halia learned her husband was killed in battle. She had two other "masters" after him, she told me, but he was the only one she loved.

After the war, the village was finally quiet. In the wartime reconnaissance photos taken over Halia's village, hydrologic forces worked like brushstrokes on the yielding terrain. The landscape swirled like Van Gogh's night sky—unfinished, raw, laid open. Within a decade, the tanks returned, retrofitted as bulldozers. From the ashes of burned villages, construction crews remastered the countryside so that it no longer followed the ebb and flow of seasonal floods. Like a hand snapping a rug straight, the magical machines shrank distances and modified terrain. The crews pulled the curling rivers into canals flowing in straight lines toward squared-off concrete dams. The canals drained the boggy land. In the newly girdled waterways, the big sturgeon no longer appeared. Commercial breeds replaced native fish. The water table dropped. The annual floods became less encompassing. As Soviet reformers toiled, the wild expressionism of the Pripyat Marshes transformed into the controlled geometry of Malevich: square fields superimposed over rectangular blocks.

And Halia was still there. Middle-aged, she walked to the market. Around her, the collective farm elongated, swelling across former swampland. Pastures silted up with thousands of cows, sheep, pigs, and goats. Solid brick structures took shape—a school, clubhouse, office building, and store. The good years of "planned farming" were aided by the farm-chemical service. New crops—sunflower, corn, rapeseed— sprang from the mineral-poor, sandy soils. The plants, enriched with chemical nitrates and phosphates, grew toward the sun as they never had before, but so too did fungi and insects. Planes dashed by trailing great clouds of sweet-smelling DDT. Small vehicles and buses appeared on freshly graded, gravel roads.[4]

Halia was almost always walking, often barefoot. As the years wore on, she ranged more widely for herbs and edible plants, having more trouble finding the water-loving plants on land dried by the network of dams and canals.[5] Birds that thrived in the swamps too were fewer. Near the new dams, the sweet, brown water crusted over with a green film of algae growing wildly on nitrate-enriched water. Midsummer the flowering heaths dried out and turned a sunburned bronze. Forest fires swept across stands of pine and birch. The infernos left sandy deserts in their wake.

Forty kilometers southwest from Halia's village, a massive construction site mounted. Crews toiled for a decade, draining 500 more acres of swamp. They heaved monumental concrete blocks onto reclaimed ground, girdled by ramparts of steel plates. As they dug, the builders unearthed spent shells, grenades, and the bodies of three World War II officers.

Finally, from enormous hourglass towers, a triumphal blast of hot steam poured forth. A city, the likes of which Polesia had never known, took shape next to the Chernobyl Nuclear Power Plant: tall apartment buildings and wide, straight boulevards lined with exotic flowers. Buses shifted nicely, like on a model train set, up and down the streets. No mud, no thorns, no swampy ground. People wore shoes and Sunday clothing every day. At a library meeting in Pripyat, a woman with an Auschwitz tattoo on her arm told young people: "We are building the first atomic station so that it brings us happiness and beauty, so that there are no more sirens and no more children's screams."[6]

The city alongside the Chernobyl plant existed for only a decade. Late on a warm spring night, the earth shuddered, and thunder rolled from a clear sky. The world might have stood still for a moment during the Chernobyl explosion. But it didn't. Fishermen kept their lines in the cooling pond as sparks lit up the horizon. That morning shepherds poked sticks at the backs of livestock as they led them to pasture. The animals hungrily tore at the sweet spring grass.

Two weeks after the accident, soldiers arrived. Halia's village was deemed clean enough to remain, but a commander ordered farmers in nearby villages to pack up. They pushed the family cow and goats into pens and began shooting.[7] Everyone cried. The soldiers, the tears, and stuffed carts reminded villagers of the war.

In the years following, Nedanchychi unraveled around Halia. A young couple packed up and left. Then another. A maintenance man shuttered the village school. At the medical clinic, the staff piled files into the trunk of a battered car and drove off. Halia and a few elderly neighbors passed as before, from house to garden, from garden to meadow, in and out of the forest.

When Halia stopped speaking, my imagination drifted back to ground level, to the dark kitchen where we had moved in the gathering twilight. For many observers, her longevity testifies to the fact that radiation can't be that bad; she lived through it. But any scientist will tell you that one person's health record is anecdotal. For me, knowing too well the dark violence of northern Ukraine in the twentieth century, her endurance for a century was just short of a miracle.[8]

"How did you do it?" I asked. "How did you survive all that?"

"Live." Halia leaned forward, looking at me suddenly sharply with her half-blind eyes. "Live! I just wanted to live, live."

//

I TOO FELT like I just wanted to live, standing in a bleak forest, sweating in a Tyvek suit and mosquito netting, while I fumbled to shut off the piercing alarm of my Geiger counter that was measuring 1,000 μSv/hr when measured at ground level. I was in the infamous Red Forest, which took the largest blast of radioactivity in the days following the 1986 explosions. Soon after the accident, the pines in the forest, very sensitive to radioactivity, turned red and died; hence, the name Red Forest. Over thirty years, the trees that were planted after the accident had worked as cleansing agents, taking up radioactive isotopes and storing them in their fibers. Normally the Red Forest sent off gamma rays measuring 50–100 μSv/hr, but a fire had swept through the forest the previous fall. The hot flames of the forest fire denatured the wood, turning it into gas and ash, and released radioactivity so that my Geiger counter was jumping in alarm.[9] And I with it. I just wanted to get through that long, sweaty afternoon as I followed Tim Mousseau and Anders Møller, two biologists who had been studying the ecology of the Chernobyl Zone since 2000. We were in the Red Forest to retrieve nylon bags of leaf litter the two scientists had left in the woods the previous fall.

Mousseau and Møller were testing a hunch. They had noticed as they walked through heavily contaminated stretches of the Chernobyl Zone that the ground sprang underneath them, the leaf litter exceptionally deep. They postulated that microbes, worms, larvae, and insects that normally break down organic matter were unable to work well in high levels of radiation (i.e., over 50 μSv/hr).[10] They stuffed leaves into 300 women's nylon knee socks and left them in places throughout the Zone where levels of radiation range over four orders of magnitude. The following spring, we came to pick them up. Later in their makeshift lab, they dried, weighed, and measured the radiation in the leaves.

I was trailing the scientists because I wanted to know more about radioactive decay in the ecology of the Pripyat Marshes. The two biologists used the Chernobyl Zone of Alienation as a living laboratory. Mousseau and Møller argue that the beauty of doing research there is that the Zone is not one uniformly radiating smudge on the planet. Chernobyl clouds missed some areas in the Zone, which today have naturally low levels of background radiation. Other spots are extremely radioactive, places like the Red Forest, where a body should not linger long.

Mousseau and Møller set up their research differently than most of the Western scientists who have worked on Chernobyl topics and who mostly engage in lab work. What happens in the wild, they wanted to know, when biological organisms are exposed to man-made ionizing radiation? This simple question is difficult to answer because of the innumerable variations on the landscape and within bodies of mice, voles, birds, and insects.[11] Many field scientists study one creature—for example, ants or barn swallows. By narrowing their field of vision they can try to account for the variables a natural habitat and genetics can serve up. In contrast, Mousseau and Møller are interested in most everything that they can count and measure that pops into view. Their field lab in an old house in the town of Chernobyl is a cluttered place: caged mice in the parlor, frozen butterflies in the aging fridge, mushrooms stacked in the hallway, fungi samples, sliced tree cores, and collections of invertebrates on the porch. When the two scientists notice something odd, they set out to study it to bring their observations beyond the anecdotal. Mousseau and Møller, in the words of the anthropologist Anna Tsing, appear to be trying to perfect the art of noticing.[12]

When he took his first walk in a forest of the Chernobyl Zone in 2000, Mousseau was surprised to notice something odd. His face was clean. He wasn't clearing spiderwebs from it. Spiders normally string webs between trees across a path so that a person walking down it gets a face full of sticky webbing. Why no webs? Mousseau went actively looking for spiders but found few. When he and Møller made a formal count over three years in seven hundred sites following line transects, they discovered that at low levels of radioactivity (at one hundred times greater than normal levels) the number of spiders decreased significantly.[13]

Next, the two scientists searched for fruit flies. For geneticists and evolutionary biologists, fruit flies are the jam to their occupational bread and butter. Fruit flies have giant chromosomes and reproduce quickly so they make perfect subjects to trace genetic mutations. In the Chernobyl Zone, the two researchers had trouble finding *Drosophila* (fruit flies). That too was startling. Most people have trouble getting *Drosophila* out of their kitchens in the summer. Fruit flies feed on fruit. Mousseau and Møller discovered that fruit trees—apple, pear, rowanberry, wild rose—in highly contaminated areas produced far less fruit. Wondering why, the biologists looked for pollinators that fertilize blossoms. They found few bees, butterflies, or dragonflies.[14] The pollinators, they realized, had been decimated by the release of radioactivity in soils where the insects lay eggs. Checking 898 points around the Zone, they found an average of a third of a bumblebee and half of a butterfly. Fewer pollinators meant less productive fruit trees. With less fruit, fruit-eating birds like thrushes and warblers suffered demographically and declined in number.[15] Frugivores, in turn, serve as seed spreaders. With a decline in frugivores, fewer fruit trees and shrubs took root and grew. And so it went. The team investigated nineteen villages in a fifteen-kilometer circle around the blown plant and found that just two apple trees had seeded in two decades after the 1986 explosion. The two trees exhibit the tapering end point of a cascade of extinction. The peril of a few species of small winged creatures magnified to threaten the entire surrounding ecosystem.

Noticing the absence of fruit flies led to Mousseau's "silent spring" moment. Rachel Carson's 1961 blockbuster, *Silent Spring*, documented that DDT, commonly sprayed to kill suburban insects, also decimated wildlife, especially birds, in communities across the United States. In her preface, Carson wrote that she noticed one day that the birds in her lush Washington, DC, suburb had gone silent. This simple act of observation sent her on a quest that led Carson to write the seminal text that inspired the American environmental movement.[16]

On a different trip, I observed the team of biologists, which included a group from a Finnish university, as they worked with voles they had trapped in parts of the Zone. After catching the micelike animals, the scientists measured and weighed them and ran them through a gamma-ray spectrometer to record their bodily levels of cesium-137. They then tagged the rodents with pea-sized crystal gamma counters and released them where they found them.

As we drove around the Zone, escorting the voles home, I practiced reading the postnuclear landscape by reckoning approximate radiation levels and matching my guess to my Geiger counter. It became a kind of game, if a dispiriting one. In the town of Chernobyl where we stayed in a makeshift hotel, the birds in June started to sing at 3 a.m. and increased in volume and intensity by sunrise to a riotous barrage. (The worst clouds of radioactivity bypassed the town and decontamination work keeps it at low levels of radioactivity so workers can live there in shifts while they manage the Zone.) In other places as we dropped off voles, I listened for birds and heard only a few shrill calls that were left unanswered. I'd check my counter to find relatively high rates of radioactive decay (30–40 µSv/hr). The composition of the forests differed from place to place. Pine trees are especially vulnerable to radioactive decay. Few pines grew in places with radioactivity measuring 40 µSv/hr and higher. When they did, they tended to have mutations.

In the Red Forest, most of the new growth was in the form of birch trees, which grow better than pines because they secrete radioactivity annually when they shed leaves. The pines that did root were more like shrubs than the straight, tall trees normally grown for board lumber. The floor of the Red Forest had little vegetation. The forests did not smell like forests with the smell of decomposition. The ground was littered with pine needles and fallen leaves that had not decomposed because the microbes, fungi, and insects that drive the process of decay also suffered from contamination. Mousseau and Møller noticed that even twenty years after the accident, trees in the Red Forest had scarcely decayed.[17] Decomposition, from death to life, is the foundational rule of forest science, a rule violated in the Red Forest. Physicists have been saying for a hundred years that time measured in unwavering increments, like seconds and years, is a human construct; that in reality time expands and contracts in unpredictable ways. If Rip Van Winkle had fallen asleep in the Red Forest and woken twenty years later, he would not have been able to tell how much time had passed. Historians seek to freeze-frame the past in order to replay it. I should have been happy to find a place where time had nearly stopped. Instead it filled me with dread.

As we delivered voles at the second drop-off spot and stepped into the forest, my dosimeter gave off a warning beep at 30–35 μSv/hr before rising to 42 μSv/hr. I did not relish the feeling of walking through woods knowing they were "hot," though I knew that by current radiation standards I was in relatively little danger unless I remained in that spot for months, ate and drank from it, and burrowed into the soil, as insects and voles did.

As we turned over the cages, the frightened voles made a dash for it. One vole with a handful of pups moved quickly to use the hay inside the cage to form a nest. We left the cage behind so she would not abandon her pups and headed back to the road where the dosimeter reading dropped in half. Because radionuclides concentrate in some biological organisms, the more plants and plant matter, the higher the readings. In contrast to everyday life, asphalt in the Zone is safer than the quiet forest.

Some biologists are critical of Mousseau and Møller's work because they say their experiments are not double-blind. Double-blind trials call for researchers to have no information that may influence their behavior while carrying out a trial. Obviously, scientists in the Chernobyl Zone know they are in a radioactive landscape and so, critics say, they look for evidence to confirm their preexisting ideas about the effect of radioactivity on biological organisms. My skittishness, for instance, under the field of a beeping monitor could manifest in skewed research results. Mousseau answers this charge by pointing to the mottled quality of contamination levels in the Zone. It would be impossible to have experimenters who did not know they were in the Chernobyl Zone, he says, but once there, scientists are ignorant of the extent of the surrounding field until the dosimeter tells them. Mousseau and Møller set up experiments where they gather information blindly. They take a census, for example, by first counting the number of butterflies or birds, and only then do they measure levels of radioactivity. A second biologist makes a second count, also blind to radiation levels, to check against the first census. They believe their science is sound. "Taking a bird census is not rocket science," Møller noted. "Not much can go wrong because so many people have done it for decades in Europe so we have validation of the method from a multitude of studies."[18]

I found in Soviet archives scientific studies from the 1980s that echo Mousseau and Møller. Brown frogs in Mogilev Province had six to nineteen times more frequent cell aberrations than controls. Farm animals developed anemia, hypertension, lung disorders, and hyperactive thyroids.[19] Animals and insects suffered a drop in fertility and more frequently their offspring had birth defects. Ukrainian scientists found that grain seeds localized radioactive alpha activity, and morphological changes occurred in a wide variety of trees. Plants, they discovered, had weak immunity against increased levels of radionuclides in soil.[20]

Even with this background evidence, rival scientists contested Mousseau and Møller's results. They contended that they did not take into account genetic diversity and that they needed more radiation readings to link their results with levels of radioactivity.[21] The pair of biologists adapted their studies to answer their critics and kept working. They published dozens of articles. Their vast body of work changed the opinions of scientists who were once critical of them.[22]

Yet, despite their findings, the idea that nature rights itself after man-made disasters is so seductive that scientists and journalists cycle back to it every decade or so. In 2015, the physicist James Smith made headlines by publishing a short letter stating that long-term census data revealed abundant wildlife populations in the Chernobyl Zone of Alienation.[23] The story went viral. Major media venues picked up Smith's two-page letter in an academic journal and repackaged it. For a few weeks, Smith became a media darling. Journalists pitted the "thriving Chernobyl Zone" story against the gloomier picture Mousseau and Møller present.

I contacted Jim Smith to ask if I could follow him on his next trip to the Zone. Though Smith had made many trips to the Chernobyl Zone, he replied he had no plans to visit in the near future. He was able to do much of his work from his desk. He did his work at his desk with data collected by Belarusian researchers or derived from camera traps and data his assistants set up around the Zone, so I met up with Smith, who had trained as a physicist, at a conference in Florida. I asked him about the records I had found in the archives. Smith asked me repeatedly about levels of radioactivity. With such a number, he could extrapolate damage from radioactivity to plants and animals. He did not need to go to the Zone. Computational studies combined with levels of radioactivity told him what he needed to know.[24]

The debate over whether the Zone is thriving is a nonstarter for Mousseau and Møller. "Every rock we turn over," Mousseau said, "we find damage." The evidence is etched in the ecosystem of the Zone, in the bodies of mice, the leaf litter of the forest floor, and the tumors that cloud the vision of barn swallows they catch. Møller noted that in twenty years of working in Chernobyl, he has encountered few other scientists. Perhaps because the work is bug-bitten, repetitive, and harmful to your health not many scientists have been willing to do it.

Mousseau and Møller are practicing the kind of science left to post-human landscapes, a science that is tedious and hazardous, as much as it is creative and invigorating. As they devise a new form of ecological literacy, their "living laboratory" is a remnant, pitted with deposits of heavy metals, chemical toxins, and radioactive waste distributed at a frenetic pace in the twentieth century.

Halia, meanwhile, was the survivor on a raft she rode through a life-time of storms. Until she died in December 2017, she was one of four survivors in her village still speaking in an indigenous dialect with a native knowledge of the forests and swamps. Her standing on that ground attests to her personal capacity to adapt and endure. At the same time, Halia's bent figure calls up the people who dropped around her, felled by the violence of a vision of technological perfection and national security.

# The Great Chernobyl Acceleration

Sleet slashed the car. A truck roared by, sending up a muddy spray. Aleksandr Komov drove blind for a few seconds, his scraping wipers struggling to clear a peephole in the windshield. For three years after the accident, he made the same trip, four hours north of the city of Rivne, to the swampy, wooded stretches of northern Ukraine. Komov was in charge from 1986 to 1990 of measuring radiation levels for Rivne Province, 300 kilometers west of Chernobyl. That job involved top-secret work. In the months following the accident, his team logged measurements and sent reports twice a day through a high-frequency connection to Kyiv.[1] They were required to include the original draft of the report because, for security reasons, data on radioactivity could not remain in the localities.

I went to see Komov because he grasped before most people how the environmentally specific qualities of the Pripyat Marshes magnified the Chernobyl disaster. He was the first person to sound the alarm that the usual, standardized computations of dose estimates had no traction in the boggy, mineral-poor marshes. Like most of the everyday heroes I came across researching this book, Komov was willing to engage in principled disobedience. When he typed his reports, he used four sheets of mimeograph paper instead of two. He broke the rules by keeping extra copies of his radiation logs for himself and his supervisor. He showed me a little orange school notebook where he entered the numbers. On top, someone had scrawled "For Office Use Only."

As he handed it to me, he laughed: "If this were still the Soviet Union, I could not show this to you." I leafed through the log. The daily record shows levels of radioactivity going up in the days after the accident and then dropping as iodine-131, with a short half-life, decayed.[2]

Komov explained to me, "They passed a law giving subsidies to people who lived in areas with over fifteen curies of cesium-137 per square kilometer. We didn't have those kinds of levels." He pointed to his notebook, "Our maximum was about ten curies, more often two or three curies."

Komov made trips to Polesian farming communities bordering the marshes. The local Polesians tended to be Pentecostalists. They had large families with ten or more children. They kept to themselves and lived from a few cows and pigs and small kitchen gardens. They foraged widely for berries, mushrooms, fish, and game. As Komov tested milk from family farms, he noticed that most samples were over the permissible limit. That was strange. With relatively low levels of radioactivity in local soils, how could 80–100 percent of the milk be too contaminated for consumption?

Komov remembered a book he had read while a student training to be a radiation monitor. He went to the library and checked out a copy. In 1974, the biophysicist Aleksandr Marei published the results of a four-year study that had an innocuous title, *Global Fallout of Cesium-137 and Man*. From 1966 to 1970, Marei and his team had traveled through the Pripyat Marshes, measuring radioactive cesium and strontium. In his publication that was censored by the Soviet military, Marei wrote the radiation they detected came from global fallout caused by U.S. nuclear bomb tests. That was only a partial truth. Soviets were also testing at the same time. In fact, in 1961–62, the two superpowers blasted the earth in a last-minute race to discharge bombs before the 1963 atmospheric nuclear test ban treaty went into effect. That last grand finale of radioactive fireworks emitted an eye-popping twelve billion curies of radioactive iodine into the Northern Hemisphere and left its mark on everything, even fine French wine.[3] One curie is equal to thirty-seven billion disintegrations per second; twelve billion curies is a lot of destructive power.

Marei's team found that the swampy, sandy soils of the Pripyat

Marshes were the most conducive of any soil type for transmitting radioactive isotopes into the food chain. Swamps in conditions of continual resaturation accumulate peaty soils that are rich in organic substances but poor in minerals. Plants searching for potassium, iodine, calcium, and sodium readily take up radioactive strontium, cesium, and iodine that mimic these minerals. Marei found that the indigenous berries, mushrooms, and herbs of the marshes showed a very high transfer coefficient of radioactive nuclides from soils to plants. His team also discovered that seasonal floods spread radioactive contaminants "in a mosaic pattern" to places where floodwaters surged. As the boggy soils delivered radionuclides to plants, grazing farm animals magnified radioactive elements in the milk they produced. For Marei, the pathway was clear: water, soil, plants, animals, milk, humans.[4]

Quizzing villagers, Marei discovered that swamp dwellers' diets consisted almost exclusively of wild game, berries, mushrooms, and milk—lots and lots of milk, for adults two liters a day. Nearly everything the villagers ate contained man-made radioactivity. Marei's team ran a thousand people through whole-body counters. The scientists recorded levels of cesium-137 in villagers that were ten to thirty times greater than the cesium-137 measured in people in nearby Minsk and Kyiv.[5] The ingested cesium-137 would still be evident, Marei wrote, in Polesians' bodies in the year 2000. Even so, Marei concluded cheerfully, 78 nCi (nanocuries) (~2.9 kBq) in a body was not dangerous. In the same years, American scientists measured 3,000 nCi (~110 kBq) in Alaskan Eskimos exposed to fallout from both U.S. and Soviet tests. The Americans, unbelievably in retrospect, showed no great alarm.[6] Marei concluded his study stating that Polesians required no protective measures. Only if cesium-137 in local soils escalated would they need protection. "And that scenario in our country," Marei brightly projected, "is highly unlikely."[7]

Marei's study ended in the same year the Soviet Ministry of Energy decided to locate the world's largest projected nuclear power plant in the Pripyat Marshes, the worst possible ecological choice, as Marei's study showed. Ukrainian party boss Volodymyr Shcherbytsky was

uncomfortable about this decision.[8] He did not trust physicists' cavalier attitude toward radioactive contamination. He had seen how they
responded when a small research reactor at Kyiv's Institute of Nuclear
Research suffered major accidents in 1968, 1969, and 1970. In 1970, several employees were rushed to Moscow's Hospital No. 6 with acute radiation exposures. One soon died. Another accident released forty curies
of radioactive iodine into the surrounding Kyiv neighborhood. The
institute's scientists did not follow up to find out what happened to their
misplaced curies. It took forty-five days for them to even notice the leak.
Nor did the staff know what to do with a mounting volume of radioactive wastewater. When the Ukrainian minister of health shut down the
experimental reactor until better safety measures were set up, Department of Defense officials in Moscow overruled him. The nation's defense
was at stake, they insisted; the polluting reactor must stay online.[9]

Shcherbytsky again watched in horror in 1972 when a team of scientists from a closed military research lab tried to use a nuclear bomb to
put out an underground gas fire in a pipeline near Kharkiv.[10] The gas fire
raged out of control for the better part of a year. Arriving to help, physicists from a top-secret bomb lab drilled a hole down two kilometers next
to the burning gas well and planted a 3.8 kiloton nuclear bomb in the
shaft. Soviet bomb designers had detonated peaceful nuclear explosions
(PNEs) in other parts of the USSR to smother gas fires.[11] They were confident that this secret "Operation Torch" would work.[12]

Soldiers showed up in surrounding villages to give notice that there
would be an explosion that morning. Villagers felt the earth rumble
violently while the ground beneath their feet liquefied. Underground,
the nuclear blast piled earth up and over the gas derrick and snuffed out
the flame. That was just what the physicists had planned. The fire was
finally extinguished.

That tableau held for twenty seconds.

And then something went awry. A scorching jet mixed with earth and
stone belched from the gas well and shot up improbably high. The blaze
rose higher than any skyscraper to pierce the summer sky. A minute
later, witnesses ducked from the force of a blisteringly hot shock wave.[13]

Radiation levels in nearby communities climbed to harmful levels. Soldiers raced in to temporarily evacuate several villages. As people piled into buses, radioactive soot floated down around them.[14]

The Ministry of Defense scientists did not tell Shcherbytsky much about this miscalculation, but he could guess. He was told that due to an "unfortunate circumstance" it would be better not to eat local produce for the rest of the summer. From Moscow, they sent the Ukrainian leadership special rations of clean food.[15] After that episode, Shcherbytsky tried and failed to halt the construction of additional reactors at Chernobyl and near other Ukrainian cities.

Once the Chernobyl plant started up, Shcherbytsky became yet more worried. The power station under normal operation was leaky, discharging 4,000 curies every twenty-four hours.[16] Soviet planners established a commercial fishery in the plant's cooling ponds. The fish grew year-round in the warm water, but KGB scientists found that the harvested fish contained strontium-90 above permissible levels. When the agents banned the sale of fish, the fishery manager sold the contaminated produce out the back door.[17] Chernobyl's director, Viktor Briukhanov, admitted he could not guarantee there would be no more releases into the pond. He already had his hands full with reactor No. 1 that was bedeviled by problems. Poorly welded fuel rods popped inside the reactor like corn kernels in hot oil.[18] They blew in September and October 1982 and again in 1983.[19] Each time, these "catastrophes" led to acutely exposed workers and lost electricity, and radioactive isotopes spilled into the surrounding forests. KGB scientists tracked hot radioactive particles to a village five kilometers from the plant. "If swallowed," the officer reported, "or breathed into lungs," the radioactive flecks "could have fatal consequences."[20]

From the start of civilian nuclear power in 1964, Soviet reactors suffered accidents every year that caused casualties, deaths, and radioactive emissions. Radioactive contamination became widespread, quotidian across the USSR. When in the 1980s a team of scientists selected a control group of citizens in Moscow for a Chernobyl study, they discovered the Muscovites in the randomly selected group registered doses on average of 100 mSv of radioactive cesium. Muscovites had somehow ingested or inhaled levels of radiation as high as Chernobyl cleanup workers.[21] Moscow medical doctors never solved the mystery of the Muscovites' whopping exposures, but it tells you something about the mantle of man-made radiation that carpeted Soviet society in the Cold War decades.

As Shcherbytsky wrote letters to the Central Committee in Moscow, opposing the plans to construct more reactors in Chernobyl, he had to tread lightly.[22] Nimbyism on the Soviet political landscape often translated into charges of nationalism. For any Ukrainian leader, being labeled a nationalist meant political death. Four decades after World War II, the memory of the Ukrainian national army fighting alongside Nazis still burned brightly, in large part because KGB propagandists kept that story alive as a way to sharpen their pursuit of political enemies they labeled "Ukrainian nationalists."[23]

Shcherbytsky's attempts to stop Chernobyl's expansion were futile. Major decisions on nuclear power were made in Moscow in the cloistered offices of the nuclear weapons industry. Republic leaders and national scientists had little input.[24] Like the RBMK reactor, built to produce electricity and plutonium for weapons, the Chernobyl power plant was scaled up not just for civilian but for military purposes as well. Chernobyl electricity fed the voracious energy needs of the secret Duga Radar system, located near the plant, to detect attacking nuclear missiles. Chernobyl plutonium could be deployed in weapons.[25]

Shcherbytsky was not the only one who worried. Behind closed doors, top nuclear scientists fretted about the safety of Soviet nuclear power plants. How much longer, they asked, could they squeeze luck from their accident-riddled network of RBMK power reactors? Coal, gas, and biofuels would be safer and cheaper, they murmured.[26] Their concerns went unheeded. Officials at the Ministry of Energy kept to their production targets, building more reactors, powering up capacity, and extending power lines. It was only after the Chernobyl accident that Soviet leaders lost enthusiasm for reactor construction. In 1987, they canceled plans for new nuclear power plants near Minsk and Odessa and in Azerbaijan, Georgia, and Moldova.[27]

In 1987, the radiation monitor, Aleksandr Komov, kept working to sound the alarm about Polesia. "Farmers in Kyiv and Zhytomyr

Provinces had subsidies and protective measures," he told me, "while our Polesians continued to live as if there'd been no accident." He sent his reports about dirty milk to Kyiv. They went unanswered. He went to Kyiv and knocked on doors. "They told me to find a place with fifteen curies/km$^2$ and they'd send a commission." He sent notes from Marei's study showing that fifteen curies in the black earth Ukrainian steppe could be safe but not in the Polesian swamps. And that is what mattered—not curies on the ground, but particular ecologies and doses that people ingested.[28]

In January 1988, Komov finally got the attention of Ukraine's chief sanitation doctor, who wrote a letter condemning the State Committee of Industrial Agriculture for failing to monitor dairies and meat factories in Polesia.[29] A commission arrived in April to take measurements. A brigade of doctors filed in to examine villagers. They found a lot of illness in the after-accident period, a doubling of the rate of birth defects, and a strikingly high incidence of infant mortality.[30] The large families in Polesia had estimated total doses two to ten times higher than considered safe. Anatoly Romanenko, Ukrainian minister of health, wrote about the "escalating radiological situation" in Rivne Province and requested that the Polesian regions be placed in the strict control zone with clean food and subsidies.[31] Nothing happened. Another year passed.

Komov and his team continued to collect milk samples and send reports. Finally, exasperated, he tried one last tactic. He loaded seven tons of radioactive milk onto a truck and sent it to Kyiv. "You test it," he told Kyiv officials, "and tell me if it is radioactive."[32] Komov's charts with scrolls of data had failed to get his message across. The canisters of milk, splashing ivory-white nourishment, finally drew Kyiv's attention. That winter, villagers were officially notified of the danger and qualified for subsidies to buy clean food.

//

I DECIDED I'D take a closer look at the swamp. The best approach into it was from the Belarusian side of the border. It wasn't so easy to get there. I traveled from Minsk to Gomel, from Gomel to the town of Ol'shany, where I bought a pair of rubber boots and climbed into a large-wheeled jeep that plowed past a control point into the Almany Swamp, the largest surviving remnant of the great Pripyat Marshes. A local forester, Ivan Gusin, was my guide.

This was a rare part of the marshes that had not been drained for agricultural purposes because in 1961 Soviet generals turned it into an air force bombing range. Exploring, we stopped at the former range headquarters. I scrambled up a rusting steel tower that generals used to observe pilots dropping ordnance on the boggy expanse. When creating the bombing range, Red Army officers evacuated ten villages to make way for the planned pyrotechnics. I met a man named Nikolai Tervonin who grew up in one of the villages moved to make way for the bombing range. I found him in his yard in Ol'shany, along a dried-up canal, weaving a rope from grass for his horse, which stood nearby twitching flies. Tervonin was born in a part of the swamp that belonged to Poland until 1939. He said they used to buy just salt and sugar. That was it. The rest they made themselves. I asked him how long his family had lived in the marshes. "I can't tell you exactly," he responded. "My grandfather and great-grandfather, for sure. How would I know beyond him? No one can tell you that."

Communities in the marshes date at least to the sixteenth century. Tsarist officials used the marshes as a destination for exile. They sent prisoners there in chains. A local collector showed me iron shackles he had fished out of the mud. A bog is an amazing natural archive: the low temperatures, highly acidic water, and lack of oxygen preserve relics. The collector had all kinds of wonders. He showed me a hollowed-out log used as both high chair and porta-potty for toddlers and another log turned into a beehive. He pointed to a hundred-kilogram anchor that he claimed came from a thousand-year-old Viking ship. "A whole vessel is down there," he said, "someone just has to retrieve it."

Tervonin lived in the marshes until 1961, when soldiers showed up and told residents they had to move. "They took our houses, barns, livestock," he said as he braided his rope, "and moved us here."[33] He missed the wetlands. He pointed to a field on the other side of his barn. "The swamp used to stop right there. Every spring it flooded. We would paddle to that field on canoes to cut the grass." Tervonin waved his hand.

"See how dry it is now. It's because of all those canals." I was surprised to learn I was standing on former marshland that used to become a seasonal lake during the spring melt.

Walking through the Almany Swamp was another story because it had not been "improved" with canals and dams to dry it out. The forester took me to a great stretch of bog, a former lake. Birds swooped down on the flowering heath. The largest spiderweb I've ever seen glistened like a disco ball with morning dew. Each step I took the earth rebounded in quaking waves. "Be careful," Gusin warned me. "Bogs don't like women." I wasn't much worried. I had been in bogs before and knew to put my arms out to stop my descent through the mat of plant matter and into the quagmire below.

Gusin drove the jeep to the site where Tervonin's village had once been. The cemetery was the only thing standing on a raised knoll, the highest ground for a mile. Large cavities pitted the site where fighter pilots had taken aim at the graves.

I noticed a strange, spindly pine tree growing from one bomb crater. It didn't look right. The pine needles were disorganized, curling in rounds rather than straight from the branch. A number of factors can cause trees to mutate, but pines are especially vulnerable to radioactive decay.[34] I'd seen plenty of pines like this one inside the Chernobyl Zone while following the biologists Tim Mousseau and Anders Møller. I asked Gusin how old he thought the tree was. About forty years, he estimated. That was ten years before Chernobyl, I pointed out. I asked him if he saw many trees like this. Gusin shook his head, looking a bit disturbed. There were a lot of other pines around. None were growing out of a bomb crater. None had mutations.

In the early nineties, after the USSR collapsed, people told secrets they had closely guarded for decades. In 1991, the politically well-connected Belarusian writer Ales Adamovich jotted in his diary that the Soviet army tested strategic nuclear weapons, the small battlefield variety, in the Pripyat bombing range.[35] I could not confirm the testing story in the archives because the records of Soviet nuclear weapons development, stored in Moscow, are off-limits to researchers. (That is the nature of state power. It can make the past go away, if it so deems.)

But the sickly pine was a clue that lent credence to Adamovich's story. I had some other evidence too. A Belarusian doctor, Valentina Drozd, told me she noticed an unusually high number of congenital malformations in the regions bordering the bombing range. The strange thing, she said, is that the spike occurred among people born before the Chernobyl accident.[36] I also noticed that the Belarusian government paid compensation to veterans who took part in military exercises using atomic weapons.[37] I wondered whether the biophysicist Marei carried out his study in the Pripyat Marshes because the air force detonated in the range small-scale, "strategic" nuclear weapons or weapons encased in depleted uranium. Maybe Marei truly went to the Pripyat Marshes to measure radioactive fallout from U.S. testing, but why there when radioactive fallout covered the entire Northern Hemisphere? Marei's specialty was nuclear accidents. Nearly everywhere in the USSR where a nuclear accident took place, Marei showed up.[38] His appearance in a village was an ominous sign, akin to the grim reaper knocking at the back gate.

More generally, thinking about that crooked tree, I was struck by the persistence of radioactive contaminants in this part of the world. Every territory on earth is host to polluting substances produced at a furious pace in the twentieth century, but here, in the midst of the marsh where I stood in rubber boots, which I did not much need because of the unusually hot, dry spring in a succession of hot, dry springs, I grasped the dramatic speed of the changes of the past half century. Until 1939 this place had been soundly cursed as lost for centuries to a primitive, unchanging existence.[39]

Staring at the crippled pine, I realized that the perforations of radioactive nuclides into the cellular structures of organisms of the swamp long predated the Chernobyl explosions and continued after the accident. Soviet propagandists and international agencies honed a public information campaign that repeatedly insisted the danger was over; the Chernobyl chapter was closed, but that is not quite right. Chernobyl was not a single event but was instead a point on a continuum; the radioactive contamination of Polesia lasted more than three decades. Chernobyl territory was already saturated with radioactive isotopes from atomic bomb tests before architects drew up plans for the nuclear power plant. And, after Chernobyl as before Chernobyl, the drumbeat of nuclear accidents continued at two dozen other Ukrainian nuclear power installations and missile sites.[40] Sixty-six nuclear accidents occurred in Ukraine alone in the year after Chernobyl blew. More nuclear mishaps transpired after the Soviet Union collapsed, including the fires in the Red Forest in 2017.[41]

Calling Chernobyl an "accident" is a broom that sweeps away the larger story. Conceiving of the events that contaminated the Pripyat Marshes as discrete occurrences blurs the fact that they are connected. Instead of an accident, Chernobyl might better be conceived of as an acceleration on a time line of destruction or as an exclamation point in a chain of toxic exposures that restructured the landscape, bodies, and politics.[42] Soviet political leaders shrank from this realization. To acknowledge the implications of the great nuclear acceleration after decades of assurances about cheap and "perfectly safe" nuclear energy, while detonating bombs in the atmosphere in the name of "nuclear deterrence," was too politically explosive. To hide it, as I describe in the next section, policy makers resorted to secrecy, censorship, counterespionage, and fabricated news.

# PART IV

—————— // ——————

# POST-APOCALYPSE
# POLITICS

# The Housekeeper

I n May 1988, foreign scientists, journalists, and the Soviet scientific elite converged in Kyiv for the first international conference on the medical consequences of the Chernobyl accident. Since 1986, scientists and politicians had called for a large-scale study of health effects caused by the disaster.[1] Everyone was eager to know the tally on human health. Soviet minister of health Evgeny Chazov gave the answer in a headline address while the cameras rolled: "Definitely," Chazov spoke into the microphone, "we can today be certain that there are no effects of the Chernobyl accident on human health."[2]

For the next few days, Soviet scientists elaborated this good news. Speaking in the dispassionate register of scientific discourse, endocrinologists presented graphs of the low levels of radioactivity they measured in children's thyroids. Pregnant women, they asserted, had no trouble after the accident.[3] Radiologists reported they discovered no milk, water, or food contaminated above permissible levels.[4] Officials described how Soviet medical personnel monitored people after the accident, not to deliver medical care (it wasn't needed) but to relieve apprehensions among nervous residents.[5] Soviet leaders did everything possible and sensible to protect the public: food testing, evacuation, agricultural innovations. The speaker's message was that science saved the day. "Even in the event of such a major accident," Moscow physicist Leonid Ilyin announced, "science-based measures allowed [us] to prevent radioactive material from spreading beyond the thirty-kilometer zone."[6]

While Soviet officials laid out these half-truths and bald-faced lies, foreign observers nodded their heads encouragingly. Hans Blix, general director of the International Atomic Energy Agency, applauded the Soviet response to the disaster and spoke of the need for nuclear power as a clean, renewable form of energy that did not emit earth-warming carbon gases.[7]

The delegates hummed on. Few people noticed a cleaning lady approach a foreign visitor, the Californian cancer doctor Robert Gale. The housekeeper had in her hand a bucket, a mop, and a small folder with papers. Before she managed to hand Gale the file, four men dressed identically with military-style haircuts fell on her. They deftly took her by the elbows, garnished her mop, and swept her quietly out the back service door.[8]

The cleaning lady was not a cleaning lady. She was a Kyiv physicist by the name of Natalia Lozytska. Acting alone, she disguised herself to slip past the guards into the restricted conference hall. The four KGB agents intercepted her as she was trying to send a message to the West, an SOS, to the effect that the Chernobyl accident had caused far more damage than Soviet authorities acknowledged.

Since the accident Lozytska had been tracking Chernobyl fallout with her husband, Vsevolod, an astrophysicist. She and her husband ran a solar telescope at Kyiv State University. Vsevolod served as the civil defense coordinator for the department. He had a radiation detector in case of a nuclear attack. Just after the accident, Vsevolod pulled out the DP-5A counter and floated the wand over the garden around the observatory. One spot had much higher levels of radiation than elsewhere. With a trowel the couple dug up the soil and placed it on a piece of paper. They separated radiating piles of dirt from clean ones until they had distilled one tiny particle the width of a human hair that measured 30 μSv/hr. Lozytska taped the particle onto a piece of paper, wrote down the location and date, and returned to measure the tiny particle daily. She and her husband found other powerfully radioactive specks in the grass, on the doormat of their apartment, on a blanket in bed. They taped the bits to paper and measured them.

As the particles decayed, Lozytska calculated the radionuclides they contained. Confronted with the media blackout on Chernobyl radiation levels, she used the tiny radioactive fragments to telescope onto the accident 170 kilometers to the north to figure out what was going on. She discovered that a whole spectrum of radioactivity existed in those globes taped to paper, a miniature, nuclear Whoville in each speck of dust.[9] From the particles, Lozytska deduced that the reactor did not blow up from a steam or chemical explosion, as Soviet authorities asserted. Rather, she determined there had been a localized nuclear blast in a thin layer of fissionable materials. Only a nuclear detonation would release the spectrum of radionuclides, she reasoned, that she found in the hot particles. Lozytska typed up her findings, but she could neither publish nor speak about them until the dissolution of the USSR in 1991. Instead, like any loyal citizen, she wrote letters describing her work to her leaders. She wrote letter after letter and received no answer. When at last the couple published their research in a small Russian-language journal, it was overlooked.[10]

Sometimes science takes time. Thirty years later in 2017, Lars-Erik De Geer, a Swedish nuclear physicist, came to a similar conclusion as Lozytska. He had directed the Swedish Defense Research Agency, whose mission was to enforce the 1963 ban on atmospheric nuclear testing. His team tracked radioactive emissions with sensitive detectors and followed global weather patterns to determine the origin of radioactive nuclides. On the Monday morning after the Saturday explosion at Chernobyl, De Geer was sick in bed with the flu. A call from his office got him up and into work. "In fifteen minutes," De Geer told me over the phone, "we did the analysis and knew it was a reactor accident."[11] Still, they could not determine what kind of accident it was, whether caused by a steam, chemical, or nuclear explosion.

Still tracking the story in 2010, De Geer learned that Russian scientists in the town of Cherepovets, a thousand kilometers from the Chernobyl plant, had detected Chernobyl-issued xenon isotopes in the days following the accident. That evidence astounded him. It showed a mysterious deviation from the known path of fallout. De Geer calculated that only a jet of radioactive gases ascending three kilometers into the air could have delivered xenon isotopes as far as Cherepovets. No steam explosion could send gases that high. De Geer computed that a nuclear explosion occurred in a part of the reactor where a two-meter-thick steel floor had vaporized in heat so tremendous that only a nuclear explosion could generate it. De Geer published his article in November 2017, resolving a dispute that had quietly played out for decades.[12] This news contradicted the long-standing assurances that reactors could not blow up like nuclear bombs.[13]

I located Lozytska in Kyiv at the address she had given in her letters to Communist Party leaders.[14] Her correspondence, written at a time when a person had to have a security clearance to check out books on radiation ecology from the library, impressed me. How did she know so much about the problems of the restricted disaster zone?

After her discovery of hot spots in Kyiv, Lozytska and her husband got in their car and drove north. They surreptitiously measured radiation levels (an illegal act) and talked to farmers and villagers. They recorded high counts of radioactivity and heard disturbing news of sore throats, nosebleeds, and dizzy and fainting children. In her letters, Lozytska pointed out that Soviet permissible limits included only one radioactive isotope—cesium-137—among dozens that were swirling in the air. Other isotopes were also perilous, she wrote. "A dissolved fraction of strontium-90 is ten times more harmful than cesium-137 in water or food. Plutonium is thousands of times more dangerous." She estimated that there were ten million hot particles in Kyiv alone. "With windy conditions," she emphasized, "particles can lodge in lungs and land on food." Referring to the special radiation-absorbing properties of the Pripyat Marshes, she warned: "There should be no production of meat and dairy in these areas. Children incorporate radioactive strontium in their bones seven to nine times more readily than adults."[15]

In the 1980s, Lozytska worked two jobs and was a mother of three young children. No one commissioned her to do this research; no one asked her opinion or provided her reports and data, which were classified. She served as a citizen scientist emboldened by Gorbachev's invitation for Soviets to criticize mistakes and participate more broadly in civil society. At night as her children slept, Lozytska read Soviet technical journals, looking for clues to piece together the complex of work on radiation medicine that scientists in closed research facilities had secretly incubated during the Cold War. She also studied her children with the close scrutiny of scientific observation. She noticed peculiar symptoms.

"After the accident," Lozytska told me over a Sunday lunch, "for a period my daughter could not get out of bed. When she stood up, she would fall right over." At the pediatric clinic, doctors could not or would not explain what was causing the problem. To get answers, Lozytska attended lectures at the medical institute and read about radiobiology. In contaminated towns, she collected snail and egg shells to test for radioactive strontium. She gathered information from local doctors.[16] And she wrote more letters: "Children in the Poliske district of Ukraine have changes in their blood-forming systems that are pre-leukemic," she typed. "Not even in towns, let alone [remote] villages, has it been possible to deliver adequate food supplies for children. The sooner people are resettled from those areas, the sooner residents of other provinces of Ukraine and the USSR will stop receiving food produced in radioactive zones," and in so doing, she concluded, "we will rescue our future."[17]

Speaking openly to the press about her concerns was out of the question. Nor could she remain silent. Lozytska resorted to a tactic deployed by Soviet dissidents for decades. If she could get a message through to the West, to Dr. Robert Gale, the famous doctor who came to the aid of Chernobyl firemen, then news of the dangers of the accident could be broadcast abroad on Voice of America or the BBC and be beamed back into the USSR. That was her idea when she disguised herself as a cleaning lady to steal into the conference hall without an invitation in the spring of 1988. Her plan failed. Without questioning her or threatening arrest, KGB agents quietly ejected her into the back alley. Lozytska picked up her bucket and took the tram to work, while the scientists at the conference stepped up to the podium and continued to deliver reassuring mistruths.

Looking back on it now, it is clear that Soviet officials staged the Kyiv conference on Chernobyl health effects to drum up support for a change of course. They were growing weary of the Chernobyl accident. With the price of oil low and foreign credit withdrawn, the Soviet economy limped at an anemic pace. Costly disasters had followed Chernobyl in quick succession: an ancient passenger ship sank in the Black Sea. Explosions rocked gas and coal mines in the Ukrainian Donbas.[18] Chemical factories suffered fires and leaks.[19] The war in Afghanistan was going poorly. Vast portions of the army proved corrupt and inept. A training brigade in northern Ukraine fired a practice missile that hit and destroyed an old woman's house.[20] Daily life also did not look good. Consumer products were in short supply. Ethnic groups clashed violently in the Caucasus. Burdened by the expense of Chernobyl, Soviet leaders sought finally to put the accident to bed. They wanted to end costly food subsidies and medical monitoring. At the Kyiv meeting, public officials presented their results to an international audience who they hoped would endorse their view that the Chernobyl chapter had closed.[21]

The invited guests were specially selected foreign experts in radiation medicine. In addition to Gale, the group included Pierre Pellerin, the head of France's nuclear watchdog agency, who was later accused in French courts of minimizing the effects of Chernobyl and concealing information. France had been late in developing nuclear weapons and in the 1980s French generals were still enthusiastically testing nuclear weapons in French Polynesia.[22] The American radiologist Lynn Anspaugh attended. He was on record saying Chernobyl health effects would be so minimal as to be undetectable.[23] Also in Kyiv was Clarence Lushbaugh, an American pathologist, who made a career in radiation-related studies at Manhattan Project sites. He had a cynical understanding of health studies: "Scientifically," he wrote in a letter to a colleague, "little 'useful' knowledge can be expected from such a study [of nuclear workers] because radiation doses have been so low." An investigation should be carried out anyway, he continued, because the negative results would reassure skittish nuclear workers, be useful in denying workmen's compensation claims, and would serve as "a counter-measure to the antinuclear propaganda." Lushbaugh understood that how a study was set up could easily determine results. He advocated blocking the nuclear workers' labor union from conducting its own study. "I believe," he wrote, "that a study designed to show the transgressions of management will usually succeed."[24]

Most of the American delegates to the Kyiv conference depended on funds from the U.S. Department of Energy (DOE), which was in charge of producing the American nuclear arsenal. At the time, worrisome reports had leaked out about how DOE scientists had exposed thousands of people in secret experiments during the first decades of the Cold War. Lushbaugh supervised a clinic at the Oak Ridge Associated Universities, where researchers covertly placed radioactive cesium and cobalt in the walls of hotel-like rooms and irradiated over two hundred patients who were suffering from leukemia, lymphoma, arthritis, and other ailments. After patients died, Lushbaugh carefully autopsied them.[25] Lushbaugh was not alone. Curious about the properties of radiation, American scientists during the Cold War sprayed neighborhoods with deadly radioactive material, injected and fed Americans radioactive substances, and zapped prisoners with neutrons. They carried out human radiation experiments with nearly every kind of American: whites, blacks, Latinos, children, pregnant women, patients, wards of hospitals, prisoners, students, scientists, and soldiers.[26] The researchers functioned behind a wall of secrecy and with generous public funding. Reflecting on that period, former Atomic Energy Commission (AEC) scientist John Gofman told an interviewer: "They gave these people a checkbook and a little wooden block with a rubber piece that said 'SECRET.' Think of the power!"[27]

In the mid-seventies, the secrecy, funding, and power began to deteriorate. Victims requested the release of records and sued for reparations. Reporters showed up on Lushbaugh's front lawn snapping photos and asking prying questions. Congress opened investigations. Lawsuits made their way to court. In the 1980s, scientists working in the closed halls of the Department of Energy felt under siege as the authority of their nuclear enterprises crumbled. Chernobyl caused an immediate drop in new orders for American power reactors.[28] Two headline-grabbing nuclear accidents in seven years—Three Mile Island and Chernobyl—undermined the claim that nuclear energy was perfectly safe. If another accident were to occur, IAEA director Hans Blix told his board of governors, "I fear the general public will no longer believe any contention that the risk of a severe accident was so small as to be almost negligible."[29]

Just a few months after Blix spoke those words, more nuclear accidents captured headlines. In October 1986, a fire broke out in a Soviet nuclear submarine equipped with ballistic nuclear missiles. The accident occurred a thousand kilometers northeast of Bermuda off the coast of Cape Hatteras, North Carolina.[30] The captain abandoned the ship and watched helplessly as the hatch holding the missiles broke open. The nuclear warheads descended slowly, fishtailing through glassy blue waters to rest six thousand meters below on the floor of the Hatteras Abyssal Plain. In 1987, in the Brazilian city of Goiania, two scavengers broke into an abandoned oncology clinic and stole a radioactive cobalt source from a radiotherapy machine. The scavengers pried open the lead capsule and found inside a powder emitting a wonderful blue glow. They shook the radioactive cobalt salt onto the floor of their hut. A child spread the glowing sand on her arms and delighted as her limbs sparkled. Neighbors crowded in to see the miracle substance. After the family started to feel sick, a garbage dealer bought the vial. He boarded a crowded tram and crossed the city, exposing hundreds more. The cobalt source held just one thousand curies of radiation but did a lot of damage. Robert Gale, in his new role as a radiation specialist, rushed off to Brazil, where four people died, including the child, two hundred people were hospitalized, and two villages were bulldozed.[31]

The career trajectories of the foreign scientists at the Kyiv conference suggested that they would not ask uncomfortable questions or raise objections. Though there is no reason to doubt the sincerity of their scientific opinions, these men had a pointed interest, given their past and the tarnished legacies of the agencies they represented, in endorsing the view that radioactivity in any but high doses caused no harm.[32] For the officials who had carefully curated the image of nuclear safety, the conference was an opportunity arising at an especially tense moment. If Soviet scientists could prove that large-scale exposures to "low" doses of Chernobyl radiation harmed only a few dozen firemen, then they could show that even the worst nuclear accident in human history had no effect on human health. And if that were true, then the fallout from nuclear testing, the seeping radioactive waste from bomb factories, the civilian reactors that daily emitted radioactivity, the widespread use of radiation in medical treatments, and the exposed bodies of workers, patients, and innocent bystanders in secret medical tests could be forgotten. At the meeting, the specially selected foreign scientists nodded their heads encouragingly as Soviet scientists delivered the headline that they had detected no health problems among the public who were exposed to Chernobyl radiation. That was very good news, indeed.

# KGB Suspicions

Soviet leaders had many reasons to downplay Chernobyl's consequences. A tempest was gathering in the USSR, a hurricane of political activity triggered by the Soviet First Premier, Mikhail Gorbachev. Unnerved by how little he had known about the catastrophic state of the Soviet nuclear industry, Gorbachev in the months after the accident stepped up his perestroika reform program. Gorbachev wanted more government transparency in order to resolve the country's economic crisis and reconstruct a crumbling, outdated infrastructure. He sought to prop up civil society so that citizens were free to voice criticisms and serve as watchdogs for corruption and waste. He encouraged journalists to engage in genuine investigative reporting. In advocating reform, Gorbachev pushed the Soviet monolith off its pedestal and got it rolling in unpredictable directions.

Gorbachev's venture was risky. Party leaders in the provinces were reluctant to upset their carefully managed police state. In Ukraine, the veteran Communist Party boss, Volodymyr Shcherbytsky, resisted perestroika at every turn.

In this contradictory climate, in 1987, Serhy Naboka, an ex-con, sentenced in 1981 for subversion, tried to organize a memorial to commemorate the first anniversary of the Chernobyl accident on Kyiv's Revolution Square (now Maidan Square).[1] Learning of the plan, KGB agents took "prophylactic measures" to prevent it. They ordered that city leaders stage a concert on that day on the square with plenty of loud dance music.[2] They planted news in the media about Western efforts to use green and pacifist causes for subversive, anti-Soviet activities. They visited dissidents and threatened them. And, just in case, KGB officers committed seventy-six people "prophylactically" to the psychiatric ward.[3] The precautionary measures worked. On the first Chernobyl anniversary, no protest took place.

But Gorbachev disturbed this carefully curated political order by issuing a mass amnesty of prisoners of conscience, the majority of whom were Ukrainians.[4] Oles Shevchenko returned to Kyiv from exile in Kazakhstan in the summer of 1987 after seven years' incarceration for human rights activism. He was on the list of people whom KGB agents considered especially dangerous because he had no job and so little to lose. In an interview, Shevchenko talked about the impact of the Chernobyl disaster on the dissident movement: "The atom, which blew the roof off the Chernobyl nuclear power plant, also blasted the foundation of the Communist system. Not just the doubters, but people who basically trusted their government saw how the authorities lied to them." Those mistruths, Shevchenko remembered, created "an atmosphere in which people lost faith. They sat silently, hesitating, angry at their government, but without the means to openly express themselves or to stand with a flag and unite with other people."[5]

Returning political exiles did not hesitate. They called each other and met up, not just in private apartments but also boldly on the street, in cinema lobbies, and most brazenly at the foot of the monument to Taras Shevchenko, the revered Ukrainian writer. Nervous about this tide of returning ex-cons, Ukrainian KGB tracked a growing camp of "Ukrainian nationalists" who worked as house painters, ditch diggers, and porters.[6] Using new perestroika laws, the ex-cons registered an "independent association" innocuously called the Ukrainian Cultural Club. Unlike Soviet dissidents before them, club members met openly in public venues. Anyone could join. Club members created groups focused on language, ecology, archeology, religion, Hari Krishna theology—"whatever anyone wanted," Shevchenko reminisced.[7]

Oleksandr Tkachuk, a university student at the time, attended the club's gatherings. He spoke about how the first meeting changed his life:

"In a totalitarian government where everything is regimented, including what you can say, when a person starts to speak differently, the effect is on the order of an explosion." For Tkachuk that moment came when Oles Shevchenko stood up and addressed the hall, not with the usual ideological "Dear Comrades!" but simply as "Ladies and Gentlemen."[8] That one phrase taught Tkachuk that personhood and belonging did not have to be regulated through a ruling Communist Party.

KGB agents circulated through the Ukrainian Cultural Club's meetings.[9] They noted who the leaders were. They dug up compromising information and fed it to a willing journalist, Oleksandr Shvets, who denounced the group as fascist extremists in the tradition of Ukrainian nationalists who allied with Nazi occupiers in the 1940s. The western half of Ukraine had been attached forcibly to the Soviet Union in 1939. For ten years after World War II, Ukrainian resistance fighters, aided by the CIA, fought a guerrilla war against Soviet forces in western Ukraine.[10] Because of this history, KGB agents were fixated on "Ukrainian nationalists" and the threat they believed they posed. Shvets named names, outing Shevchenko as a convicted felon and others as people with relatives in the Ukrainian wartime resistance. Shvets charged that the felons and fascists were pied pipers of slander and counterrevolution.[11]

KGB operatives reasoned that people who spoke about Chernobyl as a "disaster" were working for the CIA-supported Ukrainian diaspora abroad.[12] They dug through judicial records from World War II and supplied archival material on Ukrainian war criminals to Canadian journalist Sol Littman, who in the 1980s published the charges and successfully pressed the Canadian government to indict several Ukrainian émigrés.[13] KGB agents staged rallies demanding the extradition of accused Ukrainian fascists from Canada and the United States to stand trial in the Soviet Union. They cynically told diplomats the rallies were a genuine product of perestroika, "the result of democratization."[14]

Agents of the CIA and other foreign intelligence agencies, circulating undercover in the USSR, also trolled for compromising information. Foreign diplomats helped Soviet dissidents send manuscripts over the border, while émigrés broadcast this intelligence-supplied news on the BBC, Voice of America, and Radio Free Europe to Soviets and East Europeans. Radio journalists had trouble fact-checking these stories because many of their informants had to remain anonymous to protect their identities. The KGB and CIA competed quietly in this battle, planting news, managing public opinion, making careers, and ending them.

In 1988, perestroika's second year, the ex-cons at the Ukrainian Cultural Club tried again to commemorate the Chernobyl anniversary on Kyiv's central square at the same time that international journalists, scientists, and doctors like Robert Gale were assembling in Kyiv for the major conference on Chernobyl health effects. Hearing of the dissidents' plans for an unsanctioned protest, KGB agents got busy. They ordered municipal construction crews to cordon off the part of Revolution Square where the protest was to take place. Trucks and bulldozers amassed, erecting an informal roadblock. "We planned to meet in front of the clock tower," Oles Shevchenko remembered. "About fifty protesters showed up and pulled out banners. There were all these noisy trucks going." A couple of protesters shouted their prepared speeches above the roar of the machinery. Suddenly, a man with a megaphone ran up and gave an order: "Citizens with extremist views disperse!"

At that signal, several hundred young men with red arm bands emerged from the rush hour crowd. They grabbed the "Nuclear Free Ukraine" banner and flung it to the ground. Shevchenko scrambled to the top of a riser and read from an article of the Soviet constitution guaranteeing freedom of assembly. A plainclothes officer ripped it from his hand and tore the constitution to pieces. Paddy wagons pulled up. The men with arm bands drew out bully clubs and methodically beat the protesters. Shevchenko started to photograph the brawl. A man came up and warned him, "Keep taking pictures and you'll end up in a forest swinging from a rope." Shevchenko stowed away his camera. "Those were not idle threats," Tkachuk remembered. "We knew people who were beaten and murdered. They killed the artist Alla Horska with a hammer. The young poet Hryhory Tymenko just disappeared."[15]

Slugging and kicking, the police dragged the activists to the waiting vans and shoved them in. As the trucks swept away the protesters, a maintenance crew finished the operation by sweeping up the torn posters from the square. The first Chernobyl commemoration lasted no more than ten minutes.[16]

Like the efforts to liquidate Chernobyl radioactivity, this cleanup operation had only a fleeting impact. What happened next was nothing short of a miracle. Taken to the jail, the demonstrators were neither interrogated nor arraigned. Most were freed with a fine or booked for just fifteen days. That was all.

"We were ex-cons, political dissidents," Shevchenko exclaimed. "We were willing to sit in jail for years, but they gave us only fifteen days. That didn't scare us."[17]

The dissidents took the light sentence as a signal that perestroika wasn't just talk. After that first unsuccessful rally, alternative political movements gained a momentum that expanded as rapidly as the queues for groceries and winter boots in Kyiv's shopping district.

People grew bolder. Natalia Lozytska joined a group of scientists mobilizing around ecological issues. She went to meetings at the House of Culture. Cleaning up the environment was a suitably patriotic cause. The scientists supported Gorbachev's reform program as loyal, concerned citizens.[18] Lozytska held teach-ins on ecology and radiobiology. Her group received permission to stage a rally at Republic Stadium in the fall of 1988. KGB agents counted three thousand people in the bleachers, among them scientists, artists, intellectuals, civil servants, and party members. People held banners with the hopeful names of newly formed independent associations: Green World, Heritage, Community, and the Ukrainian Helsinki Committee. The speeches at the meeting strayed from ecology to human rights and politics. The science writer and medical doctor Yuri Shcherbak prodded his fellow protesters to resist new nuclear ventures in Ukraine.[19] He berated the official secrecy surrounding Chernobyl. A scientist from the Academy of Sciences said that people in Ukraine were living "inside a nuclear reactor," that Ukraine was on the precipice of ecological catastrophe.[20] Speakers called for the creation of an umbrella organization that would unite the various groups in a "national movement."[21]

A Ukrainian national front was exactly what security agents most feared. After the rally, KGB agents called in the writers who organized the meeting. Before perestroika that would have been enough to silence them, but not this time. KGB agents wrote nervously: "The organizers continued working up plans for a national front."[22]

The scale-up of people willing to openly express their political views took off in late 1988. All those kitchen conversations, hurried exchanges at work over lunch, and knowing nods suddenly erupted into the public sphere and caught just about everyone by surprise. Faced with a rising tide of political activism, KGB agents worked to counteract it. They sent operatives to infiltrate independent associations, planting people whose mission was to lead groups away from dissent.[23]

The KGB playbook was rich and experienced, but so too was the toolbox of dissidents who had scraped with KGB agents for years. Shevchenko grasped the KGB tactical advance as it took place. "Ex-cons ran most of the new political groups," he remembered. "When *Rukh*, a national front organization, emerged, we had a big fight with the KGB over who would control it."[24] That contest among activists who worked in the open and KGB operatives who functioned undercover created a climate of secrecy and distrust that ran through the perestroika years, magnifying and expanding to include not just politics but also the arts and science. These underground conflicts spilled over into debates on the medical consequences of Chernobyl as it emerged in 1989 from the state-imposed media blackout.

# PART V

//

# MEDICAL MYSTERIES

# Primary Evidence

I n the courtyard of Kyiv's All-Union Center for Radiation Medicine, the men's cigarettes flared up, one drag at a time. The patients sat like prizefighters on the backs of park benches and talked in low tones about their health problems. It was not masculine to discuss maladies, but the former liquidators broke code because they were angry. They told a film crew that their leaders called them up to battle invisible flames of radioactivity only to reward them later with contempt. The patients saved their most biting remarks for the Ukrainian minister of health: "Romanenko said we came here [the clinic] to shirk work and pretend to be sick."

"I went to the Chernobyl plant with a clean bill of health," thirty-something Nikolai Zhutkov picked up the thread. "Two years later doctors in Moscow diagnosed me with brain damage, injury to my heart, a damaged GI tract, full degradation of my bone base, and a rectal prolapse. What else could have caused that? I go to work and an hour later they have to call the ambulance."[1] Zhutkov sounded like a lot of Chernobyl-exposed people.[2] They rarely had just one disease but had instead a complex of illness swarming their bodies like a murder of crows.

Romanenko and most of the doctors at the Center for Radiation Medicine, which was controlled by the secretive Third Department of the Soviet Ministry of Health, did not acknowledge the liquidators' health complaints.[3] That was the general line from 1986 to 1989. Everyone was just fine. Romanenko reported regularly to the public on the massive medical monitoring of people in contaminated territories. He and his staff related that the medical brigades had found no change in statistical data among exposed people as compared to control groups.[4] After examination of 86,000 Chernobyl-exposed children, the Soviet minister of health in 1988 claimed that 80 percent were "healthy."[5]

I noticed as I read through reports over the years that Soviet health officials stayed on message, repeating these facts to the public until 1990, when they suddenly flipped and began to write internal reports and appeals abroad with the alarming message that one million Ukrainian citizens were at risk and that "the most dire health indicators are among people in controlled [heavily contaminated] regions." Those areas, they said, had become a "zone of catastrophe."[6] Fewer than half of the children in contaminated zones were healthy. They had suddenly reversed themselves: the same ministry, the same years, but different numbers.[7] I tried to sort it out. Somebody, at some point, was lying.

I tried to make sense of the contradictory evidence. The story of sick liquidators, fainting, listless children, and widespread public health problems emerged in the Soviet media starting in the winter of 1989, when perestroika blossomed almost overnight like crocuses springing from under the snow. In the hothouse of political fervor before the first relatively free elections in Soviet history, political aspirants linked the "crimes" of Chernobyl to their bid for office.[8] Chernobyl became a cause for all who wanted to denounce Soviet rule.[9] Opponents accused the activists of using a supposed health crisis to rattle the cup to win international aid. They also claimed Soviet doctors were poorly trained, had few diagnostic tools, and for a bribe would hand over a diagnosis of a Chernobyl illness that paid compensation. And it is true, doctors were taking bribes and public health officials did start "Chernobyl Children's" organizations in order to personally enrich themselves.[10]

Trying to sort these conflicting accounts, I contacted Alexander Klementiev, an epidemiologist who worked on the Chernobyl Registry in the late 1980s, a database set up to track health data among exposed people. Generally, Klementiev had a low opinion of Soviet epidemiology. "The system of collecting medical data," he said, "was interconnected with the political system." The Cold War superpower race included a global contest for the healthiest, most prosperous people. Proponents on each side asserted that socialism/capitalism produced the healthiest citizenry. That was the official story even if it wasn't true. In the 1930s, Soviet demographers reported a drop in the population after secret mass executions. Stalin grew so enraged over the census results that he ordered the execution of the demographers. In the 1980s, scientists would not be killed for truthful reporting, but it wasn't good for one's career to relay statistics that reflected poorly on the Soviet Union. Klementiev described supervisors who would override epidemiologists when they reported troublesome health data. He noted that regional sanitation doctors had latitude to massage unsightly facts into the proper profile of socialist progress.[11] The latitude was wide and it got out of hand. Gorbachev's demand for glasnost or "transparency" took aim at just this difficulty of distilling fact from fable in official records.

Political leaders lied to please their bosses so as to remain at their posts, while bosses knew that subordinates Photoshopped reality to satisfy them. Propaganda, in other words, was written into governance, into the warp and weave of daily life. Truckers hauling boxes of paper later shuttled the mistruths written into official records to archival repositories, where they were cataloged, filed, and ossified as "fact." Historians later unearth these documents and try to sort through a mist of facts and fabrication. That was my problem as I worked through thousands of pages in the Ukrainian Ministry of Health archives. Who was right? The people who at first said there were no health problems or the people who later said there were? And what about the individuals, like the Belarusian minister of health, who said both things?

What can a researcher do when facing this kind of controversy? In scientific debates over lead and tobacco, historians helped to solve impasses by showing how science was managed, mismanaged, or even deliberately falsified.[12] These histories pivot around what scientists knew and when they knew it. The question of timing also is important in the Chernobyl case. Before mid-1989, there was a nearly total ban in the Soviet media on Chernobyl health problems and information about levels of radioactivity. Officials could, however, write openly about health problems in documents stamped "for office use only." I figured

that, thanks to government censorship, records in classified correspondence between 1986 and mid-1989 could offer a glimpse of health consequences before Chernobyl was politicized and monetized. Because
public officials were encouraged to report only good health statistics,
if courageous bureaucrats violated this code by recording bad news, I
could give more credence to those reports.

I spread out files from the Ukrainian Ministry of Health and arranged
them in chronological order to make a time line. I could see that in 1987
Romanenko informed Evgeny Chazov, minister of health in Moscow,
that after screening people in contaminated regions, Ukrainian medical brigades logged an increase in diseases of the circulatory system and
thyroid, both organs known to be sensitive to ionizing radiation. Romanenko attributed this increase to better detection of disease among a
"[rural] population which had never before been investigated in such
a complete way." He also chalked up the spike in health problems to a
"psycho-emotional factor."[13]

A 1988 chart showing disease rates of 61,000 people with known
exposures to Chernobyl radioactivity in Kyiv Province categorized
only a fifth of children and a third of adults as "healthy."[14] Between 1985
and 1988, regional Ukrainian public health officials noticed in the most
contaminated regions of Kyiv Province an increase in thyroid and heart
disease, endocrine and GI tract disorders, anemia, and other maladies
of the blood-forming system. Doctors noticed a rise in autoimmune disorders. They watched the number of pediatric infections—tonsillitis,
chronic bronchitis, and pneumonia—climb with each reckoning.[15] They
reported a rise in cancers and childhood leukemia.[16] Leukemia, with a
short latency period, had a proven link to radiation. Four times more
workers in contaminated regions became invalids because of eye disease than in Ukraine generally.[17] Eyes, an external organ, are sensitive
to ionizing radiation.

Statistics on birth defects and pregnancies also did not look prom-
ising. Scientists knew that fetuses were especially vulnerable because
of their rapid rate of growth at critical moments in the formation of
organs. They also understood that mothers efficiently absorbed nutrients, including radioactive nuclides that mimicked nutrients, and chan-

neled them directly to offspring via the placenta. In two closely watched regions of Kyiv Province, the percentage of infants born with congenital malformations more than doubled from 10 percent to 23 percent between 1986 and 1988. Newborns were sick twice as frequently. The first reason for their illness was birth defects.[18]

Publicly officials trumpeted the fact that child mortality in Ukraine was on the decline.[19] That's true. Fewer children died in nearly every Ukrainian province except in heavily exposed regions, where child mortality levitated two to six times in 1987 and 1988 compared to 1986. Perinatal mortality (death after twenty-eight days) rose four times between 1985 and 1987. Sixteen percent of newborns in some fallout regions died within twenty-eight days of life.[20] Half of these deaths were stillborn, whereas the other half had congenital malformations "that were not compatible with life." With these prospects, many women did not have the courage to reproduce. An uncommonly high percentage of women, up to 75 percent, chose to terminate their pregnancies.[21]

Kyiv researchers kept a close eye on livestock in farms in the Narodychi and Chernobyl regions, areas that were heavily contaminated and easy to reach from Kyiv. The scientists found in the animals' lungs a pneumonitis similar to that which people develop after radiation therapy. They noticed that tumors grew one and a half times more frequently than in a control group of farm animals. A full 67 percent of tumors were malignant. Animals without tumors were sicker and died sooner than expected. Continual chronic low doses of radioactivity had a more serious impact on the animals' health than a single dose of the same amount of radioactivity. "In fact," the researchers summarized in 1988, "damage over a protracted period does not correspond with the [non-acute or low] dose, but looks like acute radiation symptoms." The researchers suggested the need to recalculate the method of extrapolating the effects of large doses of radioactivity to small doses. "The relationship might not be a straight line" or "linear."[22] This was bad news, indeed. Scientists had postulated for decades that low doses of radiation over long periods were less harmful per unit dose than one large blast.

Dying is a song the body plays, a song it eventually masters. The question is at what tempo and volume. Clearly, something was going on. In the three years after the Chernobyl accident, disease rates rose and the pace of fatalities quickened. Even Romanenko was forced to publicly acknowledge that fact. The question that persisted was whether Chernobyl was to blame. Radiation damage is hard to isolate and detect because it causes no new, stand-alone illnesses.

I wasn't sure. I kept looking. I went to provincial archives in Chernihiv and Zhytomyr and noticed a trend. Local doctors, who were not experts in radiation medicine, saw patterns in their regions of rising rates of disease in five general categories, though the weight and balance of morbidity changed from place to place. Children, infants, and pregnant women showed a yet greater intensification in the rates of disease than adults. Farther away, in Moscow and Kyiv, experts in radiation medicine who looked at larger aggregate numbers recognized the increased frequency of disease, but they dismissed the connection to Chernobyl because the illnesses did not correlate with established radiation risk estimates in charts that drove the science of radiation medicine. They were the experts. The local doctors were closer to the disaster. I kept coming to the same impasse. How was I to know which way the archives lied?

# Declassifying Disaster

What was it like to live in a community saturated with illness? Fast-moving medical brigades landed in communities, scanned a hundred patients a day, five hundred in a week, and moved on. They did not linger on personal stories.[1] I looked for ways to understand the statistics in the archives. I came across a film that provided one of the first glimpses at street level of people living through the accident's aftermath. The film eventually broke the Chernobyl story wide open, triggering a hailstorm of protest.

The documentary film *Mi-kro-fon* opens with a TV clip of a Ukrainian party boss uttering soothing assurances, then cuts to a village in Narodychi Region where a farm woman wrangles a squirming piglet. She turns the animal over to show its disfigured eye. The grotesque porcine oculus stares blindly at the camera while farmers recount the uptick of birth defects among farm animals. Other women angrily list their children's health problems and the fact that the doctors tell them to stop having kids. "I'm only thirty," a woman exclaimed. "How can a doctor say that to me?" The last scene of the film takes place at a large gathering at Republic Stadium in Kyiv. As the speakers criticize their leaders for glossing over the catastrophe, the microphones suddenly cut out. The angry crowd, daring the KGB agents to turn the loudspeakers back on, shouts in unison, "Mi-kro-fon! Mi-kro-fon!"

Most every word in *Mi-kro-fon* violated the existing censorship code.[2]

Film director Heorhy Shkliarevsky didn't stand a chance of seeing his documentary released, but somehow it passed the censors.

I sought out Shkliarevsky to learn how that happened. "We went up there to Narodychi," Shkliarevsky, in his late seventies, explained to me from a wheelchair in his Kyiv apartment. "We turned on a radiation meter and the needle went nuts. Really high numbers. And people were living there! We saw that, and the crew, together, we made a pact. We'd film this story like we saw it, with no commentary, and we'd cut none of it out."[3]

Shkliarevsky described how he submitted his film to local censors, as always, but the film was too hot to touch. The director of the Ukrainian State Film Bureau called him personally and told him to take his canister to Moscow for approval. "Let them decide."

Shkliarevsky was scheduled for a viewing the next day, in Moscow, a thirty-hour train ride away. Shkliarevsky dashed to the station and spent a night bouncing on a wooden bench. He arrived late at Goskino Studios, the Soviet Hollywood in a Moscow suburb.

Just as Shkliarevsky opened the door to the small in-house theater, another Ukrainian filmmaker, Rollan Serhienko, was being propelled out of it. The Goskino film board had just rejected Serhienko's film *Threshold*, also about Chernobyl's health effects on cleanup workers and children. Serhienko had been sharply reprimanded for his first film on Chernobyl, *The Bell* (1987), which was released after censors demanded he make major changes.[4] Serhienko's second film, also critical of the state, was no more politically correct. A secretary went off to shelve the rejected documentary in a warehouse with hundreds of other forbidden films. Shkliarevsky stepped into the auditorium. Naturally, he expected the same treatment as the projector started rolling.

When the lights went up, Shkliarevsky watched the jury members silently inspect their fingernails as they waited to see which way the political winds would blow. Finally, a doctor from the Moscow Institute of Biophysics stood up. He criticized the film's suggestion that radioactivity was causing a health crisis. A second man challenged the doctor. The two started to argue. They went out in the hall and began to punch each other. Shkliarevsky observed his first insider glimpse of the top echelons of the Soviet culture industry in astonishment. Unable to decide, the film board passed Shkliarevsky's film up the hierarchy to Yegor Ligachev, the conservative chief propaganda officer of Gorbachev's Politburo. Ligachev was known as the watchdog of communist idealism, a man who spoke in wooden tones and took few chances.

"That was it," Shkliarevsky said. "I figured the documentary would sit on a shelf forever."

Strangely, Ligachev approved *Mi-kro-fon*'s release. As Shkliarevsky tells it, Gorbachev wanted to clean house in Ukraine, starting with the party first secretary, Volodymyr Shcherbytsky, known to be obstinately resistant to reform. The documentary, critical of the Ukrainian leadership, could work like a land mine planted under Shcherbytsky's desk. The timing of *Mi-kro-fon*'s release coincided with Gorbachev's visit to Kyiv.[5] Gorbachev met with prominent Ukrainian writers in Shcherbytsky's office. They complained about the leadership of the Ukrainian Communist Party and how party leaders censored intellectual life, stifling glasnost and perestroika. Gorbachev reportedly turned to Shcherbytsky, skulking at his desk. "What is wrong, Vladimir Vasilevich? People just want to work for perestroika."[6]

Shcherbytsky offered his resignation the next day. Gorbachev told him not to rush, but the Ukrainian party boss was on notice. After that, Shcherbytsky's grasp on power slipped and he was forced to allow the writers wider latitude to organize political opposition.

No one expected uncensored reporting on Chernobyl in the USSR, so when Shkliarevsky's film appeared in February 1989, it became an overnight sensation. Working in the Soviet Union at the time, I attended elite film festivals and theater productions and believed I was witnessing the emergence of a free and independent media. Looking back on it now, I see it differently. Shkliarevsky had to submit his documentary to three censoring bodies. His film passed, while many others, like Serhienko's, were rejected and appeared in public only at the end of the USSR, although his film was no more critical than Shkliarevsky's. Authorities arbitrarily or with political strategy in mind decided what appeared in public and what didn't. Gorbachev's reforms gave oppositional voices a platform, but Gorbachev and his allies also used glasnost as a tool for

renewing the leadership. Media reform gave them a club to beat opponents while winning propaganda points abroad for apparently supporting independent thought. Regardless of motives, the appearance of the
first critical news about Chernobyl rippled across the Chernobyl territories and triggered a torrent of citizens' letters to party leaders.[7]

"Only a bureaucrat could divide a territory into 'clean' and 'unclean,'"
farmers from the Gorky Collective Farm in Narodychi Region wrote.
"Our food is contaminated. It exceeds the permissible norm two or
three times over, but we have to eat all that we grow. Our children are
often sick. They miss school. Almost all the children have enlarged
thyroids. Many have enlarged livers. Some have heart problems. More
and more adults have oncological problems. Fourteen people were
diagnosed with cancer in just the last two months." The villagers complained that because their village was not considered highly contaminated, the Kyiv doctors would not examine their children and they did
not get food subsidies. "We know that the number of sick children in
our village is greater than children in controlled areas because they get
clean food and we do not." The villagers concluded, "We are left alone
in this tragedy."[8]

Parents in another "clean" village, Rudnia Radovel'ska, went on
strike. They refused to send their children to school until a commission of doctors came to their village. A local party official wrote in panic
about the region: "Medical exams have shown that there are practically
no healthy children at all. Several times the parents demanded a bus
for a group trip to the Central Committee [in Kyiv] to solve these problems. We have held them back every way we can, but I have been to those
villages and I know what it is like. As chair of the executive committee
and as a woman, I can no longer witness the tears of mothers and even
fathers who see their children withering away in front of them."[9]

I read many such petitions. They are particular in their way, and they
are the same. Townspeople were angry about a power plant burning contaminated peat that emitted cesium-137.[10] Women prisoners who were
press-ganged into working inside the Zone cited environmental safety
laws and demanded competent medical care.[11] Everyone was incensed

that they had been living in a radioactive stew for three years with no real information about what that meant.

I wondered how the people who wrote those petitions would view the accident thirty years later. I stuffed several dozen letters in a file, rented a car, and with Olha Martynyuk, my research assistant, headed north from Kyiv to see if we could find anyone on the long lists of names at the bottom of each group petition. We rolled into the village of Noryntsi, the administrative center of what had been a large collective farm in Narodychi Region. There is not much in the way of bucolic attraction to a collective farm settlement. It is the rural equivalent to an American industrial park, a place of surveyed lines and standard-issue buildings dedicated to production. We walked up the straight, wide street, faced off by tall perimeter fences in front of farmers' houses. We knocked at a gate.

A woman looked at her signature on the petition in surprise. She didn't remember the letter or signing it. A farmer, a former cowherd, glanced at the Xerox copy and found his signature, but he wanted to talk about the collective farm, how it had once been a rich and powerful economic force, and now a foreign concern was running it to ruin. The farm used to have seven thousand head of livestock, he said. They once sent tons of milk and meat to Soviet markets plus flax, rye, potatoes, and seed. The sight of cow barns falling into the field in front of us was a raw reminder of that lost productivity.

We talked to the farm's former veterinarian. She remembered the letter. She had a sad, slow way about her. Standing at her gate, she asked reluctantly: "What do you want to know?" She looked like she didn't feel in the way of saying much. "What more is there to tell?"

There was a time in the perestroika years when Chernobyl was all villagers could talk about. They spilled forth their testimonies in letters, interviews, and meetings. They traded information with each other and made alliances with sympathetic officials and local doctors who slipped data to them. They asked questions and scoured newspapers and bulletin boards for information. They talked and talked, voices rising, crowds gathering as they tried to figure out what was happening.[12]

I asked the veterinarian about the widely reported accounts of animals born with birth defects in Narodychi.[13] She waved a hand. She wanted to talk about her own health instead.

She said a day didn't go by without pain. A sore throat started soon after the accident followed by headaches and dizziness. "And the symptoms haven't really ended," she added. "I still have a headache. My joints hurt. They are full of strontium." This was the same phrase used by the women in the village Kulykivka, and at the wool factory in Chernihiv, and from the woman on the street who gave us directions. The veterinarian said her daughter had a heart attack at thirty-four years old and her son, age twenty-eight, suffers from debilitating joint pain. A second daughter who had chronic asthma could not handle the smog of Kyiv, so she moved back to the village. The woman said she had wanted to leave Noryntsi, but her husband voted to stay. She told us she has had six operations. In one of them, she lost half her intestines. Now, she said, "almost everything is hard to eat."

The veterinarian's neighbor, a woman with a scar from thyroid surgery ringing her neck, said that she too had wanted to move away and had gone searching for a new place to resettle, but no place was as nice as their home with their river and their forest.

I'd seen that forest of pines, towering trees as quiet as a library. On the northeastern border lining the road, the red-bark pines did not grow straight as they were planted to do to become lumber. The trees split and the branches shot out like zigzag bolts of lightning. There are strange groves like this throughout the Chernobyl territories that carry the imprint of the passing radioactive clouds. I kept turning to look at those luminous, crooked limbs, impressed by the fact that trees, unlike archival records and the people who wrote them, don't lie.

"We are concerned," residents of the town of Malyn wrote, "that in the fourth year after the accident the number of children and pregnant women who are sick has grown. In 1985, 72 children suffered from anemia. In 1989, we have 1,430; that is 11.3 percent of all the children in the region." They pointed to the problem: "A steady stream of trucks from the Chernobyl Zone pass through our town. They bring in contamination and with it an increase in illnesses of the respiratory tract."

The residents of Malyn wanted sensible protections, for example, a bypass road for polluted trucks, gas instead of local wood for heating, personal dosimeters, clean produce, more soap, and a regulated radio-active waste repository. They also wanted Romanenko, Shcherbytsky, and other responsible officials to be brought to justice.[14]

Soviet bureaucrats took letters from the public seriously.[15] In 1990, Kyiv sent a commission to Malyn. Commissioners found the central hospital and clinics to be severely short of staff. They confirmed that rates of anemia and enlarged thyroids had skyrocketed; they gave the same numbers as in the citizens' letter. They added that three children had died of leukemia, a formerly "sporadic" disease, recorded in the region only once before, in 1984. The commission found that cancers among adults had doubled; 85 percent of births were "pathological" and 35 percent ended in spontaneous abortions. The commissioners summarized that the increased frequency of disease was due to better recordkeeping and more examinations.[16] With that, the commissioners went home. Case closed.

OLHA AND I kept driving on our quest to find the voices in the letters logged in the archive. We turned west and north from Noryntsi toward Polesia and Rivne Province. We stopped in Rudnia Radovel'ska, the village that staged a school boycott in 1989.

Now this was a proper Ukrainian village, not a planned collective farm conglomeration, a pretty spot, elevated above the plain with a panoramic vista. Best we could tell four people remained in Rudnia. The rest had moved away or died. We met first a middle-aged woman and her son, who was jobless, mute, and nursing a hangover. She invited us in and offered us, with little conviction, blueberry dumplings. We turned them down. Our hostess had little to say. The sadness seeped from her like a tincture. She said she took care of a bedridden woman, her neighbor, down the street. She walked us through the yard and handed us over to the fourth resident of Rudnia, Galina.

Galina had a sunburned face, a smoker's raspy voice, and a surprisingly urban look for a villager. She could have been a milk-crate merchant, hawking Chinese jewelry in the Kyiv subway. Instead, she traded in wild berries and mushrooms she picked in the surrounding forests. She wanted to tell us how she lost money because she had a radioactive hot spot in her yard right where the buyer measured her berries before purchasing them. The higher the needle went, the lower the prices she got for her goods. It was the ground, she complained, that was radioactive, not her berries. I pulled out my Geiger counter. The gamma rays from that spot were only a little above background.

"Maybe the berries really are radioactive," I said and added unhelpfully, "maybe you shouldn't be eating them."

She grew angry. "I'm fifty-two years old. I make $25 a day selling berries. Where else can I earn that kind of money at my age?"

Like everyone else we met on that trip, Galina didn't want to talk about Chernobyl. "Okay, say the mushrooms have Chernobyl," she conceded. "We still pick and eat them. And we don't look. We don't pay attention to where the radiation is. We eat everything with no limits. You go to the market and you hear, 'Oh Chernobyl this, Chernobyl that,' but we have no Chernobyl here."

She looked up at me, swatting away a fly and challenging me to contradict her. "I work," she said as she pointed around her, "I live. I carry on."

Olha and I stared back, not knowing what to say.

Galina changed the topic abruptly and told us she'd had a heart attack and two "women's cancers." She brooded for a bit and then added, "Well, maybe there is Chernobyl here, but it's all ours. A person adapts."

Driving away, I remember thinking that it was such a beautiful spot, and the living there was so lousy.

But maybe I had it wrong. Galina was saying that what mattered was the surviving. That's the victory: to be the last to hold on to that two-hundred-year-old village. Galina had it—that something that could not be extinguished by radioactive fallout or a string of fearsome diseases. An indefatigable human will. I can pay tribute to that.

# The Superpower
# Self-Help Initiative

When officials at the Ministry of Health were finally forced to admit that rates of disease were climbing in places where people were exposed, they grasped for ways to explain it. Romanenko and his colleagues might have returned to the manuals issued after the accident describing symptoms from exposures to chronic doses of radiation, but officials at the Soviet Ministry of Health had since banned the diagnosis of chronic radiation syndrome and replaced it with the less revealing, unspecific malady called "vegeto-vascular dystonia."[1] Soviet doctors might have turned to studies of thousands of people contaminated with radioactive waste near the Mayak plutonium plant in the Urals, where Soviet scientists had studied survivors for forty years, nearly as long as their Western counterparts had investigated Japanese bomb survivors. In these studies involving over 40,000 people, they found significantly increased death rates, cancer incidence two to three times greater than in Japanese survivors, a significant rise in complications at birth, autoimmune disorders, and other diseases. They learned too of the extreme vulnerability of children exposed to radiation.[2] Instead of turning to this work, the secretive Soviet realm of radiation medicine looked outward for answers.

In months when the first exchanges across the Iron Curtain were gaining momentum and Soviets were traveling abroad in greater numbers, Soviet health officials shifted westward, toward a newfound admiration of American and West European radiation medicine. They sought information about the U.S.-sponsored Life Span Study of Japanese bomb survivors.[3] They made contact with their chief adversary in the nuclear arms race, the U.S. Department of Energy.[4] As Romanenko and his colleagues developed relations with international scientists, they armed themselves with the Westerners' much sunnier understanding of radiation's effect on human health.

The Life Span Study was the largest epidemiological study in radiation medicine. Because of its scale of 200,000 subjects (survivors and offspring), it was considered a very reliable source for effects of radioactivity on humans. Even so, the Life Span Study had critics. The British epidemiologist Alice Stewart first ran afoul of it when she discovered in 1956 that X-raying pregnant women doubled the chance of leukemia in the children who were born afterward. Her finding contradicted the Japanese data, which had discovered no excess cancers in children exposed in utero. Stewart's work was heavily criticized. She was attacked and defunded for two decades.[5] Continuing to research and answering her powerful critics point for point, Stewart finally won the battle, and doctors stopped X-raying pregnant women in the 1970s. That dispute, however, concealed another important discovery Stewart had made. She had also found that children exposed in utero were three hundred times more sensitive to infections than unexposed children. She pointed out that the Japanese survivor study, initiated five years after the bombs fell, would have missed those illnesses, the deaths they caused, as well as stillbirths, spontaneous abortions, and miscarriages (known effects of radiation exposure) because no one was counting them in Nagasaki and Hiroshima between 1945 and 1950.[6]

As they began to collaborate with the Soviets, American officials looked nervously at Chernobyl. At that time, several lawsuits and independent investigations into America's nuclear legacy were gathering speed. Plaintiffs' lawyers carefully watched news on Chernobyl as they prepared their cases in U.S. courts.[7] The U.S. Department of Energy (DOE) was undergoing its own perestroika as protesters asked for records about the release of radioactive emissions from U.S. bomb production.

Moreover, the DOE had an ongoing dispute over the health of American nuclear workers. In 1965, the Atomic Energy Commission (AEC) hired Thomas Mancuso to study the health of 500,000 American bomb workers. Mancuso was considered one of the best in his field. After several years, AEC officials pressed Mancuso to make a statement that he had found no negative impact on workers' health. Mancuso refused and asked for more time because he was having trouble obtaining many of the workers' records he sought, due to obstruction from local AEC offices. In 1977, the AEC suddenly terminated Mancuso's contract and scientists in Oak Ridge Associated Universities maneuvered to take over the study, though they had no principal investigator to direct it.[8]

Undaunted, Mancuso used his retirement funds to keep working on the worker study. Alice Stewart joined Mancuso and they examined workers' health records at the Hanford Plutonium Plant. They found more cancers than predicted, "an excess," at doses "known to be too low" according to the Japanese bomb survivor data.[9] Stewart and Mancuso asked for more records and were refused access. The pair went to court to secure the release of workers' health data. Clarence Lushbaugh at Oak Ridge ordered the records Mancuso had collected up to that point destroyed. Lushbaugh was grateful that a new privacy act enabled the Department of Energy, successor to the AEC, to withhold bomb workers' records from any party, including other government agencies.[10]

This was the background in which Soviet and American scientists quietly became partners on Chernobyl research.[11] Both nuclear superpowers had a lot at stake: liability and lawsuits and the moral and financial burden of millions of exposed citizens. An unholy alliance was taking shape.

To put it another way, Romanenko was starting to get some help from abroad. The mounting list of diseases in the contaminated regions, he wrote with relief, were "not evaluated by the international community as radio-productive."[12] Nor were Chernobyl residents' doses high enough compared to Hiroshima and Nagasaki to induce even the short list of cancers that researchers working in Japan had found.[13] Since the reported rise in illness couldn't be caused by Chernobyl, Soviet public health officials concluded, the cause must be improvements in diagnostics, medical monitoring, and the recording of illness.[14] Indirect causes, such as anxiety, limited diets, and restricted time outdoors also, they speculated, must have caused the statistical increase in disease.[15] Referring to stress induced by fear of radiation, Romanenko borrowed the term "radiophobia" and used it often.[16]

It is true that people often want to attribute their health problems to a specific cause. Faced with a known but insensible toxin, people might seek out doctors more frequently, get more diagnoses, and apprehensively imagine the worst. Romanenko had a point; fear of radioactivity could cause both more stress-induced illness and more reporting of disease.

There are, however, some problems with Romanenko's arguments. Until 1989, local public health officials had no idea of trends. They did not know about health problems outside their immediate community and they had no knowledge of radiation levels. Doctors only learned to be "radiophobic" by assessing the bodies they examined.

But just look for disease, Romanenko insisted, and you will find it. He argued that mass monitoring programs led to a recorded growth in rates of illness, which had long been there, just no one had tracked it. I find, though, that medical teams were not able to carry out their herculean task to examine all children at risk in the rural areas of Chernobyl fallout.[17] Fast-moving mobile brigades missed a great deal.[18] The Soviet countryside did not have enough permanent doctors or specialists in endocrinology, pathology, oncology, or hematology. Commenting on a regional increase in cancer, a sanitation doctor wrote, "The number of cases of leukemia might be yet higher if there were any endocrine specialists in the regions."[19] That dearth got worse, not better, after the accident.[20] Young doctors refused to report to jobs assigned in contaminated territories, and nearly half of the existing doctors in contaminated regions fled after the accident, so that the number of medical personnel fell every year. The "catastrophic" shortage of medical personnel was universal.[21] Staffing got so bad in Belarus that officials in Minsk begged Kyiv to send doctors and medical equipment.[22] Kyiv replied they had none to spare.[23]

Nor was data collection consistent. Families that moved away were not tracked. Children who had been resettled got lost in the system and fell off the registries.[24] Blood drawn for exams did not make it to the lab in time because of poor roads, bad weather, or a shortage of needles and vehicles.[25] In short, the collapse of medical services in the contaminated regions led to an underreporting of illness, not an excess.

You could argue that despite the secrecy, lack of maps, and scarcity of doctors, people knew they were in contaminated regions, as the official designations "control zone" and "severe control zone" cued people in to possible exposures. They and their doctors would attribute any illness residents contracted to radiation. But what about a place where no one knew they were exposed? Such a community would serve as a good test for the idea that increased screening was finding more illness than usual.

There was such a territory 300 kilometers west of the Chernobyl Zone in Rivne Province of Ukraine, which had been declared clean in 1986. Three years later, Aleksandr Komov reversed this judgment when he proved that local milk came in routinely above permissible levels. Two northern regions, Dubrovytsia and Rokytne, had especially alarming levels of cesium-137 in milk and were belatedly categorized in 1989 as "strict control zones," where people qualified for food subsidies and medical monitoring.[26]

Because radioactive contamination had gone undetected for three years, these communities had received no special postaccident attention from 1986 to 1989 that might have caused either stress-related illness or increased detection of existing disease. In fact, the regional medical service was in terrible shape. The pediatric hospital consisted of two rooms in a peasant cabin, heated by a wood stove. The per capita number of doctors was one of the lowest in Ukraine (1.2 pediatricians for 10,000 children). Despite limited resources for diagnoses, public health officials found in 1989 that children and adults had the same growing list of medical problems as those in areas long known to be contaminated.[27] Most alarmingly, the number of tumors among children was up to twenty times higher in 1988 than in other contaminated regions, where there were efforts to provide clean food and medical treatment and to send children away for summer holidays.[28]

I had to pause to imagine the moment of detection of a rare childhood cancer in the crumbling, two-room daub-and-wattle cottage that housed the pediatric unit in the northern region of Rivne Province, a territory where roads dwindle into rutted paths then disappear, as the land gives way to the swamps and bogs of the great Pripyat Marshes. Eighty-two percent of country dwellers died at home.[29] Rural hospitals rarely had pathologists on staff, so ambulance drivers wrote most death certificates. Many cancer deaths, miscarriages, birth defects, and other illnesses from 1986 to 1990 in the contaminated areas of Rivne Province were not recorded at all.[30]

# Belarusian Somnambulists

kept returning to the health records in the archives because I was having a hard time believing them. Was there really a public health disaster among four million people that the world had overlooked? I wanted to be sure because some people get very upset on hearing that low doses of radiation are anything but good for you. I thought of the letters I was sure to receive from people employed in nuclear power plants, oncology wards, and nuclear submarines. These working people expose themselves or others for a living and some are very sensitive to suggestions that low doses of radiation might be dangerous. I thought of the scientists Thomas Mancuso, Alice Stewart, Ernest Sternglass, John Gofman, and Arthur Tamplin, among others. They delivered bad news about low doses of radioactivity and health and it cost them their research funding, sometimes their jobs, and a lot of grief.

What about Belarus? More Chernobyl fallout landed there than anywhere. I flew to Minsk and read through records in the Belarusian National Archives. I continued on to Gomel and Mogilev, the cities in the eye of the Chernobyl maelstrom. My research assistants kept busy. Olha went to Moscow during a cold snap to order documents. Katia took a train and then a bus to a tiny, unheated archive in the county seat of Cherykaw Region, where the archivists were so thrilled to have a visitor they kindly handed over records from the local hospital.

Reading through these new stacks, I noticed in Belarus a similar schizophrenic cycle of denials of Chernobyl-related health problems, until one day authorities spun around to declare southern Belarus a "disaster zone."[1] I saw a difference though. When a few weeks after the accident Moscow ordered up mass screenings of people exposed to Chernobyl radioactivity, Ukrainian officials carried out this order with all due speed. I don't see the same urgency in Belarus. The mandates to examine villagers and take protective measures arrived several months later in the fallout-saturated localities.[2] Clearly, Belarusian leaders didn't go looking for trouble. Initially they considered only five districts to be polluted with Chernobyl fallout. Later they learned they actually had fifty-four contaminated districts.[3] Anatoly Romanenko founded the All-Union Center for Radiation Medicine in Kyiv within five months of the disaster. In Belarus, the Ministry of Health stalled for three years to create an Institute of Radiation Medicine, which was then swamped with funding problems and rivalries so intense that the institute failed to carry out much work on Chernobyl.[4]

Several years after the accident, the Belarusian government issued another set of decrees to examine the exposed public. The government had to renew the original 1986 order because only a small portion of Belarusians in contaminated regions had been monitored due to a lack of staff, a dearth of equipment, an excess of secrecy, and a deficit of curiosity.[5] As Belarusian leaders came to understand the mushrooming geographical grip of the disaster, they increased the number of people under observation for Chernobyl exposures from 166,000 to 2 million people.[6] Finally, three years after the accident, the first health data of Chernobyl-exposed people flowed into the Belarusian Ministry of Health.[7]

These standardized reports were unlike anything I've ever seen. One chart showed that in hard-hit regions of Gomel Province, the population had dropped in half between 1985 and 1987. Poof, in one region, 15,000 people had voluntarily relocated. That was bad, but what sent a chill down my spine was the text under the chart: "As apparent from the table above, considerable demographic and migratory changes in the structure of the population have not occurred in Krasnopole Region."[8]

What? How could the author determine there were no demographic changes when 50 percent of the residents were gone? I suppose the bureaucrats who filed that report never imagined it would be read. I found more documents that also unhitched facts from conclusions. An analysis of cerebral-vascular and neurological health in contaminated areas showed a tripling of illness between 1985 and 1988, but then, as if no count had ever been taken, the official concluded they had found "no essential changes or a significant increase."[9]

Doctors at the Belarusian Ministry of Health engaged in other research methods that puzzled me. In a village study in southern Belarus, medical brigades used as controls for their study people living in "practically clean" areas. Those were territories with five curies of cesium-137 per square kilometer—hardly "clean."[10] I spoke to Valentina Drozd, an endocrine specialist who worked at that time as a mid-level researcher at the Ministry of Health.

"In those days," she explained, "we considered Chernobyl to be a Ukrainian problem. So, after the accident, we brought in some kids, took a look at them, gave them iodine pills, and sent them home. The files of those exams went somewhere, classified."[11]

She laughed when I showed her the reports I cited above. "That is the work of a *Sovok*," she said. "I was one too in those days."

*Sovok* is a lovingly derisive word for an unreflectively loyal Soviet citizen, bound by ideology and lacking independent thought and action. It describes bureaucrats aimlessly filing contradictory documents between long coffee breaks, somnambulists going through the motions. Told to tabulate health statistics in regions of fallout, they did. Told there were no problems in Chernobyl territories, they found no problems. And, to give them some credit, the sanitation doctors did not know anything officially about the extent of radioactivity in the villages in southern Belarus. Most did not, like Natalia Lozytska in Ukraine, get a hold of a radiation counter and take measurements themselves. They worked blindly. When public health officials met with villagers to pacify them, they could not give numbers or explain much in the way of any possible connection between health and radiation because they were denied much of the same information villagers were restricted from knowing. When villagers pressed them on the growth of endocrine and circulation system disorders in their communities, the officials stammered and spoke in "vague abstractions." When mothers asked them to clarify their answers, they were defensive and insulted them.[12]

I found in the records just a few individuals in Belarus who sounded the alarm about the catastrophe in their midst: an army general, a journalist, a university biochemist, the writer Ales Adamovich, and the top nuclear physicist in Belarus, a man who most likely believed his prestige, fame, and high-priority military contracts protected him.[13]

The physicist, Vasily Nesterenko, ran the Institute of Nuclear Energy in a suburb of Minsk. The institute had one main purpose: to build a mobile nuclear power station that was cooled not with water but with nitrogen dioxide. The mobile reactor, called "Pamir," was Belarus's largest military contract. When finished, it would consist of five enormous trucks—sixty-ton armored caterpillars that could crawl anywhere on earth—over glaciers in the polar north, onto the windswept sands of Central Asian deserts, or deep into the melting tangle of the Siberian taiga.[14] The Pamir was designed to be a self-sufficient reactor. Just start it up, and the lights would go on. Soviet generals enthusiastically funded Pamir and appointed Nesterenko, who conceived of the idea, to be chief designer. That made Nesterenko the scientific wunderkind of Belarus. Titles and awards were piled on him. He worked around the clock to build up a team of nuclear physicists and chemists to create an important research center in Minsk, a city long on the margins of the Soviet scientific world.

It didn't all go without ripples. The project got held up and missed deadlines by years. A chemist died in 1979 in an experiment. In 1985, party officials issued Nesterenko a warning for using institute workers to build his private dacha. By 1986, Nesterenko's team of a thousand employees had built two model reactors, tested them, and found they were extremely volatile, hard to control, and rocked dangerously at full power, all while nitric acid corroded the reactor from within.[15] During tests in July 1986, gases leaked from the reactor and seared the lungs of four employees. This was all business as usual in the devil-may-care experimentalism of Soviet military research and development. Despite the problems, in early 1986, Nesterenko won several major prizes.[16] He was at the peak of his career.

Then his fortunes shifted. During the interminable summer of 1986, Nesterenko dashed about southern Belarus with a radiation meter and rushed home to warn party leaders about the sea of radioactivity flooding Belarusian villages and farms. He diverted three hundred members of his staff from work on the mobile reactor to map the contours of radioactivity in Mogilev and Gomel provinces.[17] Finding eye-popping levels, Nesterenko wrote Belarusian party bosses, got no purchase, and so directed missives to Moscow leaders. He bypassed the chain of command. That was probably what got him into trouble.

At the Institute of Nuclear Energy, the personnel director, KGB Lieutenant General P. Budakovsky, wrote a letter of complaint to the Minsk City Communist Party about Nesterenko.[18] Anonymous letters preceded and followed Budakovsky's letter, though they appear to have been written with the same fuzzy typewriter by a person with the same literary ticks, one who had a great deal of insider knowledge of personnel issues and private phone calls. The anonymous letters charged that a "Zionist clan" was at the helm of the Institute of Nuclear Energy. The clan engaged in nepotism, embezzlement, and falsification of research. Just as the gang was about to be apprehended, the authors speculated, Chernobyl fell from the sky "like manna from heaven." Nesterenko, at the head of the band, used Chernobyl to deflect attention away from his crimes. He also manipulated information about the accident to "provoke residents to issue a stream of complaints about their government." Nesterenko, the writer stated, did the same at the institute. He held a staff meeting where he called for a ban on the sale of milk. "Only one sober voice," an anonymous author wrote, "rejected this display of unnecessary psychosis—that voice was Budakovsky."[19]

The accusations led to an investigation and, in no time at all, Nesterenko was on trial at his own institute. The paperwork for the investigation fills several volumes and is stored at the National Archives in Belarus. I had some trouble getting access to these files because the case got so personal and snarky that the archivists and Nesterenko's son preferred I did not copy them.

Two vice directors joined Budakovsky in attacking Nesterenko. One had recently been demoted.[20] Perhaps he wanted revenge or cov-

eted Nesterenko's job. Perhaps the men were genuinely angry that Nesterenko was, as they alleged, imperious, arrogant, and corrupt, had
trouble delegating, and grabbed credit for subordinates' work. It's hard,
however, to see any truth in the accusation that Nesterenko was part of
a "Zionist lobby."

Nesterenko wasn't a little professor who could be wiped away like a
crumb. He had a national reputation and was backed by two of the most
impenetrable and powerful ministries in the government portfolio—the
Ministry of Defense and the Ministry of Medium Machines, the department in charge of nuclear weapons. Nesterenko's close friend, Nikolai
Borisevich, ran the Belarusian Academy of Sciences. With friends in
high places, Nesterenko was safe, or so he felt. Defending himself, Nesterenko accused Budakovsky of trying to cause a scandal at the institute to silence him about Chernobyl.[21] A commission of scientists from
the Academy of Sciences and the two powerful ministries reviewed the
institute's progress and financial arrangements and generally found no
alarming problems. Instead, the visiting commissioners sharply criticized Nesterenko's enemies, Budakovsky and his allies.[22] Colleagues
testified that Budakovsky was a bully, suspicious, and had a habit of
eavesdropping.[23]

Enraged, Budakovsky wrote Nikolai Sliun'kov, the Belarusian
party boss, a four-page denunciation of Nesterenko. He held nothing
back. Budakovsky again made allegations about embezzlement, phony
reporting, sycophancy, and Nesterenko's dictatorial management style.
"All that is negative," he charged, "all that we actively fight against in
our society, is concentrated in the Institute." Upset at what he saw as
betrayal by colleagues who took Nesterenko's side, Budakovsky added,
"I have not experienced such moral despair even at the front."[24]

Budakovsky's appeal elevated the case from the city to the republic
Party Committee, Sliun'kov's domain. Sliun'kov ordered another investigation, which also didn't find much. Many of Nesterenko's colleagues
defended him. Even his rivals had to admit he was essential to the institute.[25] But in staff meetings, Budakovsky and his allies attacked Nesterenko and his allies. They passed insults and affidavits back and forth
with plenty of verbal shoving and kicking. Commissioners held meet-

ing after meeting, long and tiresome, with institute staff to discuss the director's shortcomings. Nesterenko's team lost ground.[26] His "roof" fell down around him. In Chernobyl's political aftermath, his patron, Efim Slavsky, the octogenarian director of the Soviet nuclear weapons program, lost his job. Colleagues who signed letters to defend Nesterenko were investigated and terminated.[27] Finally, the presiding party commissioner sided with Budakovsky. Nesterenko was demoted to working in a lab in the institute he had directed. He did not take this defeat lightly. He checked into the hospital with an ulcer and kept writing letters, appeals, and affidavits.[28]

Maybe because he didn't stop and showed no sign of going away, Nesterenko was punished in that special way security services use to quietly reproach their subjects. Nesterenko's son, Aleksei, told me about a series of mishaps that followed Nesterenko's termination as director. One morning on his way to work, Nesterenko noticed that someone had cut the brake cables on his car. A few weeks later, slowing for a light, he spotted an ambulance racing up behind him. Nesterenko put on the gas and accelerated to get out of the way, but the ambulance driver steered right at him. Nesterenko was injured in the collision. The driver walked away. Pursuing the case later, Nesterenko learned the driver had been hired the morning of the crash and quit at the end of that day. The last straw was the evening a few months later when Nesterenko found his dacha ransacked, papers cleared out. "At that point," Aleksei said, "my father lost it. He came pretty close to a nervous breakdown."[29]

While government officials slandered and insulted each other, people in southern Belarus were in real trouble. In Mogilev Province, more than 10,000 people lived in communities contaminated with at least 40 curies of cesium-137 per square kilometer. The highest scorer was the village of Chudiany, measuring, on average, 141 curies per square kilometer.[30] Keep in mind, after the accident, army brigades removed a scant 24,000 people from Gomel Province and no one at all from Mogilev Province. The media tend to focus on the people who returned to live inside the Chernobyl Zone of Alienation. Authorities allowed elderly people to return home after evacuation if their homesteads were not heavily contaminated. The "self-returners" or "babushkas of

Chernobyl" were supplied food by the government. Journalists often marvel that these pensioners, mostly women, are still in the Zone and are still alive. Some use the persistence of these women to argue that Chernobyl radiation was less harmful than the stress of evacuation.

In this way, commentators fall for the proximity trap, reasoning that the closer one is to the source of a radioactive event, the more dangerous it is. Because of wind patterns distributing fallout after the accident, no self-returner was exposed to as much radioactivity as the tens of thousands of people in distant areas of Belarus who lived in areas ranging from 40–140 curies per square kilometer.[31] The hospital records of the Belarusian Cherykaw Region, where Chudiany was located 300 kilometers northeast of Chernobyl, render a portrait of exposure year by year in a district of 20,500 people, 5,000 of whom were children. In the first year, medical brigades examined a couple of thousand people, half of those in the worst areas.[32] Blood tests showed that 5–25 percent of residents had "poor blood indicators" and high counts of radioactivity in the thyroid.[33] In 22 percent of samples of mothers' milk, levels of radioactivity were over the permissible amount.[34] One report said no one was hospitalized.[35] Another document counts 100 people treated in the county hospital.[36]

The next year, the indicators edged upward. In 1987, soon after the Moscow biophysicist Leonid Ilyin told international scientists that food protection measures had been successful, local sanitation inspectors in Cherykaw Region found food more contaminated than in 1986.[37] In 1988, a quarter of milk produced was above the permissible limit. All communities, even ones "relatively clean" (less than 15 curies per square kilometer) produced some portion of milk too radioactive to drink. Of 59 samples of wild boar meat, 47 were too hot for consumption. In only one forest in the region were mushrooms edible.[38]

Cherykaw Region had long had a shortage of doctors, a trend that had intensified over the years. At the end of 1987, residents in controlled areas refused the cursory exams by mobile medical units because the doctors treated them with scorn.[39] Due to the shortage of medical personnel, only a portion of the "at risk" population was examined.[40] Even so, the medical landscape looked grim.

Among adults, the number of cases of heart disease, enlarged thyroids, gastrointestinal and urinary tract disorders, cataracts, and liver and blood disease doubled or tripled between 1984 and 1988.[41] Neurological disorders exceeded averages for Belarus as a whole. In 1987, half the children had enlarged thyroids. Ten percent had anemia, four times more than the year before. Several dozen children suffered from leukopenia and thrombocytopenia, a low count of white blood cells and platelets that can cause bleeding from gums or nose, fatigue, weakness, and a high risk of infection. Of 222 births, 67 percent of the mothers had complications, twice as many as in 1986.[42] Of 103 pregnancies, 63 babies were born alive.[43] The number of babies with heart defects doubled.[44] The number of infants who died within 28 days of birth doubled in 1987 and nearly tripled in 1988.[45] Children of mothers pregnant during the accident were sicker than those born after the accident. Children with higher recorded doses of incorporated radioactivity were generally sicker than the rest.[46]

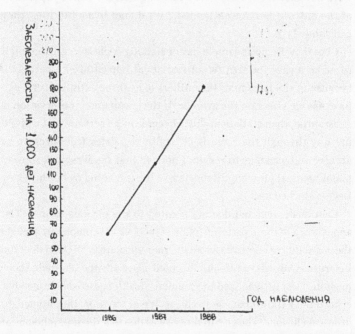

*Dynamic of anemia among children 4–6 years old living in Cherykaw Region*

Three years after the accident, 30 percent of children in the region had an anemic syndrome.[47] A regional public health official who wanted to make sure this astonishing uptick was noticed included a rare graph in her report. It shows cases of anemia in children, aged 4–6 years, living in territories with soil counts of more than 40 curies per square kilometer.[48] The graph line resembles an arrow rocketing toward infinity.

Cancer rates climbed from 1986 to 1989, as they did everywhere in the USSR, except that the cancer rate for Cherykaw Region was five times higher per capita in 1988 than for the Belarusian republic.[49] The most prevalent cancers were lymphomas, leukemia, and cancers of the thyroid and gastrointestinal tract.[50] In Mogilev Province as a whole, leukemia tripled between 1986 and 1988.[51] The increase in other cancers continued to climb in 1989.[52] And in 1990.[53]

Residents of the region wrote letters and sent telegrams. They were not answered. They held a meeting that the entire population of the city of Cherykaw attended. "We do not live a full life," they proclaimed.

Cherykaw Region's sample size of 20,500 people is small, too small to make a case for significant statistical impact after Chernobyl, because in small samples the outliers (e.g., three extra cancers) can have a large effect on the average. If the group under observation is substantial, then statistical shifts become more certain. As I worked my way through the records of health registries from archive to archive and province to region, I noticed that small regions showed major medical problems. Reports like those coming from Cherykaw Region stacked up.

Cautiously, regional doctors pointed to the obvious factor: "The appearance of these patterns [of disease] does not completely exclude the possibility of negative factors such as radioactivity."[54] Officials at the Belarusian Ministry of Health disputed especially the Chernobyl connection. They rationalized the ominous health statistics in ways that echoed their Ukrainian counterparts. It was too soon, they argued, to draw conclusions. The rise in illness was due to a "poor psychological climate."[55] The cancer rate tracked with "natural growth."[56] Thyroid problems were endemic in the region because of mineral-poor soils.[57] There was, they said, too much medicine: "The acute increase in illness...is explained by the improvement of medical care."[58] Yet they also maintained that the spike in numbers such as infant and perinatal mortality was due to poor medical care in rural clinics.[59] It was illogical to argue both ways—both improved medical attention and neglect.

The denials of Belarusian officials matched those of Ukrainian and Moscow officials. The health statistics of Cherykaw Region echoed those of Narodychi, Poliske, and Dubrovytsia regions in Ukraine and dozens of fallout regions in Gomel, Mogilev, Brest, and Grodno provinces of southern Belarus. Wherever I looked (and I was usually the first researcher to sign out the files), the evidence of a public health disaster was overwhelming, and it came from almost every possible quarter.

Even a KGB general sounded the alarm. Dr. Mykhailo Zakharash directed a specially equipped, well-subsidized KGB clinic in Kyiv. In a study of 2,000 people, he found patients' bodies had incorporated up to a dozen different radioactive elements. Cesium-137, used in dose estimates, made up only 40–50 percent of the ingested dose. Summing up four years of work, Zakharash wrote in 1990, "We have shown that long-term, internal exposures to low doses on a practically healthy individual lead to a decline of his immune system, a lowering of defensive strength, and a whole series of pathological changes and illnesses."[60] The KGB doctor estimated that 4.5 million people had been contaminated at levels above the state's mandated permissible norm, and he recommended that the Zone of Alienation be extended from 30 to 120 kilometers, which would encompass the ancient and beautiful city of Kyiv, where he lived.[61]

I was astounded to read that document in particular. I understood that the KGB's traditional role was to solve problems, silence detractors, and generally shore up a patriotic appreciation of the Soviet polity. General Zakharash's convictions must have been strong to sound this dire message. I asked around about him and learned he was still alive, still working as a doctor, by 2016, in a civilian hospital in Kyiv. I went to see him.

Between urgent phone calls and consultations with other doctors, Zakharash told me how he came to write the report that I found in the archives. "I heard about the accident on April 26th, about six a.m. I got a call. No one yet knew how serious it was."

"Three days after the accident," he continued, "I went to the site myself. I could tell it was bad. Driving past farms and towns, nothing, no one was alive, no birds, no dogs. I saw just one white stork swimming in a pond." Arriving at the nuclear power plant, Zakharash was surprised to see army generals and engineers sitting in rooms near the blown reactor with the windows open. Soldiers were camped out in the extremely radioactive Red Forest. "They had no idea what to do—how to protect themselves." Zakharash sent some of his staff to the KGB's classified library to research and write a pamphlet on radiation safety. That became the chief manual they handed around the accident site. "Scientists at the Moscow Institute of Bio-physics," he told me, "should have written that pamphlet. They should have been better prepared."

As Zakharash started to send in classified reports of the poor health of his patients, he made a lot of people angry. In 1990, Zakharash was called to Moscow, to a building near the Kremlin, black limousines winging around it. Party leaders at the Academy of Sciences gave the KGB general a verbal thrashing. They charged he was a traitor, spreading panic. Zakharash held his ground. Unlike other Soviet medical establishments, his KGB clinic had clean needles, chemical reagents, dosimetry equipment, ultrasound machines, and KGB doctors cleared to see their patients' doses. Zakharash knew his study was the best in the country. The party scientists told him he had only one case and that wasn't enough proof. "What do you want," he retorted, "to wait for another nuclear accident in order to check my numbers?"

It wasn't the KGB, Zakharash insisted to me, but the Communist Party that pressured scientists to be silent. "I could never have carried out that study in a civilian hospital. Never."

I found in the archives just the summaries of Zakharash's study. I asked him about the rest of the records. He looked surprised. "I gave all that over to the KGB archives. You should find them there." I told him I didn't find them in the archives, despite a good deal of looking.

Zakharash shrugged his shoulders. "No one listened," he said. "No one paid attention. And no one will listen now, either."[62] He got another call on his office phone, excused himself, and rushed out. There is always another emergency.

Until mid-1989, there were few motivations other than individual conscience to report the public health disaster in the Chernobyl territories. Yet, despite substantial disincentives to issue bad news, the evidence piled up. The accounts of unspecific, widespread, and chronic illness, reproductive problems, and acute increases in cancer resound like a lament across the area of Chernobyl fallout. To put it most simply, the majority of adults and especially children in the contaminated regions were sick. Residents had chronic illnesses, many suffering from several diseases at once. Many more infants were dying soon after birth or were born with birth defects.[63] Women had more difficulty conceiving and carrying their pregnancies to full term. Men quietly suffered from impotence.[64] The problem of contaminated food persisted and deepened.[65] Tests showed people had ingested a wide range of radioactive nuclides and that some bodies had achieved the levels regulated as radioactive waste.[66] Everywhere, doctors fled the clouds of radiation, so that a minority of medical staff was left to diagnose disease. Even so, the pattern of health problems kited upward.[67] The records stacked up to form a picture that is detailed and sharp. Even if you are not one to look at footnotes, you might turn your attention to this one.[68]

It is just a footnote, and it is a nightmare. The leaping, bounding, galloping rates of maladies took shape, a dark horseman riding wild across the Chernobyl territories.

# The Great Awakening

And one day the somnambulists awoke. The *Sovoks* claimed their citizenship. In September 1988 in Moscow, the Council of Ministers quietly announced that thanks to Soviet science and the heroism of cleanup workers, they had arrived at "the last stage of the accident." With that, Moscow authorities adopted a "safe residency" plan announcing that residents could receive without harm to their health a lifetime dose of 350 mSv over seventy years. They assured citizens that 350 mSv was a conservative number. It amounted to 5 mSv a year, less than a CT scan. Residents would take in this dose over a long period, which was two to ten times less harmful, they postulated, than one single, large dose.[1] With this guarantee of safety, most communities would no longer need to monitor food and health, restrict farming, or engage in expensive decontamination measures.[2] Leonid Ilyin, the Moscow biophysicist, suggested subsidies could end. People could return to their "traditional, pre-accident" lifestyle.[3]

Some residents, the policy makers admitted, up to half of the population in the contaminated territories, might get doses over 350 mSv. They would need to be relocated, but the resettlement could take place over three to five years. Ilyin and a few other Moscow scientists justified the plan based on accepted international standards. They cited several UN agencies.[4] This was the new perestroika USSR—embracing transparency, joining international conventions, meeting global standards and exceeding them. A hundred international experts at the International

Atomic Energy Agency (IAEA) endorsed the 350 mSv plan. The Ministries of Health in Belarus and Ukraine yielded to it more reluctantly.[5]

It was a cruel policy. In it, the authors implicitly conceded they had miscalculated. They admitted it wasn't safe to live in areas contaminated with more than forty curies of cesium-137 per square kilometer, yet more than a hundred thousand people had simmered in over forty curies for three years.[6] Even so, they were in no rush to move them. At the same time, the 350 mSv standard (5 mSv a year) was a questionable threshold, devised for nuclear workers clocking in and out of a job for a career, not for civilians living full-time in radioactive conditions from birth to death. The International Commission on Radiological Protection, an advisory body, and the IAEA recommended a dose that was five times lower for civilians (1 mSv a year). In asserting safety, Soviet planners repudiated their own internal reports that described their inability to decontaminate communities and supply clean agricultural produce. They also turned their backs on the alarming health statistics generated by the socialist medical system.

The timing of the announcement could not have been worse for Moscow leaders. It coincided with campaigns for elections to the Council of People's Deputies, the Soviet parliament. The Communist Party controlled these elections as usual, but some restrictions had been lifted and an election campaign like Soviet society had never known was gaining speed in early 1989. People went to election meetings and spoke their mind, criticized their leaders, and complained about their problems. Two people talking on the street would draw others and soon "spontaneous meetings" would form with hundreds of discussants. Ukrainian KGB agents counted that year 1,565 "massive events," 732 of them unsanctioned.[7] New independent groups popped up promoting causes having to do with ecology, health, and nuclear safety.[8] In this environment, the safe lifetime dose, meant to pacify anxious residents, had the opposite effect. It pulled up the control rods moderating political interactions, and society went critical.

And that is when the sleepwalkers stirred. Well, not everyone had been sleeping. It turns out that in Belarus after Nesterenko was forced to withdraw from public life, other people rose to take his place. That's the beautiful thing about human society. A lot of people lined up are hard to vanquish. While drowsy officer workers at the Ministry of Health filed nonsensical reports, across town researchers at the Belarusian Academy of Sciences quietly got to work studying the effects of the disaster.[9] In 1986, a group of pediatric hematologists selected 6,000 children from communities in southern Belarus and matched them with controls in relatively clean Vitebsk Province to begin a large-scale, long-term study of the children's blood-forming organs. They worked alongside a team of endocrinologists who investigated the thyroids of 2,500 exposed children and 500 controls. Other groups examined immune systems and liver function of children in contaminated regions, also comparing them with controls in Vitebsk.[10] A researcher tracked leukemia among children in southern Belarus. Several pathologists in Gomel dissected the organs of corpses to measure them for incorporated plutonium and other radionuclides.[11] Belarusian doctors had an ongoing epidemiology project that had begun in 1984. That study gave them a background against which to see changes after the accident.[12]

The Belarusians did not shy away from asking uncomfortable questions. Their subjects were those they identified as the most vulnerable— children and infants in areas of heavy fallout whose small bodies efficiently drank up mineral elements, radionuclides, and chemical toxins while their cells rapidly reproduced. Despite Nesterenko's well-publicized demotion, the Belarusian scientists found the courage to ask questions about possible subtle, "unspecific" effects of low doses of radioactivity on human health. "We are concerned," they later wrote, "that until this time no population in the world has ever lived with continual internal and external exposures of this size."[13]

Less than a month after the Ministry of Health endorsed the safe living plan, the president of the Belarusian Academy of Sciences, V. P. Platonov, sent to Moscow a twenty-five-page report reflecting the Academy's post-Chernobyl renaissance in the fields of radioecology and radiation medicine.[14] Not just two southern provinces, Platonov summarized, but two-thirds of Belarus had been contaminated. The contamination had a mosaic complexion with radiation levels differing ten to twenty times in areas a few kilometers apart. Even long distances from the Chernobyl plant, they found areas of 50–100 curies per square kilometer of cesium-137 in the topsoil. Because of the nature of the swampy, sandy, and clay soils of southern regions, radioactivity remained in the top five centimeters and tended to converge in plants, a concentration that focused with time. Forest trees showed a gradual increase in radioactivity levels in the three years since the accident. Soils contained cesium-134, cesium-137, ruthenium-106, strontium-90, plutonium-239, plutonium-240, and a host of other nuclides.

The Belarusians worked across registers from environments to humans. Analyzing corpses of people who died between 1986 and 1988 in the most affected provinces, Belarusian researchers discovered cesium-137 and ruthenium-106 accumulated in the spleen and muscles, strontium-90 in bones, and plutonium in lungs, liver, and kidneys. Soviet scientists had largely excluded plutonium in their dose estimates because it is not soluble, so it rarely enters the food chain. The Belarusians interrupted this thought pattern: "plutonium takes the form of a persistent, finely dispersed aerosol and its main pathway is through the lung." The Belarusian pathologists found, unnervingly, little relationship between the high levels of radionuclides in humans and the levels of radioactive contamination in the territory. All corpses in Gomel Province had inhaled plutonium, and bodies in Vitebsk, with far lower counts of radioactivity, also showed elevated levels.[15] The scientists attributed the general congregation of radioactivity in bodies to the migration of radioactive contaminants along dusty transport vectors and food pathways. Humans are not mice. They forage widely for food and trade among themselves. Soviet food distribution networks had efficiently worked to scatter contaminated food liberally.

In health studies, researchers recorded a significant increase in the average number of chromosomal mutations in newborns and children. The frequency of birth defects in southern Belarus was found to be significantly higher than the controls in areas of low radioactivity. They learned that exposed children with accumulated radioactive isotopes became radiosensitive and were more susceptible to fast-growing tumorous growths and to autoimmune and endocrine disorders. A team of pediatric hematologists found a sharp increase in leukemia, in the

provinces of Minsk, where evacuees moved, and Brest, along the Pripyat Marshes. About the cause of these diseases, V. M. Pavlova wrote,
"One cannot ignore the contamination of children by food that reached
Minsk. We have evidence of this in the cesium-137 found in the children's blood and urine, from whole-body counter measurements, and
from chromosomal aberrations in their lymphocytes."[16]

In adults, the Academy of Sciences investigators recounted a list of
diseases that corresponded with the cross section of rising illnesses that
local doctors had been reporting. Academy president Platonov acknowledged that the jump in rates of disease might be related to increased
medical attention but, he pointed out, the numbers had risen steadily
each of the three years. His researchers had found clear disorders in
bodily functions, plus a track of radioactive isotopes in sick bodies, and
similar changes in experimental animals. All of this led the Belarusians
to suspect that radioactive exposures were a factor.[17]

With their data, Belarusian scientists questioned the premise of the
350 mSv safe living dose. They argued that it made no sense to use mathematical models based on generalized exposures and extrapolations
from Japanese bomb survivors. The lifetime dose must be grounded on
the particular conditions in each locality, which, they found, served up
a different cocktail of exposures and resulted in a variable bouquet of
diseases. The Belarusian academics wanted a much lower threshold for
a lifetime dose, one already established in international regulations for
civilians (1 mSv a year for a total of 70 mSv in seventy years). Platonov
also objected to Moscow leaders' plan to end medical and environmental
monitoring. "To date, much is unknown about fundamental aspects of
the action of low doses of radiation on organs of humans and animals."[18]
If passed, he argued, the safe lifetime plan would cultivate ignorance.[19]

In Moscow, Boris Shcherbina, in charge of the Chernobyl Emergency Committee, read Platonov's report and responded a week later
demanding, as he had many times before, more action and more effective measures. Authorities in the provinces "were not doing enough."
People living in areas with extremely high levels of radioactivity (over
eighty curies per square kilometer) needed to be moved immediately,
he ordered. He commanded Yuri Izrael's State Committee of Hydro-

meteorology to distribute detailed maps to the regions so local officials knew where exactly to clean up and relocate.[20] From Ukraine, Health Minister Anatoly Romanenko shot back a sycophantic letter thanking Shcherbina for the help and assuring him he was on task with the new 350 mSv plan.[21]

But the Belarusian scientists did not back down. An anonymous report appeared at the Ministry of Health giving dose estimates for contaminated villages that far exceeded 350 mSv by the year 2000.[22] The academy's eminent epidemiologist, K. V. Moishchik, sent the Ministry of Health a cautious letter describing cancer and infant death rates in Gomel and Mogilev provinces that were much higher than those they had seen in 1984 and higher than the rest of the republic.[23] Andrei Vorobiev, the respected Moscow leukemia doctor, supported his Belarusian comrades. "Deciding to resettle people three years after the accident reflects only one fact: that the first estimates of the possible doses were mistaken, whatever the words used to express that fact, whatever the reasons for it are given."[24]

Vorobiev wrote Shcherbina that the 350 mSv dose might be safe, but mathematical modeling could not predict computed doses with reliable certainty. Residents' exposures varied so much that the only way to verify them was to check each body. As an example, Vorobiev pointed to four children in his clinic from southern Belarus with acute leukemia. Two of them had chromosomal aberrations associated with a 300–500 mSv dose of radioactivity. He expected the children's parents, with whom they shared food and housing, to have similar doses, but they didn't. Instead, the village doctor who accompanied the children had approximately 300 mSv. "All this attests," Vorobiev concluded, "to the mosaic character of damage, and confirms the need for individual, biological dosimetry."[25]

Izrael's ministry finally published maps in mid-1989 that showed just how high radiation levels were in several hundred Belarusian communities.[26] The poor-quality maps were like matches tossed into a leaking gasworks. After years of rumors, conjecture, and meaningful looks from masked radiation monitors, the maps gave concrete contours to residents' inchoate sensibility that something was terribly wrong.[27]

Armed with the maps, villagers stormed their leaders and candidates running for office. Residents living in the red spots who were told they would eventually need to move, but not yet, were especially angry. People detailed in letters their projected doses that did not match up with the 350 mSv safe living promise.[28] Villagers wrote from "clean" villages with evidence that their local food was dirty.[29] Residents sent figures on local radiation levels and health data to their ill-informed province leaders.[30] The populace, in other words, enlightened the state, not the other way around.

Looking at the new maps, local leaders, even leaders in charge of the Chernobyl cleanup, learned for the first time just how bad it was. "At five curies on the ground, we can't get clean milk," Heorhy Hotovchyts reported from Zhytomyr. This man in charge of Chernobyl liquidation saw radiation maps for his province for the first time in September 1989. That is when he grasped he would need to resettle 45,000 people.[31]

Radiation monitors from various ministries arrived in villages to investigate the claims in citizens' letters. They measured and recorded conflicting numbers.[32] Witnessing the confusion, residents asked, what kind of "science" is this? The inspectors generally found the villagers' testimonies were correct.[33] Even at acceptably low counts of curies in soils, doses from consuming local food were too high and, indeed, the disease rates were rising. More villages were added to the list of at-risk communities. The hash marks on the radiation map spread outward.[34] People learned they would be relocated, but with no prospects of going anywhere soon, a gloomy apprehension settled in.[35] If there was such a thing as radiophobia, 1989 was the year it became manifest.

AFTER 1989, there was no going back, no return to the denials and silences of the previous years. In that year, time sped up. Decades of repressed secrets poured out in just months. Journalists wrote exposés. Groups formed to take care of basic services that the state could not manage.[36] Alla Yaroshinskaya, the crusading Zhytomyr reporter who had been thwarted for years in her attempts to report on Chernobyl, finally got an article past the censors.[37] Her five hundred words about sick adults, children, and animals in Narodychi were so troubling the

Ministry of Health ordered yet another investigation.[38] Eight hundred villagers showed up to talk to the arriving commission of scientists. The experts told farmers their calves' birth defects were caused by a deficit of minerals in local soils.[39] The farmers wondered why the scientists didn't sound very scientific: a soil deficit that had manifested itself only after 1986? The farmers demonstrated a refined understanding of how radionuclides might become concentrated in the food supply and genetic lines: "What guarantee do we have that in twenty to thirty years we will not see the same defects in people?" The experts had no answer. From the hall, villagers threatened to go on strike or to start "a second Karabakh," meaning a murderous riot.[40]

Yaroshinskaya's article propelled her into elections for the Supreme Soviet. Thousands gathered for campaign rallies in Kyiv and Narodychi. Berdychiv issued a vote of no confidence in their leaders. Across Ukraine, protests broke out; then more.[41] In Crimea, loved for its resorts and beaches, citizens organized against the construction of a new nuclear power plant in an earthquake zone.[42] Thousands of residents near other nuclear power plants mobilized to shut them down.[43]

After decades of soothingly dull news on completed harvest targets, the unleashed perestroika-era media suddenly served up terrifying headlines. Reporters wrote stories of skulls disinterred in church courtyards, the bones of people executed during Stalin's regime.[44] Headlines followed about hazing in the military, drug addiction, AIDS, and official corruption. These were capitalist, not socialist, problems. Gorbachev had planned a controlled bloodletting to refresh the body of the state. He miscalculated. Rapidly losing vital fluids, the Communist Party of the Soviet Union appeared more and more anemic.

Workers shot missives at Moscow spokesmen: "We would like Ilyin and Izrael to consider us as people and not experimental subjects. If they think it is safe to reside here then let them come live with us, and they can take care of our health and of our children." Workers asked the scientists to be responsible, "as no one was in the 1930s and 1940s when millions of innocent people were sentenced." They asked, "Who was brought to justice for that?"[45]

As scientists and doctors lost credibility, Soviets turned to their television sets for help. Two TV hypnotists, Allan Chumak and Anatoli Kashpirovsky, became overnight sensations. Each had his own style. Chumak told his audience to sit back, close their eyes, and feel his healing powers across the airwaves. For the rest of his program, he would silently wave his hands while opening and closing his mouth in a fishlike way. He told viewers that if they could not be home during his broadcast, they could place a glass of water in front of the television. The water would absorb beneficial information to cure them. After Chumak's first few televised sessions, he claimed that six million letters of gratitude flooded the central TV station.[46] The other healer, Kashpirovsky, was a Ukrainian psychotherapist. His powers worked through his hypnotic voice and penetrating eyes. He became famous when from Moscow he hypnotized a woman in Kyiv who was having surgery with no anesthesia. Over live TV, Kashpirovsky chanted incantations. On a split screen surgeons wielded scalpels and the patient on the operating table sang joyful songs. The nation watched in awe. The Soviet Ministry of Health banned the men from the airwaves, but that didn't stop them. They wrote books, made videos, and filled large halls with hopeful patients. Elderly people, invalids, women, men, workers, and intellectuals—all kinds of people—tuned in to the shamans. Kashpirovsky appeared in Slavutych, the new atomic city next to the Chernobyl plant, for a special session to cure Chernobyl evacuees, a tough sell with an audience of engineers and technicians. My friend Nadia Shevchenko remembered how Kashpirovsky chanted and waved his hands in the direction of individuals who looked doubtful. "Even those who don't believe in my powers," he intoned, "will be cured."

"He was a real charlatan," Nadia remembered, "but they could not pass up the chance to make everything better for those of us from Pripyat."

TV hypnotists may sound crazy. That's how I saw it at the time. I remember trying not to look skeptical when a university dean told me that Chumak cured his wife's cancer. Now I understand it differently. The embrace of TV healers was a skillful adaptation to the ongoing disaster. Turning their skepticism of experts into a belief in miracles, Soviets created a new imaginary much richer in possibility than the meager supplies of Soviet pharmacies or the sorrowful incapacities of underheated hospitals working at half-staff. It was a smart move. In direct contrast to the aching shortages of the Soviet economy, miracles had no limits.

//

IN JUNE 1989, the first Supreme Soviet parliamentary sessions played live on television. Citizens across the country stopped for twelve full days to watch. Shopkeepers placed televisions facing outward in windows. Pedestrians stood transfixed. The impact of democracy was immediate and profound.

Emboldened by the crowds on the street, the long-submissive ministries of health in Ukraine and Belarus rebelled against Moscow's safe lifetime dose. Belarusian officials testified that eighty-five Belarusian communities could not be guaranteed doses of radiation below 350 mSv. Ukraine had a dozen more.[47] Belarusian scientists pointed to the rising volume of nitrate fertilizers, advocated by the State Committee of Industrial Agriculture to reduce the uptake of radioactive nuclides. They found that high concentrations of nitrates in rats doubled the harmful effects of radioactivity in their bodies.[48] Responding to criticism, Belarusian leaders rushed to move people from two villages that the newly published maps showed were the most contaminated.[49] Province leaders built a pretty new community with running water, gas heat, and even a grade school with a swimming pool.[50] There was just one problem. Local leaders had to resettle their residents inside their own districts. Regions like Cherykaw, blanketed with fallout, had few contamination-free sites for evacuees.[51] When party officials gathered to cut the ribbon on the new community, someone thought to pull out a radiation counter. The new town with the freshly painted houses had the same old level of radioactivity, over twenty-six curies per square kilometer.[52] A year later, it was razed.[53] After that, Belarusian leaders called the condemned Minsk physicist Nesterenko back from obscurity. He and his team of monitors had the skills to measure and, more importantly, they had people's trust.[54]

During the week that the Berlin Wall came down in the grim, hungry

autumn of 1989, Natalia Lozytska took part in a pilgrimage from Kyiv
to a nuclear power plant in central Ukraine. The fifteen activists left
Kyiv late at night. Among them were radioecologists, university physicists, radiation monitors, and, most likely, undercover KGB agents. They
planned a route through towns and villages contaminated with Chernobyl fallout. They didn't get far that first night. Just outside of Kyiv,
state troopers, tipped off about the pilgrims, pulled the bus over. The
scientists showed their letters explaining the purpose of the trip and its
authorization by the ecological group Green World. The troopers radi-
oed in to headquarters and ever so slowly sorted out the problem, which
shouldn't have been a problem in the first place. The pilgrims waited on
the side of the road as a cold, damp fog rose from the asphalt.[55] At the
cheerless morning dawn, the pilgrims were finally released. The group
spent the next two weeks visiting towns and villages and holding meetings with worried farmers and townspeople.

In the preceding three years, villagers had hosted many such visiting
scientists. Doctors had taken blood and urine samples, measured heart-
beats and lung capacity, and went away. Radiation monitors had asked
for milk and soil samples and then left. When villagers asked about measurements from the samples, the experts answered, "These results are
not for you."[56]

The Green World activists were different. They measured radioactivity and showed the meters to the villagers. They listened to the villagers' worries and health complaints. They did not scold or dismiss them.
The pilgrims answered questions in detail, trusting residents to grasp
science and assimilate honest answers. No one panicked. Lozytska and
her colleagues explained the nature of radioactive exposure, how radionuclides spread, which plants absorbed which isotopes, and where they
lodged in the body. In communities with high levels of contamination,
Lozytska advised parents they should do their best to pack up their kids
and move away.[57]

As they traveled, Lozytska asked villagers for eggshells and children's
teeth. She understood that the computed dose estimates drawn up by
the State Committtee of Hydrometeorology rarely led people to feel
secure. Lozytska had an alternative, simple, and cheap plan for estimat-

ing exposure. She started with a basic understanding of the reciprocal nature of contamination; man-made radioactive isotopes imprint on local ecologies, while contaminated environments make an impression on bodies. Testing two artifacts for depositions of radioactive strontium would give people concrete results about their individual surroundings and their particular bodies. Rather than variable and easily disputed dose estimates, the eggshells and teeth would tell them of beta and alpha energy absorbed in organisms. Mapping critical areas where people and household chickens had high levels of strontium-90, Lozytska hoped, would be the start of finding solutions to health problems. Lozytska had no idea she was re-creating the work of citizens in St. Louis in the 1950s when they collected 67,000 baby teeth to track the incorporation in children's bodies of radioactive strontium from the Soviet and American detonation of thermonuclear bombs, which grew larger and more dangerous with each passing year.[58] In Ukraine in the 1980s, Lozytska, likewise, was creating a citizen's science that held great promise.

In 1989, new leaders stepped into the void left by the deflating Communist Party. People's deputies, elected to the Supreme Soviet, derived their power not from the party but from the electorate. Residents flooded them with appeals. As the parliamentarians grasped the complexity and magnitude of the problem, they demanded actions to protect people.[59] They wanted Chernobyl emergency committees to be accountable to elected representatives, not to party organs. They demanded a deep study of contaminated zones, plus more and better hospitals. They wanted treatment of survivors "regardless of the reasons for their illness" and opportunities for relocation sooner, not later. They used words like "duty and responsibility."[60] Even deputies who had done and said nothing as appointed party leaders suddenly found their voice and courage as elected delegates. The party boss of Zhytomyr Province, Vasyl Kavun, arranged to run unopposed in Berdychiv.[61] After he was elected, he became a new man. For three years he had suppressed news of Chernobyl problems. Suddenly he was writing letters opposing the 350 mSv plan.[62]

This was the point when Chernobyl became politicized and monetized.[63] Budding politicians, like Kavun, grasped that showing concern for Chernobyl victims bought them votes. Republic leaders realized the safe lifetime dose implied a retraction of federal funds. In a rare joint appeal, the leaders of Ukraine, Belarus, and Russia begged Moscow to extend subsidies to an additional 140,000 people.[64] Republic ministers drew up their own thresholds for lifetime doses that were five times lower at 70 mSv over seventy years. They wanted to regulate at the republic, not federal, level.[65] They wanted more power in their own hands.

Kavun was a decorated Communist Party boss who surfed the turbulent wave of democracy all the way to the beach. Others crashed with arms and legs cartwheeling in the frothing breakers. At the end of that eventful year, Gorbachev returned to Kyiv to open the newly elected Ukrainian parliament. He gave the final nod to Volodymyr Shcherbytsky, who after seventeen years ruling Ukraine submitted his resignation. Anatoly Romanenko was shifted out of his job at the Ministry of Health that same autumn. In retirement, Shcherbytsky began to smoke several packs of cigarettes a day and walk with a cane. He died two months later, on the day before he was to testify before a parliamentary commission about his role in covering up the Chernobyl disaster.[66] Some say it was suicide.[67]

In late 1989, dissidents formed a group called "Trust" and passed around a petition to have American specialists arrive to judge the safety of nuclear reactors.[68] Assurances and myths about the miracles of science and technology were sinking fast. KGB agents worried that even Soviet scientists had lost confidence in Soviet science.[69] A quicksand of distrust and accountability emerged in the three postaccident years in which Yuri Izrael slow-walked radiation measurements and Leonid Ilyin computed the sunniest possible health outcomes from those numbers. Gorbachev's handlers did their best to stifle discussion of Chernobyl problems at the televised parliamentary sessions, but it was becoming clear that Chernobyl required something more than old-school repression.[70] Something proactive was needed.

# PART VI

—————— // ——————

# SCIENCE ACROSS
# THE IRON CURTAIN

# Send for the Cavalry

Seeking to calm the unruly public, Soviet minister of health Evgeny Chazov called for backup. He asked the World Health Organization (WHO) to send a delegation of "foreign experts" to assess the 350 mSv safe lifetime dose. Wasting no time, WHO officials put together a commission of physicists—not physicians—who were already on record minimizing the Chernobyl accident.[1] That fact didn't matter because the phrase "foreign expert" rang like a church bell across the Chernobyl territories, clear, sharp, and humane. Residents of Navrolia, the closest Belarusian city to the smoldering Chernobyl plant, announced a general strike until the foreign experts came to their town. They had a list of demands; the first was a study of Chernobyl health consequences. The foreigners, they were sure, would help them.[2]

Honoring the strikers' wishes, the commission of well-dressed scientists took a detour to Navrolia, on an itinerary that included a number of suffering communities. In each town, the visitors spoke to packed auditoriums with the cameras rolling.[3] They sped through Minsk, where they met with Belarusian scientists, who told them about the wide range of health problems they were finding. The experts moved on to Moscow.[4]

After their whistle-stop tour, the WHO consultants came to the conclusion that any association between the reported rise in noncancerous diseases and Chernobyl fallout was a mistake of "scientists who are not well versed in radiation." They suggested instead that health problems were due to "psychological factors and stress."[5] They wrote that

350 mSv "would present only a small risk to health, comparable with other risks to human life."[6] Using as a baseline the findings of the Life Span Study of Japanese bomb survivors, they recommended raising the safe lifetime dose two to three times higher. This was a remarkable statement, given that there was no accepted understanding at the time of how bodies responded to chronic, low doses of radiation. The WHO consultants were proposing a threshold ten to fifteen times higher than the recommendations of the IAEA and the International Committee for Radiation Protection.[7]

After the foreign experts boarded planes home, Minister of Health Chazov prepared a circular to party leaders. He singled out the rebellious Belarusian scientists for censure: "foreign experts had an exceedingly grave impression of the level of radiological competency of some specialists of the Belarusian Academy of Sciences."[8]

Unfortunately, slander is one tool in the cultivation of knowledge. When villagers said they were sick from Chernobyl fallout, they were derided as frightened and ignorant. When Belarusian scientists who had spent the previous four years studying the effects of Chernobyl exposures said people were ill, they were dismissed as poorly trained and incompetent by experts who visited for just a few days. The president of the Belarusian Academy, V. P. Platonov, wasn't having it. He aimed a reply in a letter to Nikolai Ryzhkov, Gorbachev's second in command. Chazov's statement, he said, was "insulting" and ill-informed. The 350 mSv safe lifetime dose was "scientifically unsound."[9]

The world's leading health agency's hasty assessment greatly disappointed Soviet green parties, citizens' groups, and activists.[10] The World Health Organization report was also an embarrassment for international science. Radiologist Fred Mettler, U.S. delegate to the UN Scientific Committee for the Effects of Atomic Radiation (UNSCEAR), said of the WHO mission, "No one received what they said very well. They put in no time, no real resources."[11] With the cameras following, the WHO mission was more media spectacle than investigation. Science performed for political theater.

The controversy over the Chernobyl diagnosis caused a rift. Ninety-two Soviet scientists signed a joint letter to Gorbachev supporting the

350 mSv safe lifetime dose. They justified it by citing the Life Span Study of Japanese bomb survivors. They also argued that the lower lifetime dose Belarusians and Ukrainian leaders sought would displace a million people and cause a lot of anxiety, which would be far more damaging, they argued, than 350 mSv. The signatories included the roster of nuclear physicists and experts in radiation medicine in Moscow and Leningrad and their collaborators in Kyiv.[12] Lined up against them were scientists and doctors who worked in territories directly affected by the disaster. The Ukrainian Academy of Sciences and the entire Union of Soviet Radiobiologists sided with the Belarusians.[13] Belarusian scientists cited the work of Moscow and Leningrad researchers who opposed them, scientists who apparently no longer openly followed the science they once charted in closed institutes.

The dissident scientists' point was simple and essential. One could not generalize the estimated radioactive exposure from Japan to Ukraine and Belarus, from a single bomb blast (as it was conceived in the Life Span Study) to an eviscerated reactor. Denying that "residual radiation" was a factor, scientists in Japan calculated that the atomic bombs delivered one large dose of gamma radiation to Japanese survivors, which passed through their bodies in less than a second. They failed to take into account the buildup of radioactive fallout in the food chain and environment of the bombed cities.[14] That estimation of a single-shot exposure, the Belarusian scientists contended, differed essentially from Chernobyl's slow drip of beta and alpha particles ingested in contaminated food and dust to accumulate in human organs and flesh over many years. The Life Span Study began five years after the bombing of Hiroshima and Nagasaki, whereas Ukrainian and Belarusian doctors got to work within weeks examining people exposed to Chernobyl radioactive contaminants. Grasping these differences, the Belarusian and Ukrainian scientists argued they saw more subtle, immediate changes to health and had a better grasp of how radioactivity at low doses incorporated into bodies damaged the health of infants, children, and adults.[15]

While central Soviet leaders tried to control the situation, Belarusian and Ukrainian diplomats went rogue. They sent emissaries abroad as if they were independent countries. They asked for foreign aid and for their own foreign experts to review the safe residency plan.[16] Meanwhile, Belarusian scientists kept working. Even the drowsy Ministry of Health got involved. In Minsk, Larisa Astakhova, assistant director of the recently opened Belarusian Institute of Radiation Medicine, gave a newly donated ultrasound machine to her assistant, Valentina Drozd. Astakhova scanned the recently published maps of radioactive contamination and realized that many thousands of Belarusian children had received high doses of radioactive iodine to their thyroids. Astakhova told Drozd to go to southern Belarus and check children's thyroids.[17]

Yuri Spizhenko, who replaced Romanenko as Ukrainian minister of health, wrote to Chazov doubting the assurances of Moscow scientists: "Many meetings with people and questions from deputies have convinced us that our system of public health is expected to firm up demands for effective radiological protection." Spizhenko worried especially about strontium-90. "At first we were told there was not enough of it for concern, but from year to year the sanitation services find the biological availability of strontium is on the rise. This is a growing phenomenon."[18] Natalia Lozytska, the physicist who disguised herself as a cleaning lady, wrote to Spizhenko's ministry previously that year about strontium-90 and its dangers. Apparently, her message and that of other scientists got through finally to the top ranks of power.

I was struck by the force of this Lozytska Effect. She and other concerned citizens—people who were shut out of information, censored, threatened, watched, bodily removed from conferences as Lozytska was in 1988, and had microphones silenced while they spoke at rallies in Kyiv—mobilized their expertise and moved a state as large and elephantine as the Soviet monolith.

I visited Lozytska and her husband, Vsevolod, several times at their university observatory and at home in a Brezhnev-era high-rise. Natalia gave me a Ukrainian peasant shirt embroidered with sky blue thread because she was grateful that after three decades someone finally answered her letters.

Vsevolod showed me around the circa 1953 solar telescope he and Natalia managed in an observatory, hidden in a courtyard garden of flowers and fruit trees in central Kyiv. "Everyone has satellite telescopes now," he explained, running a hand over the missile-shaped barrel of the scope. "No one uses telescopes like this anymore." The couple dedicated their careers to studying the eruption of plasma jets on the sun. Vsevolod was obsessed with the topic. He returned to it at every opportunity, speaking almost mystically about plasma jets. His ever rational wife would gently change the subject.

One evening Natalia was out, and Vsevolod could dwell on his favorite topic unhindered over a meal of stuffed peppers. He said the solar telescope taught them not just how the sun affects the earth, but also how the earth and the actions of humans on it impact the sun. He showed me charts of sun eruptions over 150 years of observation. The biggest jet occurred in 1946, which he interpreted as right after the bomb fell on Hiroshima. More eruptions followed in quick succession during the years of atmospheric nuclear testing. I listened skeptically. Later I found signs not of the bomb's impact on the sun but on the planet's stratosphere when American and Soviets exploded high-altitude nuclear bombs in the early 1960s with the specific purpose of changing the electromagnetic fields and radiation belt surrounding the planet.[19] The bomb, Vsevolod read into his graphs, was a mistake, a huge error that altered the heavens itself. "Before we turn the earth into something miserable for human life," Vsevolod wanted me to know, "we have to recognize that mistake."

# Marie Curie's Fingerprint

For many the bomb wasn't a mistake. Plenty of people believed then and now in the great promise of nuclear chain reactions to deliver national security in the form of deterrence and economic security as renewable nuclear-powered electricity. At a time when somber crowds were forming human chains around Soviet nuclear power installations, Alexander Kupny applied for a job at the Chernobyl Nuclear Power Plant and moved with his wife and children into the brand-new town, Slavutych, put up to replace the ghost city, Pripyat.[1] In 1989, Kupny got hired as a health physics technician at reactor No. 3, once paired with reactor No. 4 and still functioning after the accident. He volunteered because the wages were good, Slavutych's shops were specially supplied with furniture and sausage, and he believed it was his duty to help out.[2]

Kupny was also intrigued professionally. The smoking Chernobyl plant had levels of radioactivity like nowhere else on earth. "Chernobyl," Kupny told me the first time I met him, "was the Klondike of radiation fields." Because of the accident, Kupny had a chance to measure radioactivity at levels few others could experience. Speaking about the entombed reactor No. 4, he said, "I didn't look at the Chernobyl sarcophagus with fear—you can't study something you fear. I saw it as opportunity."

As the years went by, Kupny wanted better ways than a ticking needle to grasp the tremendous power erupting from a nucleus. Daily he passed the sarcophagus, which had two doors covering cavelike openings that workers dug out to access the crushed control and machinery rooms inside. After staring at those beckoning voids for years, Kupny's curiosity overcame his good sense.

He and a friend, Sergei Koshelev, put on hazmat suits and gas masks, strapped on cameras, and, when no one was looking, slipped down the rabbit hole into the blown reactor.

Kupny showed me photos he snapped during those expeditions. The fire generated by the eruption burned at over 2,000 degrees Fahrenheit. The tremendous heat melted iron, steel, cement, machinery, graphite, uranium, and plutonium, turning it all into running lava that poured down through the blown floors of the power plant.[3] The lava eventually cooled into stalactites, black, sparkling, and, at first, impenetrable. One stalactite is called the "elephant's leg" for its thickness, gray shade, and deep furrows. In the months following the accident, scientists estimated (because it was too hot to measure directly) that the elephant's leg emitted 100 Sv/hr. A few hours' exposure could serve a fatal dose.

"We went there as partisans," Kupny recounted. "We took the risk on ourselves. The fewer people who knew about it, the better."

Kupny was cavalier about the danger of their expeditions. They had half an hour, forty minutes max, to stoop, crawl, and wriggle into the underground chambers, take pictures, and get back above ground before they were overexposed. After a few trips, the men did not bother with a detector. "The radiation fields do not change much," Kupny remarked. "I knew what spots recorded single digits, and I knew where the numbers spiraled into the tens and hundreds."

Even without exposure to radioactivity, the excursions were risky. The caverns under the reactor are former rooms—the control room, a pump house, the turbine halls—but they are no longer in the same places or in the same order they were when the reactor was operating. The basement of the turbine generator hall was filled with water. Planks thrown over it served as flooring. Spilled oil, slimy and slippery, seeped everywhere. The men had to step lightly around cables, potholes, and ankle-trapping crevices. "Just walking," Koshelev noted, was "a hazard."[4] It would have been easy to trip and pitch into a hole or have heavy doors swing shut and jam, trapping them below. Flashlight batteries are unreliable in fields of radioactivity, and they could have given out, not gradually, but suddenly with no warning. Pitched into darkness, the explorers would have had to feel their way out of the crypt. The hazmat suits were hot and cumbersome, and gamma rays still penetrated them. Trip after trip, a body accumulates doses.

I asked Kupny again why he went down there. Was it the same motivation that drives people to climb Mount Everest?

Kupny bristled. "I don't do it to put a flag down and beat my chest. I went under to figure out what happened. I see nuclear energy as a natural force. I wanted to know that force. Unless you go there, you can't understand it."

Kupny is not your usual disaster tourist, capturing photos of ruin as a metaphor for human folly. Squirming into the spleen of the burned-out reactor was as close as Kupny could get to entering a mushroom cloud. His life work had been to visualize and map the scarcely sensible phenomena of radioactive decay. He placed his vulnerable body before a barrage of gamma rays because he sought to grasp what he considered was the elemental force burning at the center of the universe. Kupny is not unusual. He comes from a long line of scientists and technicians who have been fascinated with the energy released from splitting atoms.

And that tradition has a particular history. It grew roots at the end of World War II and flourished in the heavily fertilized soil of the Cold War arms race. Without the Cold War, civilian nuclear power reactors like Chernobyl would never have made sense. The technology for nuclear power generation was borrowed from bomb-producing reactors, yet even with free, army-issue designs, the reactors were pricey to build and risky to operate. The rationale to construct expensive power reactors at a time when oil was flowing cheaply from the Middle East makes sense only if you factor in the Cold War. Bomb-producing nations sought a peaceful atom as an antidote to the skin-melting horrors that nuclear war presented. "Peaceful" nuclear power made for good public relations.

After Hiroshima, American propagandists glossed over the deaths and acute radiation illness suffered by Japanese bomb survivors.[5] Soon after the war ended, Hollywood distributors sent MGM's film *Madame Curie* to occupied Japan, a biopic about the first scientist to distill radium. The film shows the Curies in their ramshackle lab at the moment of discovery. In a soft-focus shot, Pierre Curie looks at the empty space behind Marie's shoulder and evokes the limitless metaphysical horizon, "If we can prove the secret of this new element, then we can look into the secret of life."

In 1953, as the public grew anxious about global fallout from accelerating nuclear bomb tests, the U.S. National Security Council resolved that "economically competitive nuclear power" must become "a goal of national importance."[6] President Dwight Eisenhower unveiled the "Atoms for Peace" program, dedicated to using nuclear power for medicine and to generate electricity "too cheap to meter." Eisenhower offered to share American nuclear technology with other countries.[7] With that salvo, a race began between the Cold War superpowers to outproduce each other not only in first-strike capability but also in civilian reactors for electric power and nuclear medicine.[8] American companies, prodded on by generous federal subsidies, geared up. In 1948, the newly created U.S. Atomic Energy Commission gave away to research institutions 2,000 shipments of radioactive isotopes, while investing a million dollars in research for the use of isotopes to treat cancers.[9] By 1957, U.S. officials had plunked down twenty-nine small research reactors abroad, some in countries that had trouble providing general education and medical care to their populations.

Soviet leaders gamely joined the race with the Americans. In 1954, engineers plugged in the world's first nuclear power generating reactor in Obninsk, a closed city near Moscow. The reactor produced all of five megawatts for the grid, but the propaganda value of juxtaposing the peaceful Soviet atom against the martial American A-bomb was immeasurable. In the years to follow, crews built reactors outside major cities of the USSR and East European allies. Soviet society welcomed nuclear power and accepted the idea, blasted from billboards, that the "Soviet atom was a worker, not a soldier."

For Americans, selling the idea of nuclear power in Japan was not seamless. In 1954, the wind shifted during the Bravo test of a hydrogen bomb in the Bikini Atoll and contaminated a Japanese fishing vessel, the *Lucky Dragon*, with a thick coating of black fallout. By the time the vessel reached port, the crew was suffering from radiation poisoning. Japanese panicked on learning that the *Lucky Dragon*'s radioactive catch was making its way through fish markets. A few months later, the antinuclear film *Godzilla* hit Japanese theaters. Nine million Japanese paid to see the sci-fi horror film depicting a deep-sea, prehistoric monster wakened by oceanic nuclear tests.[10] The confused and angry beast pulverized Japanese cities with his radioactive breath. As protests erupted in Japan against nuclear testing, Japanese and American leaders quickly settled in closed-door meetings on a deal to transfer U.S. nuclear designs and fission products to Japan. Among Japanese leaders, who were eager to secure an independent power source for industrial expansion, the choice between the two competing visions of nuclear power—that of *Madame Curie* or *Godzilla*—was easy. The antidote to nuclear fear was the beacon of nuclear power.

President Eisenhower proposed the creation in the United Nations of an international nuclear regulatory agency. The UN International Atomic Energy Agency (IAEA) started as a relatively toothless branch of international governance.[11] It had no power to inspect or control the bomb-producing superpowers and could only regulate countries seeking to build nuclear reactors. For many nonnuclear countries, the IAEA's lack of oversight over nuclear arsenals and bomb testing was troubling. A Norwegian geneticist drafted a resolution in 1955 noting the "already demonstrated" genetic damage nuclear weapons tests had caused and called for an international study of health effects from testing.[12]

The body poised to carry out that study, historian Jacob Hamblin points out, was the UN Educational, Scientific, and Educational Organization (UNESCO). Lewis Strauss, chairman of the U.S. Atomic Energy Commission, vehemently objected to an international organization staffed with independent scientists who would meddle in the private concerns of the U.S. military. In the same years, Strauss and Pentagon leaders were fighting off Soviet proposals to end all nuclear testing. Soviet leaders and European protesters argued that testing spread toxic radioactive fallout and encouraged the proliferation of larger, more powerful bombs.[13] Eisenhower, John Foster Dulles, and Strauss masterminded a new UN body to work alongside the emerging

IAEA. Brushing UNESCO aside, Dulles and Strauss pushed to create a new, high-level UN agency to which political leaders, not scientists, would elect the scientists to serve. The new agency would appear to be an independent scientific body but could be deftly controlled by strategic political appointments. The other nuclear powers—Great Britain, Canada, the Soviet Union, and Sweden—enthusiastically supported the new UN Scientific Committee for the Effects of Atomic Radiation (UNSCEAR).[14] In tandem, American diplomats shaped the two new bodies that would govern the international use of nuclear power: the IAEA, a lobby to promote peaceful uses of nuclear energy, and the politically controlled UNSCEAR to evaluate the environmental and medical impact of exposures to radioactivity. It fell to UNESCO to take up studies to help populations psychologically adjust to their "irrational fears" of nuclear power.[15]

Alexander Kupny came of age in this intellectual and social current. He grew up in the shadow of civilian reactors. His father, a nuclear engineer, directed nuclear power plants in the Russian Urals and southern Ukraine. Kupny trailed his father, working a series of blue-collar jobs as a radiation monitor.

The photos Kupny snapped inside the sarcophagus stray far from the utopian promises of nuclear power. They look more like an episode from *Planet of the Apes*. Wrecked machinery and frozen control room dials rest amid wires dangling from fuse boxes and cages of twisted steel. Scattered everywhere are cement blocks tossed on end. Yet the most haunting aspect of Kupny's photos is the snowfall of tiny crystalline flakes floating in deep-sea darkness through his scenes of silent ruin. The tiny orange flecks are not aberrations in Kupny's film. As Kupny snapped his shutter, the effervescence of the decaying reactor fuel jeweled the atmosphere in the dark chamber into a spidery chandelier. These points of light are not representations. The photons of radioactive energy swarming inside the sarcophagus imposed their image on Kupny's film. They are energy embodied. The specks are none other than cesium, plutonium, and uranium self-portraits.

Henri Becquerel first discovered in the nineteenth century the power of irradiated objects to take selfies when he accidentally burned an image of a copper cross onto a glass plate he had left in a drawer with a rock containing a uranium sulfur compound. In 1946, David Bradley, conscripted to monitor radioactivity for Operation Crossroads, reproduced that experiment by slicing open a puffer fish he caught in the warm waters of the Bikini Atoll a few days after American generals detonated two Nagasaki-type bombs on the island. Bradley pulled the fish from an irradiated lagoon, cut it open, and placed it on a photographic plate in a dark room. Returning several hours later, he found the fish's contaminated bones, organs, and final meal had burned their images onto the photographic paper to create what art historian Susan Schuppli calls "a new kind of photosynthesis." Schuppli believes that radioautographs undercut the idea that humans have a monopoly on image making. They show that matter can write its own history.[16]

But that can only happen when bodies switch places with specimens of uranium. A radioautograph is the opposite of an X-ray in which radioactive energy emanates into rather than out of a body. Behind the self-portrait of the puffer fish's skeleton and the raw, pure crystals of light in Kupny's photographs is the work that goes into retrieving uranium from underground depths, isolating it, refining it, and setting up the conditions for controlled or uncontrolled chain reactions.[17] Marie Curie knew that work intimately. Over the course of five years, she and her husband, Pierre, distilled eight tons of a uranium compound known as pitchblende down to a few grams of radium. They did so by boiling and condensing the pitchblende in huge vats. The indefatigable Marie Curie would stay at the lab late into the night to stir the mixture with poles taller than she. She breathed in the vapors and heaved buckets, spilling compounds on her hands, which blistered and festered. Over time, her body and that of the isotopes in the pitchblende joined.

The film *Madame Curie* made a big impression on Japanese audiences. It triggered a widespread interest in nuclear science and in the scientist herself. After Curie died, Japanese collectors purchased some of her papers, which ended up in a Tokyo library. In the early twenty-first century, manga artist Erika Kobayashi went to take a look at one of Marie Curie's notebooks. It was not exceptional; just a sketchpad of notes and observations. But in the library, Kobayashi pulled out a Geiger counter and held it close. She watched the needle rise. Kobayashi was amazed to find that seventy years after her death, Marie Curie's radioactive fingerprint still registered on the page. Curie never visited Japan, but the touch of her finger emits energy in Japan, where her life work played out in both productive and disastrous ways.[18]

Curie wasn't thinking about catastrophe and destruction when she left her fingerprints on that notebook. Her dream was to create a source of energy that would heal wounds and relieve human want and misery, of which she had seen plenty growing up in the Polish dominions of tsarist Russia.

What Curie began, the hopes she invested in achieving a pure source of radiant energy, Kupny carried along. Kupny believed that a lot went wrong at the Chernobyl Nuclear Power Plant, but what he had in mind when he discussed mistakes was not the accident but its aftermath. He thought the remaining Chernobyl reactors should never have been shut down in 2000. "They were," he asserted, "perfectly safe." Kupny was convinced that nuclear power plants offer an acceptable and minimal health risk.

Certainly, exposure to radiation doesn't seem to have been a problem for Kupny personally. At age sixty-eight, he was past the average male life expectancy in Ukraine and presented as ten years younger: lithe and agile, with a brisk step and quick mind.

"As it does with food," he told me with a smile, "radiation works on me like a preservative."

Kupny's intellectual influences followed a straight line directly from the nuclear industry where he spent his career. Officers, especially, in nuclear navies, where sailors live in close proximity to submarine reactors, frequently refer to "hormesis," the idea that exposure to radiation in small doses has favorable biological effects. Scientists tried for decades to prove this notion in studies. They poured lots of money into the endeavor to no avail.[19]

The debate between scientists promoting hormesis and geneticists' warning that no dose is safe echoed in 1989 more loudly than ever

before. A month after the World Health Organization's unconvincing
"triple it" assessment of the 350 mSv safe living dose, the Soviet Ministry of Atomic Energy asked the IAEA to do another "independent"
assessment.[20] In Vienna, Abel Gonzalez, in charge of reactor safety for
the IAEA, was both surprised and delighted at the Soviet invitation.
He set out to resolve the debate over Chernobyl health effects once and
for all.

# Foreign Experts

Abel Gonzalez grew up on the pampas in Argentina in a village with no electricity. Like Alexander Kupny and like most of his colleagues, Gonzalez was dedicated to nuclear power as a solution to hunger, inequality, need, and illness. A nuclear physicist, his was a noble profession with a lifelong mission to bring light to the dark corners of the earth.

In June 2016, I made an appointment with Gonzalez who kindly fit me into his busy travel itinerary between Malaysia and Moscow. I met him at the UN's Vienna International Center, a complex of glass, steel, and convenience. Inside the sleek, midcentury modern buildings, men and women in fashionable suits and impractical shoes glided on polished floors that reflected the pools of sunlight spilling in through glass walls. Behind the international diplomats trailed the scent of cologne and good health.

The UN complex is a long way from both the Argentine pampas and the wash-bucket, cigarette-stubbed corridors of Soviet institutions where in the 1990s researchers in polyester suits and gum-soled shoes hustled in and out of tiny offices, cheap doors banging behind them. At the time, few people doubted which societal choice—socialist or capitalist—was superior. The wash of consumer goods presented the tally clearly to almost every commentator. Capitalism was ascendant. The angel of history flew in triumph across the Berlin Wall in November 1989. East Germans, climbing onto the concrete blocks that had sealed them in, marveled at the beauty and freedom of the West, marveled that they were standing on that wall and not getting shot for standing there. That night in November 1989 opened new possibilities for citizens across Europe.

For decades, intellectual collaboration across the Iron Curtain had been intermittent and managed by government officials. After the Berlin Wall came down, the floodgates opened. Citizens carried along by decades of pent-up economic and social desire started dashing across the Iron Curtain like never before. In the packed Pan Am flights, missionaries with suitcases full of Bibles practiced how to say "free gift" in Russian. Businessmen skated around Moscow looking to sign contracts as they eyed the great, untapped Soviet consumer market. I recollect a man trying to sell lobster tanks to restaurants at a time when it was hard to buy a single salted fish, let alone a fresh one. Another entrepreneur had the mission of marketing diet sweeteners long before Russians considered body fat as anything but a healthy reserve for the next famine.

Those years were a high point of faith in civil society. It was a time of sister city delegations, tele-bridges, and telethons designed to connect citizens who would dismantle the Iron Curtain, one faxed manifesto, one airplane ticket, at a time. Hundreds of charitable organizations, religious groups, and start-up nonprofits scrambled to find Soviet partners with the simple idea that people could engage in diplomacy, trade, and cultural exchange and bypass their warring governments. In this atmosphere, the field of Chernobyl charity quickly grew crowded. In the first months of 1990, competing Soviet ministries and independent associations besieged international charities and the United Nations with requests for help.[1]

European and North American charities answered the call. Loosely organized groups like Green World in Ukraine became small United Nations, fielding scores of guests from Europe and North America. In Paris, UNESCO officials planned to establish a Chernobyl research center to collaborate with the Academy of Sciences on analyzing Chernobyl data.[2] In Moscow, officials of the World Health Organization negotiated with the Soviet Ministry of Health to join forces on a long-term epidemiological study of the low-dose effects of radiation, a study planned to dwarf the Japanese Life Span Study.[3]

With this hurricane of goodwill, the judgment of Chernobyl health effects could get out of control. Gonzalez understood what he was up against. His office tracked the alarming media coverage of Chernobyl health problems. Competitors were cropping up who might offer alternative evaluations of the Chernobyl disaster. Taking on the Soviet request for a new assessment, Gonzalez needed a strategy to head off rivals. It had long been thought that either the World Health Organization or the UN Scientific Committee for the Effects of Atomic Radiation (UNSCEAR), not the IAEA, would assess Chernobyl medical effects.[4] The IAEA was compromised in this one respect: the agency was already on record predicting that Chernobyl would deliver no detectable health problems and that the 350 mSv dose was safe.[5] Even inside the United Nations, the IAEA was seen as a nuclear lobby. UN Assistant Secretary-General Enrique ter Horst described the IAEA's approach to a Chernobyl assessment this way: "The IAEA has its own institutional interest for the promotion of peaceful uses of nuclear energy and would like to allay the fears of the public as to the actual and potential damages of the Chernobyl accident."[6]

Thinking creatively, Gonzalez put together an interagency committee of UN organizations, including WHO and UNESCO, and called it the International Chernobyl Project (ICP). The committee rarely met but presented an image of a collaborative UN project, while Gonzalez actually directed the assessment from IAEA offices.[7] At a May 1990 meeting, the committee members agreed there would be no free market of scientific studies or independent action by UN agencies: "Sending one [UN] mission after another," the minutes read, "without careful preparation would impact negatively on the credibility of each organization." Gonzalez cautioned that agencies in the UN family needed to work together "to avoid premature and unverified conclusions." He suggested that the IAEA should take the lead to "harmonize" the various Chernobyl projects and medical studies to follow the "IAEA-led assessment." Proposing to send in a hundred international experts to make a "snapshot" of health and ecology, Gonzalez asserted that the IAEA's evaluation would be important in the future for recommendations that would be "fed into longer-term programs."[8]

That was the mandate. The IAEA would be in charge of Chernobyl health and environmental studies, and the IAEA-led assessment would be the blueprint in determining how UN agencies funded future Chernobyl relief programs.[9] The IAEA would prepare and write the Chernobyl assessment but, the American delegate stipulated, the International Chernobyl Project, rather than the IAEA, "should be the stated author of the report."[10] With that agenda established, the interagency committee took a long hiatus. It did not meet again for over a year.[11] Gonzalez appointed Itsuzo Shigematsu, head of the Hiroshima-based Radiation Effects Research Foundation, to lead the committee, which would use the bomb survivor Life Span Study as a template.[12]

In Kyiv, KGB agents also worried about the uptick of international activity. They were anxious about the inquisitive foreign scientists who materialized in Ukraine and mixed with Soviet citizens.[13] KGB agents had a practice of taking a daily count of the number of foreigners in Ukraine. Irina, a former Intourist employee who worked with foreign students, told me that KGB agents circulated in places where foreigners dined and socialized. "Nearly everywhere you went," she told me, "someone was nearby keeping an eye on you." As the number of visitors scaled up, KGB agents feared CIA counterespionage, sleeper cells, and American interference in the upcoming elections to the Ukrainian Council of People's Deputies.

I was one of those foreigners in the KGB daily tally. I worked in the Soviet Union in the perestroika years for a Soviet-American educational exchange. We placed American students in universities across the Soviet Union. When our students had problems, it was my job to solve them. I got a call one day from an American student I will call Bob, who was studying in Kyiv. He called me in Moscow and asked me to come to Kyiv right away. He didn't sound right. I boarded a train for Ukraine with a Soviet colleague. Bob met me at the station. He was pale, thin, and shaky. He told me his girlfriend had encouraged him to go to Tashkent on a holiday. Bob did not have a visa to leave Kyiv. She suggested he use a friend's Soviet passport. Bob did, and was nabbed at the airport for committing a border violation. KGB agents threatened him with serious jail time unless he informed on fellow Americans in Kyiv. Bob agreed. After that, Bob would be walking down a street and a black sedan would pull up, a finger gesturing for him to get in and spill his guts. After a few months of this, the twenty-year-old student was a nervous wreck.

We negotiated an end to the harassment with the university president, and Bob went on to marry the girlfriend who led him into the KGB trap. Reading in KGB archives thirty years later, I noticed that KGB agents were convinced that the American CIA sent operatives disguised as students to the USSR to gather intelligence.[14] Meddlesome outsiders, KGB agents reasoned, incited crowds on the streets, but by 1989, Soviet citizens needed little provocation. A loud and angry mob stood each day in front of the Supreme Soviet on Kreshchatik Street, Kyiv's Broadway. Protesters picketed the entrance to the Chernobyl plant. They broke into the territory of another nuclear power plant, demanding the reactors be shuttered. Nuclear power workers talked of going on strike. Chernobyl patients in the radiation medicine clinic did go on a hunger strike, demanding aid and medical help for Chernobyl victims.[15] Hunger-striking students camped the winter of 1990 on Revolution Square in Kyiv. Pale and emaciated in white headbands, their shivering forms embodied "revolution." In Lviv joyful demonstrators held an elaborate funeral for the Communist Party. Students threw volumes of Lenin's essays at one of Lenin's statues, but he didn't flinch so they climbed up the pedestal and took Vladimir Ilyich down. The fallen leader wasn't debased enough, so they chipped away at his jutting chin and high forehead.[16] A student offered me Lenin's brow, just a heavy block of stone. I turned it down.

In 1990, challengers rose up to resist the destruction of all they saw as sacred. Conservatives championed the restoration of law and order and the preservation of the federation in the face of a growing tide of separatist movements throughout the USSR. New associations with traditional names sprang up—"Shield," "Union," and "Fatherland." Starting in 1990, Gorbachev wavered, sometimes acting like a democrat, sometimes like a hard-liner. Solidarity and unity had been the most important myths of the Soviet state. Suddenly social harmony dissolved into a brawl. Parliamentarians in the Congress of People's Deputies, who had in the Communist past agreed on everything solemnly and unanimously, turned on each other. The televised shouting matches heralded the approach of reality TV. Crowds on the streets battled with police and each other. Shots rang out in what were once peacefully dull Soviet cities. In those days, I did my best to steer clear of the swaggering, muscle-bound men in track suits who worked for shadowy "biznesmen." They used the English word, because Russian had no equivalent to describe an entrepreneur who obeys no laws and protects his wealth with mercenaries.

In this political maelstrom, KGB agents rushed to defend the country's scientific interests. "The democratization of society," a KGB officer grumbled, "leads to the problem of the defense of intellectual property."[17] The agents understood that Americans sought to use the growing public disenchantment with the economy and the Communist Party to ignite a "social explosion of immense strength."[18] KGB agents worried that Soviet scientists who went abroad with no hard currency were financially and morally dependent on their hosts. They suspected that foreign security forces carefully watched and possibly recruited Soviet scientists. Using this vulnerability, Western spies had scored some serious intelligence victories. Americans from the Department of Energy and the National Academy of Sciences openly approached researchers at the Kyiv All-Union Center for Radiation Medicine. "They want to work with the absolutely unique data base at the center. They especially want the data sets for the study of indirect and distant effects of radiation to a person's organism," the KGB chief Nikolai Golushko wrote. "We have fixed on persistent attempts to talk the Soviet side into making accessible the preliminary scientific results of our work and also the raw materials of the database." They will use this scientific intelligence on Chernobyl, Golushko feared, "for military purposes and to reduce expenditures for similar research at NATO institutions."[19]

To get at this material, Golushko speculated, Westerners were using international organizations, especially the IAEA and the World Health Organization. They took note of Gonzalez's interagency committee meeting. Instead of seeing that Gonzalez was trying to limit and control information about Chernobyl health impacts, they construed that he was calling for various UN organizations to span out broadly to gather Chernobyl data. Most of their interest, Golushko wrote, "centers on ecological and medical-biological effects."[20]

As usual, Golushko's report ended with a plan of action: "We need to take measures that in case of international contact we do not lose authorized information on the actual problems of liquidating the accident."[21] KGB agents rarely spelled out what kinds of "measures" they had in mind.

# In Search of Catastrophe

I n the spring of 1990, the IAEA mounted a major effort to assess Chernobyl damage, and Abel Gonzalez managed it. He recruited two hundred volunteer scientists categorized into groups—dosimetry, history, health, and social consequences.[1] The first thing IAEA investigators needed to know was how much radiation was out there. So, in the summer of 1990, small IAEA teams dropped onto Chernobyl contaminated lands on a hunt with no quarry. The international experts sought the measurements of bygone jets of elusive radioactive energy. The experts complained that their Soviet partners, especially the Belarusians, withheld information and maps with radiation levels, doses, and calculations.[2] Soviet science managers were under strict orders not to release data to visiting foreigners. Even sick children were considered "biological material" and not to be shared.[3] For their part, Soviet scientists complained that the IAEA experts, in their quick, two-week trips, one parachuting after another, landed at the same institutes, in the same small towns, asking for the same information with no apparent knowledge of earlier missions.[4] The IAEA teams, which Gonzalez created to avoid inefficiency and duplication, were inefficient and redundant. Apparently, many international scientists were enthusiastic about a junket to the closed Chernobyl regions to see what was billed as the world's most radioactive zone. The foreigners were less interested in writing reports or returning for multiple visits to cold Soviet hotels with cranky staff and bad food.[5]

The IAEA teams visited Kyiv's All-Union Center for Radiation Medicine and the Moscow Institute of Bio-physics, Leonid Ilyin's institution, which had carefully managed and controlled Chernobyl data since 1986. Everywhere the IAEA consultants went, KGB agents accompanied them, posing as interpreters. Gonzalez laughed about the surveillance between the two camps of scientists. He described to me his "KGB-interpreters." He reported, "They were smart guys. One told me that a scientist in our group was working for Israeli intelligence. I found out later he was absolutely correct."[6]

KGB agents in Ukraine especially suspected the IAEA and WHO as covers for foreign spies. And that is how Soviet security used UN agencies themselves, as hosts for undercover agents abroad. The mutual suspicions put a damper on scientific exchange. The IAEA missions were looking for dose measurements drawn from Chernobyl-exposed people in the years after the accident. The Soviets had compiled a federal registry of over a million Chernobyl survivors. The secretive Third Department in charge of radiation medicine at the Ministry of Health was the clearinghouse for most of the dose data, but it was well known that the department did not share information.[7] It was the KGB's job to prevent foreigners from getting hold of this material and they kept it locked up in safes.[8]

These were the closely watched parameters of joint Chernobyl research across the Iron Curtain. KGB agents posing as scientists sat down at formal banquets to share mayonnaise-fortified salads with Western spies masquerading as scientific administrators. Each eyed the other knowingly. Maybe they had a secret handshake.

IAEA teams flew to Belarus, where researchers at the Academy of Sciences described the full range of health problems they saw after Chernobyl exposures.[9] They told the visitors about the growth of health problems related to the thyroid and blood systems, about sick children and pregnancy complications.[10] The IAEA teams were doubtful. These reports did not track with the existing science in the West. What, they asked, about doses? There could be many reasons for rising rates of disease, but if the increases in health problems matched doses of radiation in a linear line pointing upward, that would turn a correlation into a significant case for the Belarusians' claims that radiation damaged multiple human organs and caused disease at much lower doses than ever known before. The Belarusians had an IBM computer with a registry of individual doses of over 100,000 people, including evacuees, liquidators, and 34,000 children.[11] The foreigners asked for the registry once, twice, many times. They were put off.[12]

And then something strange happened. Soon after the first IAEA scientists began arriving in the USSR in the summer of 1990, "hooligans" broke into the offices of the Institute of Radiation Medicine in Minsk. In what looked like a professional job, the thieves knocked out a security guard with a blow to the head, took his keys, stole two computers and a pile of floppy disks, and sped off in a waiting car.[13] The computer files contained the data on radiation exposure for 134,000 Belarusians.[14] The police recovered the computers, but the hard drive was erased and the disks were gone.

That unique information, the only database of its kind in the world, was never recovered. And that wasn't the only incident. Again in the summer of 1990, files with dose information about villagers hospitalized in Moscow went missing from a computer at a Moscow institute. Even the notebooks disappeared.[15] At about the same time, a rich store of chromosomal dose data vanished inexplicably from an institute in Bryansk in western Russia.[16] I could not get access to the KGB case files, but I suspect they were behind the missing files—one of those "measures" KGB agents deployed to deter "Western spies" from getting Soviet intellectual property.

Instead of the full records, IAEA teams used dose estimates supplied by Moscow scientists but without supporting data to explain the calculations. According to these estimates, several scores of villagers would have to be relocated because people were getting doses higher than the 350 mSv lifetime limit. IAEA teams took some of their own measurements to compare against the Soviet data.[17] The Soviets' tally was higher than the IAEA count. The Soviets, they acknowledged, had a sophisticated understanding of regional ecological factors.[18] They also had a good idea of local habits. Nonetheless, the IAEA often dismissed information Soviets gave them and made their own guesses about what people ate, the time they spent outdoors, and work patterns. They assumed that people were eating clean food from stores and that "changes in agricultural practices had been effective in minimizing doses."[19]

Ukrainians and Belarusians told the experts this assumption was not true.[20] Local scientists explained to IAEA consultants that Polesians, living in northern Ukraine and southern Belarus, ate large quantities of mushrooms and berries between July and September, 200–300 grams a day.[21] For some reason, IAEA scientists calculated a diet of 300–600 grams of mushrooms a year and omitted berries from the estimates as insignificant.[22] Villagers in Polesia drank on average two liters of milk a day. IAEA teams estimated less than a liter. These omissions were important because mushrooms, berries, and milk were the most radioactive food products. Scientists now know and Soviets knew it then that most of the Chernobyl dose came internally from ingesting food.[23] American researchers had learned the same insight in secret studies of Marshall Islanders living on atolls saturated with fallout from American bomb tests, but they did not share this information.[24] In 1990, IAEA scientists attributed most of the dose, conversely, to external exposures.[25]

The IAEA consultants did not figure in their dose estimates the fact that families burned radioactive wood that contaminated cookware and indoor air.[26] Nor did they consider that villagers used radioactive ashes and manure to fertilize gardens. Foreign experts on a separate UN mission to contaminated areas noticed the empty store shelves, the vibrant kitchen gardens, the people walking along forest roads with baskets of berries and mushrooms.[27] But these signs of exposure were either missed or ignored by the IAEA when calculating how large a dose of radioactivity people received.

Consistently, the IAEA teams selected the most optimistic numbers.[28] These estimates were three times lower than those of Belarusian scientists and two times lower than those of the Moscow group.[29] This transnational computation of doses worked like a game of telephone. Researchers in Minsk and Kyiv reported numbers to Moscow.[30] Moscow scientists recalculated, reduced, and passed their numbers on to IAEA teams, who downsized them again.[31] The Chernobyl doses, the IAEA teams concluded, were two orders of magnitude lower than

Japanese bomb survivors, which meant there was no ground for worry about radiation-induced health problems.[32] The IAEA reporters admitted their numbers were uncertain, but the uncertainties were lost in the solid appearance of the final figures they produced.[33] (Several years later, an international group of scientists came up with dose estimates closer to the Belarusian numbers.[34])

The numbers were not innocent. They had agency. IAEA physicians and physicists used dose estimates to figure out the impact of the accident on human health. Gonzalez appointed Fred Mettler, a professor of radiology and an American delegate to UNSCEAR, to lead the health section of the IAEA assessment. I contacted Mettler.

Mettler had a good memory for names, dates, and details. He was very helpful. He told me how he invited specialists in radiation medicine from around the world to participate in the IAEA assessment. These were top scientists, but missing were some prominent figures in the field who dissented from established opinion on the relative safety of low doses of radiation. Mettler solicited over a million dollars' worth of donations of medical supplies and equipment. He drew up the protocol for the investigation quickly in the spring of 1990 with no peer review. He designed a case control study, picking villages near one another with very different levels of radioactivity in soils, though likely the same food networks and so similar exposures from food pathways, which the study did not acknowledge. The contaminated villages would be the case study and neighboring "clean" villages would serve as controls.[35] In a half dozen communities, Mettler's teams randomly selected twenty residents, adults and children, who were then given thorough medical exams. They did the same for the six control communities in neighboring areas that the dosimetry group determined to have low fallout levels. They examined a total of 1,600 people.[36]

What were the IAEA scientists looking for? Mettler told me over the phone: "We looked for everything: cancers, disease, birth defects." Mettler had no baseline on which to evaluate the data his teams collected. There were no long-term studies of people exposed to chronic low doses. Western practitioners could only see a definitive "marker," or a sign of radiation damage, on bodies at acute, very high doses over 1,000 mSv.

For Mettler and other IAEA experts, ignorance of the effects of low-dose exposure was not an obstacle. Using knowledge of doses they estimated Japanese bomb survivors received, they extrapolated to the Chernobyl case.[37] As I noted above, these were very different exposures. Researchers in Japan calculated a single, high-dose, external blast at Hiroshima and Nagasaki, while Soviet scientists understood Chernobyl contamination to be chronic, low doses of ingested or internal exposures.[38]

Belarusian scientists pointed out the differences between A-bomb and Chernobyl doses. Much of the danger, they told visiting IAEA scientists, came from ingested radioactive isotopes, some in the form of inhaled hot particles, which they found caused more damage than external exposures.[39] On hot particles, and IAEA scientist reported that no calculations had been made since there is "no official method."[40] The Belarusians also saw a problem with the selection of controls. It was unlikely that people in the "clean" regions Mettler chose were really control cases. Belarusian researchers supplied information to IAEA teams showing that people outside contaminated zones had ingested nearly as high levels of radioactivity as those in contaminated areas because of the exchange of food across regions, a deliberate strategy Soviet officials used to contain the disaster.[41] The Belarusians told the visitors they suspected that the ingestion of isotopes had a lot to do with the sharp jump in disease rates in Belarus, increases of 100–400 percent.[42]

On hearing this information, Mettler did not change course. He had a low opinion of Soviet medicine. He found the data they showed him inconsistent and incomprehensible. Most Soviet doctors, Mettler told me, had little training. I asked him what he meant by that. He described how his Soviet counterparts did not know how to operate an ultrasound machine, new technology at the time. He found them sharpening steel needles with a knife because they had no disposable needles. They took small X-ray images to economize on film. Like many Westerners, Mettler conflated Soviet poverty—a lack of disposable needles and equipment—with incompetence.[43]

Mettler told me he detected a lot of disease in both contaminated and control communities, with no real difference between them, and so nothing he could associate with Chernobyl radiation. "The doses were too low," he said. "The evidence was not there." The wide range of symptoms Soviet doctors associated with Chernobyl fallout made no sense to Mettler because scientists had found nothing like it in Japan. Sure, there were differences. The bomb survivors' dose was calculated on a single, high-level exposure for less than a second and the Chernobyl survivors had taken in low doses for long periods. And, yes, Soviet researchers had begun their exploration a few weeks after the accident whereas the Life Span Study in Japan began five years after the bombings, but the models generated by the Life Span Study predicted that there would be a linear relationship between exposure and symptoms. As the doses dropped, so too did the prediction or risk for disease. At the low doses IAEA teams calculated, Mettler expected to see, at best, a few excess cancers in a very large population over many years. Leukemia and lymphoma would appear, if at all, within a few years. All other cancers would manifest after ten years.

I took a closer look at Mettler's study protocol to try to understand what exactly his study had the power to find. What does it mean to examine 1,600 people from a possible 4.5 million exposed? According to contemporary understandings of low-dose radiation, the probability of detecting risk to an aggregate population would require that a study look at hundreds of thousands of cases.[44] A random examination of 800 exposed and 800 controls would turn up statistically significant effects only if people were dying in the streets, tragic effects so great they would be difficult to miss.

And, indeed, as one might expect from the study design, the IAEA teams found nothing they could associate with Chernobyl radioactivity. Without mentioning the study was designed to detect only catastrophic effects, the final IAEA report stated that the international scientists found no increase of leukemia or thyroid cancer and no statistically significant growth in fetal anomalies or infant death. Among the people in the study, the IAEA teams saw a lot of disease but, they reported, "no health disorders that could be attributed directly to radiation exposure."[45] The UN press release suggested that poorly trained Soviet scientists had erroneously produced the impression of adverse health effects.[46] Psychological stress, the IAEA group asserted, was the biggest

Chernobyl health consequence. The anxiety from evacuation and fear of radiation caused, they speculated, the high rates of disease Mettler's group had noticed.[47]

With these results, the IAEA made recommendations. The consultants determined that measures such as relocation of people on highly contaminated ground and food restrictions for them were not necessary.[48] The one note of caution in the optimistic IAEA report, which few commentators noticed, was that there might be a statistically detectable increase in thyroid cancer among children in the future.[49]

Mettler told me when he wrote his report that he understood that childhood thyroid cancers were inevitable because some children had received very high doses of radioactive iodine to their thyroids. He said that when he first wrote that prediction into his report, the editors in Vienna had taken it out. "I had to fight very hard," he told me, "to get it put back in."

I was left with the impression that Mettler had a very good memory for names, places, and dates and was a scientist with integrity—and that he prided himself on these qualities.

# Thyroid Cancer: The Canary in the Medical Mine

I n the spring of 1991, Abel Gonzalez rushed a summary of the IAEA's Chernobyl assessment into print. His Belarusian partners were furious because they did not have a chance to read a translation in Russian before it was released to journalists. Once they saw it, they were angrier still. Foreign Minister Piotr Kravchenko complained that the IAEA assessment of "no health effects" was "overly optimistic" and did not take into consideration Belarusian and Ukrainian research.[1] Worse, the report threatened to undermine the success of an upcoming United Nations pledge drive to raise $646 million ($1.1 billion in 2017 dollars) for a long-term epidemiological study on Chernobyl health effects and for relocation of several hundred thousand people from contaminated territory.[2]

In May, the IAEA held a conference on the results of the Chernobyl assessment at the agency's Vienna headquarters. The conference proceedings make for dramatic reading. Scientists backed by the IAEA, Fred Mettler among them, calmly reasserted that their case control study had found nothing and that doses were too low to predict any but a tiny percentage of future cancers.[3] Scientists from Ukraine and Belarus noisily rejected the IAEA's dose estimates. People in the study's control group in reportedly clean areas, they argued, ate food that was contaminated. Even as far away as Zagreb the amount of strontium-90 in people's bones doubled after Chernobyl.[4] They charged that IAEA investigators overlooked a lot: hot spots of radiation, the resuspension of plutonium particles kicked up in dust, and the ingestion of radioactive particles. People's doses were much higher, they insisted, than the IAEA estimated.

"The main factor governing radiation dose now is contamination by cesium," Gonzalez directed the conversation. "It is not contamination by strontium; it is not resuspension."[5] One manifestation of power is being able to determine parameters and definitions. Gonzalez had that muscle. He steered the group to concentrate on the collective dose instead of individual doses. The concept of a collective dose grew out of technocratic planning of the 1960s when scientists used cost-benefit analysis to calculate the risks of nuclear testing against the benefits to the nation as a whole of increased security.[6] A collective dose was a number that described the exposure of no real person but to a population as a whole. It was an estimate, a number that included guesswork and uncertainty.[7]

Soviet researchers at the conference had more immediate concerns about actual doses to actual individuals who were particularly vulnerable—fieldworkers, tractor drivers, pregnant women, and children in contaminated territories. The Belarusian and Ukrainian scientists showed slides and charts laying out the increase in health problems and autoimmune disorders they had found in the previous four years. The international scientists repeated that there could well be a surge in health problems, but the doses were too low to attribute them to Chernobyl.

The Ukrainians and Belarusians asked why the report did not include fourteen Belarusian children and twenty Ukrainian kids who had thyroid cancer.[8] Thyroid cancer in children is extremely rare. Belarus before the accident had fewer than two cases a year. The increase, if verified, would be telling. In their report, IAEA scientists stated that "most of the reports of thyroid cancer were anecdotal in nature."[9] At the conference, they said they did not know of the alleged thyroid cancers. They accused their Soviet partners of withholding this information from them.[10]

When I talked to Fred Mettler over the phone, I asked him whether he noticed thyroid cancer when he was carrying out his study in the contaminated zones. "Thyroid cancer is very difficult to diagnose," he told me, "and it's easy to get wrong."

I asked Mettler about a 1994 BBC report stating that a Soviet scientist had given him twenty histological slides from Ukrainian children diagnosed with thyroid cancer tumors to verify. He sounded surprised. "The BBC reported that? I was never given any samples. I first heard of those thyroid cases at the meeting in Vienna."

The BBC story could have been, I reflected, another example of sensationalized reporting on Chernobyl. Still, I puzzled over his words. I realized that truths that appear clear in the present were not so easy to discern in the past. Today, childhood thyroid cancer is recognized as the major medical outcome of the Chernobyl accident, but it wasn't always the case. In the 1990s, scientists fought bitterly over the alleged outbreak of childhood thyroid cancer and its tie to Chernobyl. The controversy hinges around timing; what did scientists know and when did they know it? I tried to piece together a time line.

BY THE TIME Mettler arrived in the Soviet Union in 1990, Soviet doctors were well aware of a new trend in thyroid cancers among children.[11] In 1989, Valentina Drozd, a young pediatric endocrinologist, turned on a Japanese-donated ultrasound machine to examine several thousand children in Gomel Province. She expected to find not cancer but thyroid disease from exposures to radioactive iodine. She went to some of the same towns Mettler visited for his IAEA study. Like Mettler, Drozd discovered that about 1 percent of the children she examined had nodules on their thyroids, rare for children. She knew the nodules could be malignant.[12] Mettler did not follow up, but Drozd did. She performed needle biopsies and discovered to her astonishment six children with thyroid cancer. She knew right away this was a significant medical event. She kept going. In the heavily contaminated Belarusian town of Khoiniki, she diagnosed five more children with malignant tumors. She reported this news to her boss, Larisa Astakhova, who passed it up the chain of command. In 1990, the number grew to thirty-one cases, fifteen times higher than background.

Drozd kept working. She gathered more information. She typed it up, checked her results, and checked them again. A few months later, Drozd suggested to Astakhova that she report on the cancers at an important

Soviet conference about Chernobyl health effects. Astakhova wasn't sure. "What if we are wrong? We should confirm our numbers again." At the time, Astakhova, as assistant director of the Minsk Institute of Radiation Medicine, was pressed on all sides. Foreign scientists were asking her for Chernobyl data that she had no authority to release. Moscow pressured her to say less. Belarusian diplomats, appealing abroad for humanitarian aid, wanted her to say more. The in-house KGB agent would call her into his office. She would return in tears. Drozd, no longer a submissive *Sovok*, convinced Astakhova to report the cancers by reminding her of the children who would go untreated if they waited.[13]

At the conference in August 1991 in Chernihiv, Astakhova approached the stage to take the podium. She showed slides and gave Drozd's figures to the hall of scientists, mostly men, many from the Moscow Institute of Bio-physics and some from the American National Cancer Institute (NCI) and the World Health Organization. As Astakhova spoke, the sound of grumbling began to vibrate through the hall. Before she had finished speaking, a voice shouted from the audience, "Get off the stage, you little fool [*durochka*]!" More voices joined in a round of heckling. Right there, at the podium, Astakhova burst into tears. "They took her," Drozd recalled, "and made a cutlet out of her."

Astakhova dropped down from the stage. As she passed Drozd, sitting in the front row, she hissed, "I'm going to kill you."[14] Emotions blew hot on all sides. Leonid Ilyin, the biophysicist in charge of Soviet health assessments, called the Belarusians "traitors" and vowed never again to set foot in the republic. But a couple of American scientists from the NCI approached Drozd and told her they did not doubt her research. The Americans had seen similar epidemics of pediatric thyroid cancers in the Marshall Islands and in Utah, near American nuclear test sites.[15]

Not that the Chernobyl thyroid cancer epidemic was a secret. Lots of people knew of it. Why did Soviet scientists shout down Astakhova? Gender might have had something to do with it. Pediatric endocrinologists were rare and they tended to be women, while medical research was generally a male realm. Women in Soviet medicine served as assistants but rarely as directors, as doctors more than researchers. Rushing down halls crowded with pale children, women did much of the heavy lifting in treating patients. A female researcher who broke ranks from established opinion made an easy target. Also radiologists were used to thinking of radioactive iodine as a diagnostic tool and a beneficial medicine to treat Grave's disease and cancers.[16] They had a hard time assimilating the fact that the same medicine that cured cancer could also cause it.

To sustain faith in the benevolent qualities of radioactive iodine a person had to overlook the history of nuclear weapons development. Leonid Ilyin and other scientists at the Institute of Radiation Biology knew to expect thyroid cancers and leukemia after Chernobyl because such cancers had been among the mushrooming diseases that grew after radioactive rainfall at Soviet bomb sites.[17]

American scientists too had a lot of data on radiation-induced thyroid cancer. In the 2000s, anthropologist Holly Barker and historian Martha Smith-Norris drew on declassified American records to reveal that U.S. officials used Marshall Islanders in human radiation experiments after exposing Islanders in tests of nuclear weapons. American officials called the exposure of Marshall Islanders in the Bravo detonation an "accident." Smith-Norris finds that inexplicable because they relocated Islanders for earlier, less powerful tests, while they left the Islanders in place for several days as a gray snowfall of fallout blanketed their homes after the Bravo test of a much more powerful hydrogen bomb. Once removed to safer ground, exposed Islanders were given regular medical exams (but no treatment) by researchers from the Brookhaven National Lab. The scientists set up a control group for comparison. The existence of a control group suggests the intent of a medical study. The Americans were interested in the rates of ingestion and retention in human bodies of harmful isotopes such as plutonium and strontium-90. In the top-secret studies, the American scientists recorded thyroid cancers and thyroid disease among 79 percent of exposed Marshall Islands children under age ten. Anemia in the group was rampant. They also learned that Rongelap women exposed in the Bravo test had twice the number of stillbirths and miscarriages as unexposed women.[18] The investigations of Marshall and Bikini Islanders offered a much better analogy to the Chernobyl case than the Hiroshima/Nagasaki Life Span Study because American researchers recorded the Islanders' initial exposures in the blast, plus exposures from

fallout that rained down on them afterward, much like Soviet researchers recorded the emissions from the nuclear explosion at the Chernobyl
plant and the subsequent fallout as the reactor burned in the weeks to follow. American scientists also tracked the radioactivity in food the Islanders ingested in subsequent years. They began to examine the health of
the Islanders within days, not after five years as did researchers in the
Life Span Study and Mettler and other IAEA consultants in their small-
scale study of Chernobyl-exposed people. The secret studies of Marshall
Islanders took in the full range of exposures much like Soviet researchers'
work in their classified investigations of the Chernobyl-exposed.

In the early 1990s as this Chernobyl thyroid story broke, U.S. officials
did not openly speak of their findings in the Marshall Islands. The study
was still classified as secret most likely because the Brookhaven investigators violated basic laws protecting human subjects in medical studies. At the time, Marshall Islanders and downwinders of nuclear tests
in Nevada were pursuing their case in U.S. courts.[19] For decades, U.S.
officials had stated that medical examinations of the Marshall Islanders
had shown "no aftermaths of fallout" and that the Islanders' "general
health is satisfactory."[20] Closer to home, scientific administrators at the
National Cancer Institute and the U.S. Public Health Service were in
the same years sitting on studies indicating that children directly downwind from the Nevada test site had three to seven times more cases of
leukemia and thyroid cancers.[21] These were the liabilities—the known
facts—that Chernobyl threatened to lay bare.

Joseph Lyon, a Utah epidemiologist, told me over the phone about his
political difficulties in the 1990s carrying out research on fallout from
the Nevada tests. After a court battle, Lyon finally won funding from the
National Cancer Institute and the Department of Energy for a study of
test site downwinders.[22] Bruce Wachholz, a DOE legal adviser in radiation
damage claims against the US government and involved in the AEC's medical studies of Marshall Islanders, left the DOE in 1983 for the NCI the year
the study began. He was put in charge of managing the NCI fallout study.[23]
According to Lyon, Wachholz badgered the Utah scientists, sending com-
missions every few months to grill the investigators and demand extra
paperwork. Lyon described to a Senate hearing how NCI officials piled on

bureaucratic procedures to slow the work down. Over the phone, Lyon told me, "I don't think they ever wanted to produce anything."[24] Someone at the DOE leaked Lyon's findings to a newspaper and called his numbers into question so that the *Journal of the American Medical Association* embargoed publication of his results. At the same time, the NCI started its own rival study, an elaborate dose reconstruction of nuclear test fallout across the United States.[25] Wachholz seemed to slow-walk the study for fifteen years. He dragged his feet so long that one colleague quit the study and another formally protested the delay.[26] Robert Alvarez, a senior official in the Energy Department recalled discovering in 1997 the NCI fallout study, which though finished five years previously, had not been published. He asked why the study was not yet published and was told that Wachholz's office was waiting on the results of the Chernobyl study. One of the disgruntled former employees told Alvarez that "Bruce [Wachholz] was convinced if they just sat on the fallout study, it'd never get out".[27] NCI director Richard Klausner issued an apology over the delay. When I reached Wachholz on the phone to ask him about the case, he refused to speak to me, and the next day the public relations officer of the NCI directed me not to contact him.[28]

The NCI study discovered doses of radioactive iodine to Americans across the country at much higher rates than previously known. U.S. military leaders detonated a hundred nuclear bombs in Nevada, which sent 145 million curies of radioactive iodine wafting across the United States. The blasts blew fine radioactive particles high into the troposphere and stratosphere where they skated eastward across the continent on rivers of air to gradually silt down with rainfall. Proximity to the bombs mattered little. Estimated doses in Tennessee were nearly as high as those in Utah and Arizona near the test site.[29] NCI statisticians later estimated that Nevada fallout caused an extra 11,000 to 200,000 thyroid cancers among Americans.[30]

These devastating results have not been fully assimilated into American history, which has focused mostly on soldiers and bomb workers, not suburbanites and school kids encouraged to drink milk for strong bones.[31] Fallout of radioactive iodine from atmospheric detonations of nuclear bombs in Nevada dwarfed Chernobyl emissions three times over.[32] Yet American leaders hardly monitored the radiation as it spread north, east, and south across the continental United States. Had American leaders restricted the sale of fresh milk from pastured-fed cows in the weeks after each test, they could have prevented tens of thousands of cancers and cases of thyroid disease.[33] That course of action was politically impossible at a time when U.S. politicians made soothing statements about the relative safety of radioactive fallout and when they sent American soldiers to blasted islands in the South Pacific, telling them they were receiving no more radiation than a dental X-ray.[34]

In short, the U.S. Department of Energy, successor to the Atomic Energy Commission, was on the line for having exposed millions of Americans without taking measures to protect them, while the U.S. Public Health Service and the National Cancer Institute had played along in covering up evidence of detrimental health impacts from testing. In the 1990s, as Chernobyl made headlines, Senator Ted Kennedy proposed a Radiation Exposure Compensation Act, and thousands of plaintiffs were preparing to sue U.S. government contractors for their exposures.[35] The number of dollars to pay in reparations was nearly as colossal as the volume of spilled radioactive curies. Chernobyl was the disaster that, if examined too closely, could expose all other nuclear incidents to a heap of lawsuits.[36]

To head off a free-for-all of open-ended questions on the impact of the Chernobyl accident, American officials focused on damage control. Seeking to stifle a multitude of Chernobyl studies, the U.S. Department of Energy (DOE) sent out a circular mandating that government-funded Chernobyl relief aid and research be channeled through the DOE.[37] The U.S. Nuclear Regulatory Commission (NRC) published a study saying Chernobyl could never happen in the United States. Internally, however, one of the five NRC commissioners, James Asselstine, argued the same accident could indeed occur in the United States and that the NRC was not prepared for it. His concerns dismissed, Asselstine left the NRC that month.[38] In the UN, American delegates voted consistently to limit international investigations into Chernobyl health effects, while in 1988 officials quietly created "Working Group 7.0" in back-channel negotiations with Soviet scientists to do joint studies of the Chernobyl-exposed.[39] As administrators at the National Cancer Institute put the brakes on research on Utah downwinders, American researchers set to work on a Chernobyl thyroid study that took two decades to complete.

You can see where this is going. "A nuclear accident anywhere in the world," Dr. Robert Gale noted, "is everywhere in the world."[40] UN officials repeated that Chernobyl was of "unprecedented dimensions," but it wasn't.[41] The U.S. government was just one of several parties that had exposed earth-dwellers to a chronic blanketing of exposure to nuclear fallout. The total emissions from nuclear tests were a thousand times greater than emissions from Chernobyl. The Soviets, British, French, Chinese, Indians, and Pakistanis also made nuclear bombs, a messy process, and blew them up, exposing people to radioactive emissions from 520 atmospheric and 1,500 underground detonations of nuclear bombs (underground tests also vent radioactivity into the atmosphere). From 1945 to 1998, military leaders blew up bombs in deserts, in polar regions, on tropical islands, underground, under water, and at high altitudes. They exploded nuclear bombs on towers and barges and suspended them from balloons. Nuclear weapons tests make up the primary man-made contribution of radioactive exposures to the world's population. Globally, atmospheric tests released at least 20 billion curies of radioactive iodine alone.[42] Chernobyl issued far less at 45 million curies of iodine-131.[43] Three-quarters of fallout from nuclear testing landed in the Northern Hemisphere.[44] That was one problem with Chernobyl health studies; a generation of residents had already been exposed from nuclear testing, Soviet researchers had shown. Background levels of radioactivity and cancer rates in the Chernobyl territories had been on the rise for a decade before the Chernobyl acceleration.[45] By 1986, there was no longer a "natural" level of radioactivity to use as background. With the Cold War in remission, officials had a hard time using "national security" as a reason to keep secrets about emissions from nuclear tests. Citizens learned the extent of their exposures and their governments' denials of it. Lawsuits mounted, so did resistance to nuclear reactors and nuclear weapons.[46] Chernobyl was nothing short of a catastrophe for the nuclear defense establishment.

But, if someone could show that Chernobyl, billed as "mankind's greatest nuclear disaster," caused only the death of a few score firemen and no other health effects, then all those lawsuits, uncomfortable investigations, and recriminations could go away.

# The Butterfly Effect

Valentina Drozd walked into the secretary's office of the Institute of Radiation Medicine in Minsk one day in the fall of 1991, soon after the failed Soviet putsch. The secretary handed her an envelope with a foreign stamp.[1] The letter, from the World Health Organization (WHO), announced it was holding a meeting in Germany about Chernobyl health effects and wanted a report about what was going on in Belarus. Drozd's bosses were out of town. She took it on herself to write a reply, saying she had made some important discoveries about cancers in children's thyroids, the small butterfly-shaped organ on the throat that regulates hormones. She wanted to share them.

A few months later, Drozd and her boss Larisa Astakhova arrived in the Bavarian town of Neuherberg, near Munich. The women described to the group of European administrators the rise in childhood thyroid cancer from a baseline of 1.9 cases annually in Belarus before the accident to 54 cancers in the year 1991. Most of the children's cancers were in advanced stages and were extremely aggressive.[2] The most cancers issued from the most highly contaminated areas in Gomel Province.[3] Children in northern Belarus, where Chernobyl fallout did not rain down, were selected as controls. They showed less significant increases in thyroid nodules and cancers.[4] As proof, the women brought biopsies on slides made from window glass they had cut themselves.[5]

At the meeting, WHO officials, who had been informed about the cancers, pretended it was news to them.[6] They expressed skepticism about the existence of a total of 80 new thyroid cancers in two years among 2.25 million Belarusian children. They judged the women's evidence as patchy. A WHO official criticized the poorly prepared slides.[7]

The Belarusians' homemade slides did not dissuade Keith Baverstock, a scientist from the European office of the World Health Organization. He was impressed with Drozd's well-documented data based on 250,000 measurements. Her research showed that children in Belarus had incorporated extremely high amounts of radioactive iodine in their thyroids, three to ten times more than adults.[8] Baverstock understood that the poor, sandy soils of the forested terrain surrounding the Chernobyl plant were naturally low in iodine. Children's thyroids would have hungrily soaked up radioactive iodine as a replacement for a deficit of the mineral in its stable form. Baverstock also knew of the excess of thyroid cancers among children in the Marshall Islands and Utah. The only question was whether the eighty reported cases were really thyroid cancer. European and American scientists tended to have little confidence in Soviet diagnostic abilities. Baverstock suggested that scientists from the WHO and the European Commission (also at the meeting) form a fact-finding mission to Belarus to look at the cases on-site.[9] The mission was set for May. American scientists from the U.S. National Cancer Institute agreed to join them.[10]

Just as the scientists were preparing to head to Belarus, the mission suddenly dissolved. Baverstock received a letter saying that the European Commission no longer supported the trip. The Americans also withdrew. "Everyone knew of the thyroid cancers," Baverstock told me over the phone. He assumed the cancellations were political. "There was a dispute in the United States over the doses of thyroid cancers at the Hanford [plutonium plant] and the Nevada Test Site. They didn't want the issue raised in the US."[11] The European Commission included the European Atomic Energy Community, an agency with a mission of promoting nuclear energy in Europe, a goal that was increasingly difficult amid popular protests against nuclear power.

Wilfred Kreisel, director of the WHO Division of Environmental Health, told Baverstock that he should also pull out of the mission or risk losing his job.[12] Undaunted, Baverstock found a Swiss donor, recruited two world specialists in thyroid cancer, and went to hungry, crises-ridden Minsk. Six months later, the number of cancers had grown. At a Minsk clinic, Baverstock's team saw eleven children with surgical dressings on their necks. They looked at tumor samples, X-rays, and echograms of the child patients. They studied the histological slides of 104 cases and agreed that 102 of them were cancer. The case was clear. There was no way a country the size of Belarus could produce that many cancers at one place at one time without an external factor. "We believe," the scientists wrote, "that the carcinogenic effect of radioactive fallout is much greater than previously thought."[13]

In early September, Baverstock and his collaborators published their findings in *Nature*, accompanied by letters from Belarusian scientists.[14] The scientists announced the unexpectedly early and large spike in cancers in children from the most contaminated regions and underlined that these cases were aggressive and so would have been detected with or without screening.

In the next several issues, scientists wrote letters to *Nature* refuting the link between the Belarusian cancers and Chernobyl radiation. In a barrage of articles, Valerie Beral from Oxford University, who frequently worked with WHO, Elaine Ron from the U.S. National Cancer Institute, Itsuzo Shigematsu from the Radiation Effects Research Foundation in Hiroshima, and J. W. Thiessen of the U.S. Department of Energy argued the increase in cancers were most likely due to intensified screening.[15] They doubted the cancers were caused by radioactive iodine from Chernobyl. The authors called for a suspension of judgment and for further study. At the same time, Fred Mettler published an article stating he found no difference in thyroid nodules among children in contaminated and noncontaminated areas of the Chernobyl Zone. He advised against any further studies.[16] Repetitive and dismissive, the letters read like a pile on.

Baverstock's *Nature* article caused a feud in the UN family between the World Health Organization and the International Atomic Energy Agency. From WHO, Kreisel wrote to Baverstock's supervisor in Copenhagen reprimanding him for the mission to Minsk and for the publication in *Nature*: "While the increase in thyroid cancer, as established by the mission, is consistent with data that have been available to WHO for some time, the publication of the findings without prior consultation with HQ [Geneva headquarters] causes concern." UN agencies involved in nuclear issues had a practice, Kreisel maintained, of "inform[ing] the other members of major developments with respect to Chernobyl." Most ominously, Kreisel continued: "One of the members, the IAEA, has been questioning, at the highest level, WHO's attitude in this instance."[17]

According to Baverstock, Kreisel drafted a press release with officials at IAEA, which he presented to Baverstock. Kreisel insisted he sign it and withdraw his name and WHO's association from the *Nature* article.[18] He told Baverstock he would be sacked if he did not endorse the retraction. In a phone call, Kreisel told me he did not recall the press release but did remember the disloyalty to WHO of an employee who caused "lots of problems, which in my view could have been avoided had this person been released from WHO."[19] Kreisel blocked funding for Baverstock's independent thyroid initiative.[20] In this controversy, Baverstock again stood his ground. He didn't withdraw his name from the *Nature* article and he continued to organize forums to publicize and explore the thyroid epidemic.[21]

The conflict in the pages of *Nature* sparked a firestorm of medical studies that duplicated and competed with one another.[22] Baverstock initiated a thyroid project in Belarus with a cancer research institute in France that rivaled WHO's slow-moving thyroid study. The IAEA pursued its own investigation of thyroid cancer, violating a prior agreement that the World Health Organization would handle Chernobyl medical issues, while IAEA took care of technical problems.[23] The U.S. National Cancer Institute worked up a protocol for a third study, which advanced far more slowly than tumors in children's glands.[24] Japanese, French, British, and German institutions all funded independent investigations.[25]

In one month, German, Japanese, and French doctors crowded into the small city of Klintsy, each team taking measurements and blood until the residents got fed up with the poking needles.[26] Western agencies treated exposed villagers as did Soviet researchers—as experimental subjects. Despite promises to return to inform individuals of test results, they often failed to do so.[27]

Overcome by this confusing mashup of thyroid research, WHO officials traveled from meeting to meeting insisting that they should have a monopoly on the topic for the sake of efficiency and to "avoid duplication."[28]

IAEA and WHO officials wrote letters back and forth about Baverstock and his meddlesome pursuit of Chernobyl issues.[29] They fought among themselves over control of Chernobyl studies. Officials at WHO argued that they were the UN agency specializing in health. IAEA officials replied they were the experts in radiation.[30] With a grant from Japan, WHO officials organized a pilot study of five possible Chernobyl health effects: thyroid cancer, leukemia, oral health, brain damage in utero, and mental health.[31] Twenty-five years later, the IAEA's Abel Gonzalez remembered the WHO study as a personal act of betrayal.[32] Gonzalez was furious that WHO officers pursued this program because he believed his IAEA assessment, quickly carried out with no peer review, had conclusively resolved the question; there were no health effects. WHO, he believed, should fall in line with the IAEA's judgment. Referring to the 1991 IAEA assessment, Gonzalez wrote to the World Health Organization: "The IAEA has . . . the one documented study on Chernobyl which has been peer reviewed [sic] internationally and it should therefore be the major reference base for any international Chernobyl related initiative." Gonzalez continued, "WHO should tailor its activities" to IAEA's "recommendations and conclusions." He called the World Health Organization "adrift" and its projects "scientifically unsound."[33]

Gonzalez need not have worried. The goal of the WHO studies, officials wrote to each other, was to pacify a nervous public.[34] They did not expect to find increases in leukemia ("doses were too low"), dental problems, or mental retardation ("since the doses were not of that magnitude"). Nikolai Napalkov, assistant director general of WHO, justified the Chernobyl investigations as "important to provide an answer, albeit negative, because the population is very concerned about the possibility of mental effects."[35]

In the mid-1990s, Abel Gonzalez and Barton Bennett, scientific
secretary at the UN Scientific Committee on the Effects of Radiation
(UNSCEAR), traveled widely and attended meetings where they dis-
couraged medical monitoring as bad science: "fishing in a pond and seeing what fish come out."[36] They fought journalists critical of the IAEA
assessment and disparaged scientists who spread "misinformation surrounding the Chernobyl accident."[37] They consistently refused to consider possible health outcomes outside the narrow list prescribed by the
A-bomb survivor Life Span Study.[38] After Baverstock's announcement of
the thyroid cancer epidemic in Belarus, several UN agencies—including
UNESCO and the Food and Agriculture Organization—drew up plans
for Chernobyl relief in their fields.[39] Bennett at UNSCEAR referred to
the IAEA "no health effects" assessment as he advised UN agencies to
cancel and defund these Chernobyl projects. When the childhood thyroid epidemic hit the press, Bennett refuted it.[40] The IAEA, he said,
had shown that radiation levels and food were safe and relocation was
not justified, nor were any other urgent measures required. The main
humanitarian needs in the region, he insisted, were "unrelated to the
accident."[41]

The major UN donor, the United States, curiously failed to take a
leadership role on Chernobyl relief.[42] U.S. diplomats, when they wanted
to, had a powerful influence on funding initiatives. They raised three
times more in donations than they originally sought to rebuild the
crumbling sarcophagus.[43] Yet officials in the George H. Bush administration actively blocked fund-raising programs for health, resettlement,
and UN-directed research.[44]

Bennett and Gonzalez coordinated their work from UNSCEAR and
IAEA offices located in the same Vienna complex.[45] Bennett, in charge of
UNSCEAR, had a small budget and borrowed employees and technical
expertise from the IAEA.[46] Staff cycled between the two agencies. They
cooperated so closely that a U.S. delegate suggested saving money by
placing UNSCEAR within the well-funded IAEA. Bennett and Gonzalez
strongly objected to this plan. The two agencies had incompatible goals,
they argued. "The IAEA supports and promotes development of nuclear
energy," Bennett wrote, while "UNSCEAR reports from an independent,

objective standpoint on the effects and risks of ionizing radiation. The reputation of UNSCEAR would be severely jeopardized by an association with the IAEA." Enlisting the help of Fred Mettler, Bennett and Gonzalez successfully quashed the proposal to merge. Meanwhile, the two agencies continued to cooperate closely, the independent UNSCEAR dependent on the richer IAEA for staff and research help.[47]

Addressing the reported psychological factors, the IAEA published a manual in Russian to teach Soviet leaders how to talk to an anxious public about nuclear issues. Public relations specialists, they counseled, do a better job of informing the public than nuclear physicists. The guide advised public relations specialists to be careful in their language: "Don't tell the public not to eat vegetables, and then tell them if they do eat them, nothing terrible will happen." The manual advised against trusting citizens with complicated scientific data and truthful statements: "In discussions with the 'man on the street,' it is more important not how exhaustive and scientifically correct your message is, but how it is accepted and understood."[48]

In 1994, UNSCEAR initiated a major review of Chernobyl outcomes. Bennett set the parameters. He directed scientists to use the IAEA's very low dose estimates. Then, pointing out that Chernobyl doses were so low, he guided an investigator toward the results he expected: "This [the dose estimates]," he instructed, "should give you some lack of expectations for your epidemiological studies."[49] He coached a Polish scientist who was drawing up a study of hormesis, the theory that low doses of radiation are beneficial to human health: "We must not take a strongly evident radiological protection point of view . . . I know your intention," Bennett winked. "With the right words we can greatly limit the amount of discussion needed."[50]

UNSCEAR delegates reviewed research on Chernobyl health effects that included the work of Soviet researchers whose results pointed to widespread, chronic health problems associated with Chernobyl exposures.[51] UNSCEAR delegates characterized these investigations as "unverified," "sloppy" with "poor quality control," and to be "treated with caution."[52] They found that Soviet and post-Soviet research did not adhere to standardized study protocols used in the West. They did not publish in English, the language of international science, and their findings did not match with results from the A-bomb survivor Life Span Study. Fred Mettler and two leading Moscow radiobiologists volunteered to edit the first draft of the UNSCEAR Chernobyl report, a 600-page compendium, which included Ukrainian and Belarusian research investigating the pathways and biological mechanisms for the rising disease rates in the Chernobyl territories. Mettler and his Moscow colleagues, probably Angela Gus'kova and Leonid Ilyin, returned a year later with a 300-page revision that cut out Ukrainian and Belarusian research and highlighted the conclusions of the IAEA assessment: the doses were too low to cause health problems. Psychological damage and economic hardship, they concluded, were the most likely culprits for a rising incidence of disease. Like the IAEA before it, the 1996 UNSCEAR report recommended against follow-up studies for a number of reasons, including the "presumably low level of risk."[53]

I elaborate on the inner workings of these UN agencies to spotlight how far I believe deeply embedded preconceptions and scientific stubbornness can obscure the truth. In my view, administrators of international science unwittingly shrank from the conclusions that would have turned their scientific views and recommendations on their heads. Like many before them, they were misled by the conditioning effect of careers in research that hitherto had pointed to different risks, different outcomes.[54] The inadvertent effects were seemingly rich and varied: classify data, limit questions, reject investigations, block funding for unfavoured research, sponsor rival studies, relate dangers to "natural" risks, draw up study protocols more likely to find nothing but catastrophic effects, extrapolate and estimate to produce numbers that avoid uncertainties and guesswork and cast doubt on known facts so that scientists must pursue expensive and duplicative investigations to prove what to others is clearly evident. Experts working in the United States had long laboured under the same habits of mind to question scientific evidence and survivor testimony about the harmful effects of exposures to low doses of radiation.[55]

The general effect of this instilled scientific closed-mindedness was to deny damaging evidence and accept only what could not, in the face of overwhelming evidence, be repudiated. For these reasons, after each new nuclear event—for example, after discovery of leaking, burning radioactive waste in St. Louis in 2010 and after the meltdown of three reactors in Fukushima in 2011—scientific administrators announced that little was known about the health effects of low doses of radiation and they needed to study the problem.

While international administrators bickered, the number of cancers reported in Belarus and Ukraine doubled and doubled again, reaching the same order of magnitude in eight years as the total number of all cancers ascribed to the atomic bomb explosions in Japan over forty years.[56] This was a curious fact after international consultants had estimated that Chernobyl doses were much lower than those that A-bomb victims received, so presumably the number of cancers would be fewer. No one stopped to explain the puzzle. In retrospect it appears that researchers in Ukraine and Belarus were correct; doses people received were greater than the IAEA initially estimated.

Nor were all the thyroid cancers "easily curable," as IAEA sources attested. The Belarusian mother of Nina Kachan wrote desperate letters to officials in Minsk about her daughter who had a thyroid tumor removed, but the cancer had spread to the girl's lymph nodes. The family had no money to relocate from their radioactive farm in southern Belarus, and the Belarusian government had no funds either to move the family to safety.[57] The international politics of Chernobyl medical investigations were angry, competitive, and jealously guarded, and the children in crowded, poorly lit clinics were lost in the scrum.[58]

NOW IT IS clear that most of the "foreign experts," global leaders in radiation medicine, were wrong. In 1996, WHO, UNSCEAR, and IAEA conceded that, seven years after Ukrainian and Belarusian officials announced the problem, the still skyrocketing increases in thyroid cancer in children were due to Chernobyl exposures. Mettler, Wachholz, and other scientists conceded that their models failed them: "There is a major discrepancy," they wrote, "between estimates of thyroid cancer predicted using dosimetry together with standard risk-projection models and the magnitude of the increase that has actually been seen."[59]

So they were late with this recognition. What difference do a few years make? It turns out, a great deal. The denials meant that programs aimed at treatment and screening children were slow to start, and so aggressive cancers were caught too late. The IAEA's refusal to recognize the epidemic of thyroid cancers also crashed international aid. In 1991, the UN General Assembly had been waiting for the IAEA's assessment of Chernobyl damage before holding a pledge drive to raise $646 million for a large-scale epidemiological study of Chernobyl health effects and for relocation of over 200,000 people living in areas of high contamination. Coming on the heels of the IAEA's "no effects" report, the pledge drive netted less than $6 million.[60] The big potential donors—the United States, Japan, Germany, and the European Community—begged off, citing the IAEA report as a "factor in their reluctance to pledge."[61] The American delegation, especially, emphasized that in light of the findings of the IAEA's assessment, population resettlement was unnecessary.[62]

After the failed pledge drive, the UN Secretary-General created a Secretariat for Chernobyl Relief, but the Secretariat was footballed from agency to agency within the UN, five directors appointed in five years. Chernobyl was a UN hot potato no one wanted.[63] Repeated appeals for aid rarely raised more than a million dollars.[64] UN officials shrugged and mumbled about "donor fatigue."[65] The IAEA and UNSCEAR's erroneous insistence that Chernobyl produced no health problems continued to strangle fund-raising. "No conclusive scientific proof of disease from Chernobyl exposure," a diplomat wrote in 1995, by which time few doubted the thyroid epidemic, "led to a reluctance among the international community to offer decisive and meaningful assistance."[66]

If IAEA scientists had reported the unexpected cancers when they learned of them in 1990, advocates would have had more leverage to demand funding for a large-scale study to determine if there were other health problems that specialists in radiation medicine had also neglected to forecast. In 1996, when the IAEA finally recognized the thyroid cancers as a Chernobyl health effect, Angela Merkel, then German minister of the environment, again called for a long-term epidemiological study of Chernobyl effects on a mass scale, equivalent to the

A-bomb Life Span Study. No study ensued, again, for lack of funds and a shortfall of leadership.[67]

From the Chernobyl morass, the IAEA with its UNSCEAR sidekick emerged triumphant.[68] These two agencies had long recommended no action on Chernobyl because they identified no problems other than fear and ignorance. The World Health Organization, overtaken by infighting, proved a disappointing leader in the crowded Chernobyl field. In 2003, WHO scientist Keith Baverstock attempted to put together an International Chernobyl Research Board to try again to ask open-ended questions on Chernobyl health impacts.[69] Gonzalez preempted Baverstock's initiative by creating the Chernobyl Forum, an umbrella organization representing seven UN agencies, with the IAEA again at the helm. Instead of a major investigation of health effects, the Chernobyl Forum reviewed investigations in Western literature—most of them repetitive examinations of pediatric thyroid cancer—and issued a "comprehensive report" that ran in the same well-worn track of earlier IAEA/UNSCEAR reports of 1987, 1988, 1991, 1996, 2001, and 2002.[70] With an insistent repetition, the Chernobyl Forum blamed the rise of health problems on psychological trauma and called estimates of fatalities in the thousands "exaggerated and incorrect."

Fred Mettler, the author of the 1991 IAEA assessment and editor of the 1996 UNSCEAR Chernobyl report, also took charge of the Chernobyl Forum report published in 2006.[71] Mettler was arguably the single most influential voice in producing judgments of Chernobyl-related health damage that are today cited in the media as fact (fifty-four fatalities and six thousand cases of childhood thyroid cancers). He wasn't involved in UN politics or the larger skirmishes between pronuclear and antinuclear lobbies and so did not share their political concerns. He was just a university scientist doing his job, which meant focusing on the scientific evidence before him. How did he miss the childhood thyroid cancer epidemic in the 1990s? Why wasn't he the one to break the news of the unexpected early spike in thyroid cancers to a surprised scientific community? The simple answer is that Mettler didn't know about them, as he told me over the phone.

But, in fact, he was honest enough to call me back a few days after we first talked to tell me that he had misspoken.

After our conversation in which he refuted the BBC's claim that he had taken home biopsies of thyroid cancers in Ukraine, he looked at the proceedings of the IAEA's 1991 meeting in Vienna on the Chernobyl assessment. In the transcripts, he is quoted as saying that he had been given twenty slides, which he delivered to his lab in New Mexico and verified.[72] He read me the passage. On the line, we both grew silent for a minute.

"So you forgot about the slides?" I asked.

Mettler replied, "Yes, I suppose I did. It says here that I received them and the lab verified them."

I then asked, "What else other than radiation could have caused those cancers?"

Mettler answered, "Nothing (pause). It was a mistake."[73]

Mettler later disavowed this exchange to a *New York Times* fact-checker. But after that conversation, other evidence that linked Mettler to those slides and his knowledge of the thyroid cancer came to my attention. First, a Greenpeace memo written about the 1991 conference stated that an endocrinologist in Ukraine, Olga Degtyariova, told Greenpeace staffers about twenty childhood thyroid cancers.[74] (Since Degtyariova mentioned twenty thyroid cases, and Mettler verified twenty thyroid biopsies, it seems likely that Degtyariova or someone from her institute gave the twenty slides to Mettler.)[75] Degtyariova lost her job as deputy director soon after that.[76] Second, Valentina Drozd, the doctor who first recorded thyroid cancers in Belarus, told me that she had spoken at length with Mettler about thyroid cancers during a meeting with the IAEA delegation in Gomel's run-down Tourist Hotel. She had met Mettler earlier in a village where they were both working. According to Drozd, Mettler recognized her and invited her to sit next to him. Drozd recounted to me that she spoke to Mettler through an interpreter for two hours. She described to him the childhood cancers she had found. The cancers had emerged after just two years, not, as the A-bomb Life Span Study predicted, after ten years. She remembers his reaction as that of a sympathetic listener. "He was very interested," Drozd recalled, "and I was happy to talk to him." The next day, Drozd's boss reprimanded her for violating the chain of command in speaking to Mettler.[77] In short, two people took personal risks to convey the news of the thyroid epidemic to Mettler—information Mettler then forgot about.

IT IS NOT surprising that Mettler, like most people, had trouble recovering events thirty years before. Human memory is a wild liar. And that makes oral history a notoriously vexing practice. Subjects asked about events in the past rarely have a precise grasp. The Life Span Study, which is so important to the radiation risk estimates Mettler used, was based on data that had been collected by questioning Japanese survivors five years after the bombing about where they were standing when the blast occurred. Women were asked to give secondhand information on the location of their husbands' exposures. All this made for a very "messy" situation, as James Neel, a leading geneticist on the project, recognized.[78] Scientists are not immune to the same fuzziness of memory.

The problem here is not that Mettler forgot thirty years later about having seen and verified the slides and about having spoken with Drozd. The problem is what he did with that information at the time. First, very soon after he verified them, he noted the twenty thyroid cancers in his IAEA Chernobyl technical report but concluded, illogically, that his teams had found no accident-related health effects, even though, as he admitted to me, nothing but radiation could have caused that spike in thyroid cancers.[79] Second, a year later Mettler published a major article about thyroid nodules among children in Chernobyl territories, but he did not mention the twenty thyroid cancers he had verified.[80]

Such omissions riddled the international drama over Chernobyl health effects. Mettler excised thyroid cancers from his memory just as officials at WHO, UNSCEAR, and IAEA deleted them from scientific journals, press releases, and international assessments. Mettler and his colleagues, who had spent careers in radiation medicine, were evidently unable to suddenly recalibrate their models and acknowledge the evidence that a cancer epidemic could occur so much earlier and at lower doses than they had calculated. For Western radiologists to seriously consider the full range of evidence Soviet doctors presented would have called into question an entire medical infrastructure and wiped out a lifetime of publications and assurances that the public was safe from exposures to fallout from bomb tests, radioactive waste, medical treatments, and daily emissions from neighboring nuclear power plants. The thyroid cancer story might easily have slipped permanently from memory, lost in dismissals of poor Soviet science, had not Olga Degtyariova and Valentina Drozd risked disclosing it, and had not Keith Baverstock gone rogue and persisted in bringing international attention to children crowding into oncology wards.

When I asked Mettler what else might have been overlooked in the Chernobyl medical story—what about the Soviet doctors' reports of a range of illnesses—he replied that nothing else was possible. "The doses were too low." He referred me to a host of UNSCEAR documents on the topic. I pulled them out.

They are wonderful to look at. After wading through, as Mettler must have, thick volumes of tables and charts of health statistics generated by Soviet agencies, with confusing, sometimes conflicting data in various calibrations and measurements, the UNSCEAR charts felt like meditation. They were simplicity itself, soothing and lulling; the only thing better than the sunny lucidity of the charts' risk estimates is the promise of mathematical certainty amid the vast chaos of data that Chernobyl presents. Feed the UNSCEAR charts an estimated dose from one isotope, cesium-137, which itself is a gross generalization, and the charts tell the reader the increased probability of cancer in a given organ. Lost in the risk estimate's wonderful magic trick of making dozens of other harmful radioactive elements disappear are the bodies that ingested them and an accounting of how they fared. I too wanted to believe in the charts, to dissolve into them and make those sick kids in the contaminated regions go away.

# Looking for a Lost Town

I n late 1989, authorities ordered the town of Veprin and two neighboring villages in Mogilev Province of Belarus, 400 kilometers from Chernobyl, to be evacuated "immediately."[1] Radioactivity in local soils ranged from 25 to 143 curies of cesium-137 per square kilometer.[2] The evacuation didn't happen immediately. After the failure to drum up international aid, villagers lingered in extremely radioactive regions for four, five, eight years after the collapse of the USSR. Veprin remained for several more years, as local officials struggled to find funds to relocate families.[3]

Nine hundred people lived in Veprin. It was the administrative center of a large collective farm. The town had farm offices, a grammar and high school, a store, health clinic, even a music school—rare for a town that size. In 1999, the town was finally abandoned.

I rented a car in Minsk in 2016 and drove south to find Veprin. As I entered Mogilev Province and passed the town of Krasnopole, I saw fewer and fewer villages lining the road. Many miles went by with no sign of life. I could see low-lying trees, wild grasses, flowers, vapors rising from swampy lowland, and the ribbon of asphalt road running before me.

Stopping for a break, I walked down a grassy trail. I was surprised to find the ruins of a village materializing on both sides of the path. Vines and saplings covered cottages, sheds, and tumbled barns. I got a little nervous, having left my Geiger counter on the seat of the car. I headed back to the safety of asphalt. Driving again, I got better at noticing the abandoned villages, which dotted the road every five to ten miles.

Finally, I saw a road sign for "Veprin 6 km." I turned off and drove, six, ten, twenty kilometers. No sign of the community. I pulled out my phone and checked Google maps. Satellite images from above showed that most of Belarus is rural, heavily agricultural, with many small hamlets. In the large territory where I was, between Chechersk and Cherykaw, I could see forest, overgrown fields, and the outlines of former communities. On the map, the thirty-kilometer Chernobyl Zone of Alienation is marked, but no signs distinguished these abandoned stretches of Mogilev Province, far outside the Zone, as an imprint of nuclear disaster. The unannounced emptiness made me wonder how much the Zone of Alienation, which journalists and tourists regularly visit, serves as a distraction, a way to flag people's attention, while the catastrophe plays out elsewhere.

Backtracking on the empty road, I noticed a cracked drive leading to a leveled field. I turned in: Veprin, at last. The former town was an orchestrated ruin. In the center was a large cement pad, surrounded by mounds of bulldozed rock and brick, ceramic pipe, rusting tools, and shards of porcelain dishes. I climbed the pile and looked around. In the near distance, another ridge of debris undulated like a fault line. Maybe humans a thousand years from now will notice these earth formations striking up from the flat Belarusian plain. Maybe those humans will wonder about the mound people and their religious ceremonies. Or maybe they will detect the layer of plutonium in the soil and guess at the truth.

As I explored, I spied a fruit tree. A few geraniums and irises thrust up through cracks in the cement. They brought to mind the hands that cultivated them. Tracking the former gardens, I wandered the erased streets of Veprin and visualized the log cottages with brightly colored shutters before they were demolished.

I had been in dozens of rural, post-Soviet towns like Veprin over the years. It was easy to reconstruct it in my imagination from the immense silent erasure surrounding me.

I could conjure the patched-together gates and carefully tended gardens. I could see the children dressed formally, as they did in the USSR, like miniature adults. They carried small violins, heads bent, rushing through the square and up the steps of the music school. Getting your child into music classes in Soviet society was a real accomplishment. Teachers only took on children who showed natural talent. They worked their pupils with exacting rigor, preparing them to go far, maybe all the way to the Bolshoi.

I pictured warm lights coming from the music school on a winter night, the snow falling thick, tinting the ground a pure white that illuminated from below the velvet-black night. Inside, children prepared for a recital. The tiny maestros sawed away at their violins. Their brows, fixed in concentration, reflect the strained attention of the adults, whose aspirations centered on those darting hands and feet. Musical ability was a good way out of the village, out of a job at the collective farm, and on to higher education. Down the hall, in the dance studio, a plunking piano sounded metronomically. Boys and girls in saggy beige tights worked at the barre. A voice pitched high called out, "plié 1st, plié 2nd." A boy walked to the center of the studio. Dark circles ringed his eyes. He arranged his thin body in position, while the pianist tapped out a melody. Rising on his toes, the boy pirouetted, wobbled off balance, and fell. He went down not slowly as in the movies, but hard and fast, dropping like a board.

In 1986, a wind had kicked up, blowing suddenly hard. Raindrops pelted the earth. Only a few villagers still believed in black magic and "unclean forces," but everyone agreed nothing was ever the same again after the evil wind passed through in the spring of 1986.

Of the seventy children living in Veprin in 1990, county medical examiners characterized six as "healthy."[4] The rest had some kind of chronic illness. On average, Veprin children had 8,498 becquerels per kilogram (Bq/kg) of cesium-137 in their bodies: 20 Bq/kg is considered safe.[5] The children's flesh was as radioactive as the bodies of steers that were rejected at the Chechersk packing plant down the road.[6] As late as 1999, Belarusian officials still struggled to relocate people from dangerously contaminated land.[7] In making the decision to do nothing year after year, Soviet party bosses, post-Soviet politicians, and international diplomats took a pass on fixing a problem in a region where several hundred thousand people lived in a stew of contaminants. The years that villagers took in persistent doses of radioactivity created the conditions for what Soviet doctors once called chronic radiation syndrome.[8] "Acute radiation illness is an accident," the Russian hematologist Andrei Vorobiev wrote in his memoirs. "Chronic radiation syndrome is a crime."[9]

Before I got back in the car, I climbed the rubble and took one last look around. Vacated land often suggests loss, but emptiness also calls up the prospect of infinity on a limitless horizon. The pile of debris evoked the day that Veprin's families were finally released from their prison of worry and illness that comes with living on contaminated ground. Their departure signaled a fresh start, the potential of living more hopeful lives elsewhere.

# Greenpeace Red Shadow

I looked around for other international scientists who worked on Chernobyl health problems to see how investigators who were not representing politically sensitive UN agencies judged Chernobyl's environmental and health impact. In 1991, half a dozen humanitarian-medical programs were working in Chernobyl territories.[1] I found records of an International Red Cross fact-finding mission in late 1990. A large party of international scientists toured contaminated territories to talk with Soviet researchers and villagers. Belarusians told the Red Cross team, as they had other visitors, about the public health disaster from Chernobyl radioactivity. The Red Cross consultants, among them a World Health Organization official, returned home and wrote a report.[2]

In it, they echoed the verdict of WHO and IAEA consultants: "People attribute all their complaints to radiation, clinging to this explanation. Many doctors seem to support their patients in their suspicions that their symptoms are due to radiation and appear to lack knowledge of scientific facts on matters of radiation protection." Told that thyroid disorders had increased, Red Cross investigators replied that was impossible because the doses were far too low; there was "no direct radiation risk." They explained away the increase in disease as caused by a screening effect and stress.[3]

When local doctors appealed to the Red Cross for donations of ultrasound and MRI (magnetic resonance imaging) machines to reduce their patients' exposures to X-rays, the Red Cross team determined such donations would be expensive and exposures were too low to worry about a few extra X-rays. Western charities generally sent used equipment to Soviet hospitals. Soviets did not want the broken goods but had to keep them lest the charities charge them with corruption. The items piled up in storerooms.[4]

The real threat, the Red Cross reporters summarized, was not radiation but ignorance. They had a remedy for that: "A Red Cross worker armed with counseling skills, a Geiger counter, and appropriate publicity material could do much to help the population affected by the Chernobyl disaster come to terms with their new situation."[5] In other words, more enlightenment was the answer. I'd seen that before.

The International Red Cross sounded almost exactly like the UN officials who circulated through the same hotels and airport lobbies at that time. Science is by nature conservative, which is generally a good thing. Scientific collectives negotiate and curate standardized protocols in order to come to agreement on results. Scientists protect the parameters of their fields by judging each other in formal venues and informally in conversations. It is important for scientists to maintain a reputation for objectivity, integrity, and careful and thorough judgments. On this status-conscious landscape, it can be easy to corral others into conformity. A scientist who blundered outside the lines and asked the wrong kind of questions could be labeled as "sloppy" or "political." Those words could kill a career.

I noticed that Greenpeace had opened the first Western lab and a Chernobyl children's clinic in Ukraine. I found that interesting because data from Greenpeace's lab and clinic could supply a snapshot of the Chernobyl medical landscape created by Western scientists using Western methods but without the baggage of the bigger international organizations and government agencies. I went to Amsterdam to read through the archive.

In the spring of 1990, Greenpeace chairman David McTaggart swept into Kyiv just as Abel Gonzalez was putting together the first IAEA teams of scientists to tour Chernobyl territories. Short and powerfully built, McTaggart had been fighting what he called the "nuclear mafia" for a decade. In 1972, McTaggart, a retired Canadian businessman, sailed his yacht to Mururoa to protest French atmospheric nuclear bomb tests in the ecologically vulnerable atoll. Anchoring in the bombing range and refusing to move, he stopped the test. McTaggart's actions against French testing concluded in a court victory in 1974. Soon after this victory, the French ended atmospheric testing.[6] The risk-taking McTaggart quickly became the force that built up Greenpeace from a small nonprofit to a global environmental watchdog.[7]

With Chernobyl, McTaggart took Greenpeace in an entirely new direction. He wanted to beat the IAEA at its own game by gathering data to independently determine Chernobyl's impact on health and the environment.[8] McTaggart took a two-week tour of contaminated villages in northern Ukraine in the spring of 1990. He brought with him four doctors and, uncharacteristically, asked for no media involvement.[9] In Polesia they saw a barricaded train car full of radioactive meat. Told it was dangerous, McTaggart approached the refrigerated wagons anyway to measure the gamma rays, which were impressive. The Greenpeace delegation traveled on roads lined with fences that walled off radioactive territory. The group was shocked to see that inside the barriers farmers plowed the fields. At a clinic, the Canadian doctors looked at 350 children and took home 100 blood samples. They promised to inform parents of the results of the blood tests in contrast to secretive Soviet practices.[10]

After the trip, McTaggart drew up an action plan. The Soviet government and other Western organizations had tried to correct Chernobyl problems but had almost no impact. "They came, they saw, they left," McTaggart quipped, "but rarely did anything of lasting benefit to the people here." McTaggart set out to change that. "We decided on a three-pronged effort: monitoring, treatment, and relocation."[11] Monitoring would be in the form of a tractor-trailer equipped as a mobile lab to measure radiation and chemicals in air, water, and food. Staffers would also gather blood samples and other medical data. A Greenpeace clinic would examine child "victims" in the field and treat them in Kyiv. In addition, McTaggart planned to design and build fifty eco-houses as a pilot project for resettling 200,000 people from contaminated territories.

These were outsized ambitions, more than Greenpeace had ever attempted anywhere else in the world.[12] Greenpeace normally special-ized in big media campaigns aimed at focusing public attention on envi-ronmental problems. Never before had Greenpeace pledged to actually solve those problems. McTaggart was aiming high.

McTaggart returned to Kyiv for meetings with Yuri Shcherbak, leader of the Ukrainian ecology movement, Green World. Shcherbak suggested that Greenpeace work with the Ministry of Health's special clinic for radiation protection.[13] Shcherbak drove the Greenpeace delegation out of town to a broken-down, one-story brick building, a tsarist-era tuber-culosis sanatorium built for "poor and Jewish people." The clinic had a dirt floor and not much in the way of plumbing.[14] This would be the new Greenpeace Children's Hospital. McTaggart loved it. He could build on five adjacent acres a small resettlement village with Greenpeace eco-housing. He was not daunted by the fact that Greenpeace would have to wholly renovate the ruin. Shcherbak recommended a construction firm. McTaggart, who made his fortune in real estate, estimated four months for renovations.

Shcherbak set up a meeting with Ukraine's Prime Minister A. V. Masol and the deputy chair of the Council of Ministers, Kostiantyn Masyk. Masyk took a keen interest. McTaggart and Masyk did not speak a com-mon language, but through an interpreter, they hit it off, drinking John-nie Walker late into the night, trading stories and jokes. McTaggart did not know that KGB agents had tasked Masyk with the mission of pro-tecting the secrets of the Soviet nuclear power industry from meddle-some Westerners.[15] McTaggart was just grateful for Masyk's high-level access. Elevating up to the halls of power, Greenpeace quickly received permission to register as the first foreign nongovernmental organiza-tion (NGO) in Ukraine. The same day, the two parties drafted proto-cols for an agreement between Greenpeace, Green World, and Masyk's Ukrainian Council of Ministers.[16]

McTaggart continued on, spinning around Kyiv like a whirling der-vish. Looking for a building to serve as headquarters for Greenpeace Ukraine, McTaggart toured several sites. He picked a historically sig-nificant nineteenth-century villa on Andriivskyi Descent, arguably the most beautiful street in Kyiv. The villa also would have to be gutted. It was all very exciting and left McTaggart a bit hungry. He jotted down: "Went out to forage for food, most unsuccessfully."[17]

McTaggart asked Shcherbak about existing data on children who had high doses of radioactive iodine to the thyroid. Perhaps Shcherbak could help them get access to those records. Shcherbak was not sure, but he introduced McTaggart to Volodymyr Tykhyy, a Green World scientific consultant. Tykhyy was not known among Chernobyl researchers or among scientists in Kyiv generally, but he did speak English, and Shcherbak spoke highly of him.[18] That was good enough for McTaggart. He drew Tykhyy into the project.

Funding appeared to be ample with donations from the Soros Foundation, plus a million rubles from the sale of a rock album.[19] At the end of a breathless week, McTaggart flew home to Italy. I imagine he felt satisfied with his work. His big plans had come together seamlessly.

Perhaps too seamlessly.

The renovations of Greenpeace property stalled out soon after they began. The contractors disappeared, so did building supplies. Deputy Minister Masyk offered to help out by having the contractors arrested, but before that happened he stopped taking calls from Greenpeace staff.[20] Making do, the staff lived and worked in cramped hotel quarters. McTaggart planned to send to Kyiv twenty-five world-class doctors to work at the new hospital, but Greenpeace managed only to hire three physicians—it wasn't easy to find Western doctors willing to sign on to the hardship conditions of the late Soviet Union.[21] The temporary clinic was a bare room with a table and a few chairs.[22] To attract doctors, Greenpeace had to pay ample salaries and be less choosy; none of the doctors who agreed to go had training in radiation medicine or epidemiology.[23] Two physicians flew in and out, making short trips.

One, a young, cigar-smoking pediatrician from rural British Columbia, Clare Moisey, staffed the Greenpeace Chernobyl Children's Hospital. At first, many children came to be examined. Moisey found that the children were mostly healthy. After a month, the number of children dwindled to fewer than twenty-five a week.[24] When I asked Moisey why he had so few patients, he replied, "They all thought we were offering a radiation vacation in Canada. As soon as they realized we were not taking children to Canada, the kids disappeared and a lot of the cooperation was withdrawn."[25]

That didn't sound right. Doctors struggled in Kyiv hospitals to treat twice as many children as they could handle. Three hundred children with cancer were on a waiting list for a bed in the city's one pediatric oncology ward.[26] Instead of opportunism, a former Ukrainian Greenpeace employee suggested a more disturbing reason why so few children showed up at the Greenpeace clinic. "We wondered," he said, pointing to KGB interference, "whether the children were being filtered."[27]

With few children to examine and not much to do, a Greenpeace technician, Anne Pellerin, collaborated with Ukrainian doctors, who asked her to run tests for them on Greenpeace diagnostic equipment.[28] Pellerin was happy to help out, although politically, it was risky to collaborate. The Ukrainian bosses didn't like it. Moisey resented the fact that his lab techs were working for other doctors.[29] Pellerin went ahead anyway. She built up a network of contacts and a database of hematological data on over 2,400 exposed children.[30]

I was happy to learn this news—finally a cache of records created according to Western standardized protocols that I could use to check against Soviet data. I looked in the archives for Pellerin's records. No luck. I called up former Greenpeace staff. No one knew where the records were. Volodymyr Tykhyy said Greenpeace was disorganized and probably lost the records. I noticed, though, that Greenpeace staff also tried to locate the same files soon after Pellerin left her job in Ukraine in 1991. A month before the records went missing, Anatoli Artemenko, who worked intermittently with Greenpeace, told a staff member that the KGB was closely monitoring Greenpeace offices in the Hotel Kyiv.[31] A week later, the floppy disks that held the children's medical records failed. Staffers were mystified that the data on the original and backup disks, one after the other, were stripped clean. The missing Greenpeace data make, according to my count, three sets of computer records containing Chernobyl health records that disappeared mysteriously in the course of that first year of Soviet-Western collaboration.

The Greenpeace medical program, which took a year to set up, collapsed out of inertia six months after it began.[32] Moisey, a descendant of Ukrainian immigrants, came away with a low opinion of Ukrainian doctors and Soviet medicine. He said he found basically healthy children "but a very sick medical care system. Something died in the medical system," Moisey said, "with the communist revolution."[33] His Greenpeace colleagues held similar views. Soviet medical care was, they told each other, "20 to 40 years behind the US."[34] Moisey said the lack of respect was mutual. Kyiv doctors were not interested in guidance from Moisey. "They wanted our equipment and our money rather than our expertise," Moisey remembered.[35]

For Westerners, Soviet medicine was like the bloodless, late socialist economy, which was like Soviet hotels and Soviet restaurants. They were bad and poorly organized with incompetent and corrupt staff. Soviet science long had a poor reputation abroad. Western commentators often focused on Trofim Lysenko, a Stalin-era geneticist who rejected evolutionary science and argued that plants and animals acquired characteristics in their lifetimes that could be passed on to their offspring. Historians judged that Lysenko slandered famed Russian geneticists to climb to the top of the scientific hierarchy. In the decades that followed, Lysenko became a synecdoche for the politicization of Soviet science, which was said to corrupt most fields of research, including that of Chernobyl radiation medicine.[36]

Historian Hiroshi Ishikawa argues that Soviet science was not entirely undone by the politics of Lysenko. Geneticists who were purged from biology moved into physics departments in the early fifties where colleagues welcomed them as they puzzled over questions of radiation and health.[37] Rather than view genes, as geneticists did in the West, as impervious to ecological influence, Soviet scientists influenced by Lysenko's ideas came to understand intergenerational transfer of induced genetic abnormalities caused by environmental factors.[38] With this science, Soviets gained a sensitivity to exposures to radioactivity and other toxins.[39] They grew critical of Western risk estimates about radioactive fallout. The nuclear bomb designer Andrei Sakharov led Soviet scientists in backing an international campaign to end nuclear testing, which he argued, with ever greater urgency, damaged global human health.[40] Lysenko, in other words, was wrong about many things, but not all the way wrong. In the 1980s, scientists in the West came to understand that although a person's genetic sequence changes little, how and when the sequence is expressed can determine important bodily functions, and those traits can be passed on epigenetically.[41] Children can even inherit a parent's trauma by way of cell communication.[42]

In pushing aside evidence of a public health crisis in Chernobyl contaminated lands, foreign scientists, even those hired by Greenpeace, worked within a frame they had trouble seeing around, which had been created by seven decades of science split between East and West. According to this worldview, the West and capitalist democracy naturally prevailed over the East and socialism in all fields, including science and medicine. That ordering appeared to be natural so that even as the Iron Curtain rusted away, it continued to cast a shadow.

# The Quiet Ukrainian

After the collapse of the Greenpeace Chernobyl Children's Hospital, Greenpeace staffers redoubled their efforts to build up their monitoring and eco-housing programs. There too they ran into obstacles. No progress was made on building fifty prefab eco-houses. Shcherbak had originally specified the houses should be made of wood. When the American architect submitted drawings for wooden houses, the officials at the Ministry of Construction said wood was no good. "We have no timber left to harvest in Ukraine, and any lumber we have here is radioactive."

"How," a Greenpeace staffer asked, "did we get this far without knowing this?"[1] The architect would not change his design. He insisted on the ecological value of timber-frame houses to Ukrainians who had been building sustainable, timber-frame houses for centuries. With that, the eco-housing project stalled out and was apparently forgotten.[2] When Ukrainian officials asked a few months later when the fifty houses were arriving, a new Greenpeace staffer did not know anything about it and believed the Ukrainians were making it up. "How bizarre!" she wrote.[3] In this way, Greenpeace failures in communication and staffing rebounded into judgments of Ukrainian incompetence.

The Greenpeace monitoring program, billed as "the first independent survey of environmental contamination of food, water, and soil ever in the Ukraine," developed at a snail's pace.[4] Greenpeace had trouble finding Ukrainians to work with them.[5] The first scientist they hired was

Volodymyr Tykhyy. He was charged with drawing up a plan to monitor radiation levels in villages.[6] He had the use of equipment donated by the city of Munich, but for some reason Tykhyy and his colleagues did not go to villages to measure. Months passed.

Six months into the program, an office manager wrote that "what are we doing here?" was the most common phrase said among the foreign Greenpeace staff. "It seems that an awful lot has been promised," he observed, "and yet GP doesn't really have the capability to fulfill."[7] Faced with these frustrations, employees quit and went home to countries where a person could easily buy toilet paper and groceries. Greenpeace had trouble replacing them. New staff members were yet younger and less experienced.

Amid the struggling Greenpeace project, the Soviet economy sank. Inflation soared. When Kyiv banks ran out of cash, the staff pawned the office curtains to tide them over.[8] As the tax laws changed, so did the language of business from Russian to Ukrainian.[9] The Western Greenpeace staff did not speak either Russian or Ukrainian. They were dependent on interpreters for all interactions. Soviet citizens assumed that interpreters were undercover KGB agents. That made for distrustful and guarded relations. Communication with headquarters in Amsterdam over an early internet connection was patchy at best. The Western Greenpeacers became depressed, isolated, and a bit paranoid. Staff continued to turn over about every six months. The learning curve was steep for each new arriving employee. One director called the job in Kyiv a career-killing "kamikaze mission."[10]

In January 1991, McTaggart returned to Kyiv to try to work out these problems. He had a tense meeting with his Ukrainian partners, Green World leader Yuri Shcherbak and Valery Gruzin, director of the Soros-sponsored foundation Renaissance. The two men were furious. Shcherbak lit into McTaggart in broken English about Greenpeace's lack of progress: "Since spring of last year we have spoken of mobile lab, nothing."[11] Because of Greenpeace's failures, Shcherbak charged, he was being pushed out of his own organization, Green World. The Ukrainians accused Greenpeace of being out of touch. "Information," Gruzin instructed, "is the muscle of politics." The political field was changing, he hinted, "and there is a fierce competition of international groups coming in with much more energy (Tykhyy made the money gesture) than GP."

McTaggart parried back, also angry. He complained their architect delivered plans for eco-housing and they flew people in, "and nothing was done."

Gruzin laughed at McTaggart: "This is an ancient country," he said, "seventy-four years and nothing has happened, and he wants action in six months?"[12]

The men turned to the stalled-out radiation monitoring program, which was just about to finally get going. After a lot of efforts to ship in equipment and hire staff, Shcherbak suddenly suggested they change direction.[13] "We know some previously secret info," Shcherbak disclosed, about government environmental studies that had been classified but were coming to light. "We do not need to repeat what has already been done." He told McTaggart that they should study chemical toxins instead. McTaggart tried to get more information on this secret radiation data.[14] Shcherbak remained in conspiratorial mode: "I need an additional meeting with people who have top-secret info because they will deal with Green World only, not GP." Shcherbak suggested rethinking the whole program: "Maybe it would be wiser to expand study beyond Chernobyl zone."

That threw a spanner in the works. Greenpeace staff had just spent nine months setting up the radiation monitoring program. McTaggart, accommodating his colleagues, did not point out that Greenpeace came to Ukraine specifically to focus on Chernobyl, nor that it was they who had recommended the contractors who walked off the job and stole materials. Gruzin and Shcherbak repeated that they had a long line of potential collaborators begging to send them humanitarian aid. Greenpeace needed them, they emphasized, not the other way around. "Without our support," Gruzin jabbed, "our bureaucracy will swallow you in one month."

I imagine it was frustrating to have such partners, who stalled, steered them toward lame business contacts, failed to deliver on promises, derailed plans, and then, after being critical of Greenpeace's slow pace, laughed that the foreigners expected anything done quickly in this "ancient" land.[15] That was the entire record of environmental data collected by Greenpeace that I could find in the organization's archives.

After running through several young Greenpeace directors, imported from abroad, McTaggart appointed Volodymyr Tykhyy to manage the lab and office.[16] I met with Tykhyy, whose name in Ukrainian means "quiet," in a noisy Kyiv diner. He was friendly and talkative. Tykhyy told me that when McTaggart hired him, he advised Tykhyy to fire the other Ukrainians working for Greenpeace "because they were all KGB." Tykhyy told me he did not in fact terminate anyone, though he estimated "about half were KGB."[17] I spoke to several Ukrainians who worked in the Greenpeace office during those years. All but one were sure their office was infiltrated by undercover KBG agents. No one confessed they were among them. A Ukrainian staffer described KGB tactics as subtle. "They were clever not to prevent too much," he said. "They tried to be very supportive and friendly. Being involved [in an organization] was the best way to get information from inside."[18] "The KGB's role was not only to stop and arrest," Irina P. remembered while reflecting on her years as a Soviet tour guide for foreigners. "It was also to disrupt and disorient."

Greenpeace administrators wondered whether Tykhyy, the son of a Ukrainian dissident, was the right person for the job of director.[19] Tykhyy's father had been imprisoned as a Ukrainian nationalist in the 1970s and died in a Mordovian prison in 1985. Soon after, Tykhyy became involved in Green World activities. That was risky. As the son of a dissident, Tykhyy was vulnerable. Parents who were party members could protect their children if they got mixed up with security forces. The offspring of convicts had no roof to shield them, and they were watched closely.

I asked Tykhyy if he had KGB handlers. He recalled many visits from KGB agents: "They approached me and said, 'Well, look, you have contact with your father and with his contacts. We advise you to convince your father to write a confession and then you can help us by talking to the other people.'"

"What did you do?" I asked.

"I told them I would never do this," Tykhyy replied. "They approached me for quite a long time."

Tykhyy recounted how a KGB major, or a captain, summoned him to headquarters and offered him compensation for his cooperation. "He said, 'We know you are a good physicist. We can assist you. You can travel to foreign countries. You can be registered to live in Kyiv.'" The agents also threatened Tykhyy if he didn't help. "They told me, 'You will never find a job in your specialty as a nuclear physicist.'"

I was confused. Tykhyy did get a good job in nuclear physics in a classified institute. He did receive permission to live in Kyiv and he traveled abroad a great deal in the late 1980s. He went to the best university in the country and even learned to speak English very well, rare for Soviet scientists, but not for KGB agents because the KGB had an excellent language school. "How," I asked, "did you manage to get all those benefits without cooperating with the KGB?"

Tykhyy has an easy laugh, and he deploys it often. "At the time it was just a job for them." He chuckled. "The KGB were not bloodthirsty. They did not think you had to destroy everyone. Besides," Tykhyy added with a giggle, "I wasn't involved in political activities."

Again, I didn't understand. Tykhyy joined Green World from the beginning, an organization closely allied with the Ukrainian independence movement, *Rukh*. KGB agents suspected that members of *Rukh* were plotting an armed insurrection against the USSR.[20] "Wasn't that political?" I asked.

"Ecological causes," Tykhyy cheerfully replied, "were important even for party officials. At the time there was a joke that the Communist Party was red on the outside, but green on the inside."[21] He laughed at his witticism.

That was news to me. In archival documents, I saw few signs that party members were interested in ecology. Ukrainian KGB officers, on the other hand, were alarmed by the run of environmental disasters in their republic—chemical spills, mine explosions, accidents at nuclear power plants, and the wanton use of agricultural pesticides as part of the Chernobyl decontamination effort.[22]

Tykhyy was very helpful. He sent me his private archive from those

years. I read his plan for environmental monitoring with a mobile van. I
was astounded at how much it differed from Greenpeace goals. "People
no longer believe government information," Tykhyy wrote. "Four years
after the accident even confirmed good news coming from official agencies is perceived as untrustworthy." Tykhyy proposed creating an "Info
Lab" to restore public confidence in government information by carrying out independent measurements to show that existing official data
were correct.[23] McTaggart also wanted a mobile lab to address the "confidence gap," but he sought to use that data to demonstrate just how bad
the accident was. Tykhyy planned to use it to pacify worried citizens.[24]

These opposing goals never became apparent because very little
environmental monitoring ever took place. As a Greenpeace employee,
Tykhyy worked slowly, accomplishing little.[25] A year passed and no measurements were taken. Explaining the lack of progress, Tykhyy emphasized supply-flow problems, communication shortcomings, and poor
science in Ukrainian labs.[26] A Greenpeace administrator noticed that,
according to Tykhyy, "The delays were always someone else's fault."[27]
Rather than measure for radioactivity, Tykhyy, a nuclear physicist,
wrote a long report on chemical toxins.[28] Instead of taking the van out
to contaminated villages, he busied himself with visas, customs, public
relations, and building a costly steel garage to store the mobile lab.[29]

Tykhyy also prepared for the arrival of a large shipping container
outfitted as a sophisticated chemical and radiological lab.[30] The container was to be housed in the parking lot of the Institute of Hygiene,
from which it would draw water and electricity. Tykhyy organized the
arrangement. After many problems with customs, the container finally
arrived in Kyiv. Martin Pankratz, a German Greenpeace employee,
delivered the container and wrote an angry report from Kyiv. "All information I received in advance about electricity and water supply was
wrong. I had a number of problems to solve which, according to Volodymyr Tykhyy, had already been resolved or did not exist." When Pankratz tried to deal directly with technicians at the institute to connect
the lab, Tykhyy blocked him. Tykhyy, Pankratz charged, deceived him:
"I received false information on numerous occasions from him and he

keeps information to himself which is vital for us to know. This behavior hindered my work immensely."[31]

It took another six months for the container lab to open. Once going, the lab, which cost $100,000 a year to operate, did not work "to its full capacity" and lacked political direction.[32] When the lab was finally in place, Tykhyy proposed abandoning the monitoring work altogether. After investing a million dollars in lab equipment, Tykhyy thought that Greenpeace should do no science at all.[33]

I asked Tykhyy why he rarely used the radiological and mobile labs. He explained they needed the expensive van as a taxi to pick up Greenpeace staff at the airport or drive to Germany for supplies. When he did go out, he added, he did not find much. He told me a story about measuring food for radioactivity in villages. "When it came to giving milk to authorities, the villagers would all go to this one guy and take his radioactive milk and give it to authorities."

"Why, to get a subsidy?" I asked.

"Yes. It happened every time."[34]

That anecdote might work on some people, but I knew from the records of the State Committee of Industrial Agriculture that a lot of milk was contaminated a majority of the time. Besides, a Greenpeace lab technician told me she took milk directly from villagers' cows in order to be sure of its origin.[35] Again, I was confounded. It seemed incongruous that a former Greenpeace director would want me to go away with false information about the Chernobyl disaster.

I had sought out the Greenpeace International archives to corroborate data I found in Soviet archives. I did not find data, or hardly any. Nearly three years and a million dollars after the optimistic Greenpeace founding in Ukraine, Tykhyy filed one study of one village in northern Ukraine. His team asked questions of twenty-five families and took ten samples of food products from other families.[36] That was the entire record of environmental data collected by Greenpeace that I could find in the organization's archives.

McTaggart had a reputation as a shrewd, even ruthless, businessman. He was suspicious of KGB covert operations. I can't tell if he knew he was being played. It's clear he was seduced by the excitement of operating behind the Iron Curtain and loved the thrill of riding the quaking Soviet political terrain. In August 1991, when eight military and KGB leaders announced they were taking over the Soviet state in a coup d'état, McTaggart hopped a plane to Kyiv. He went straight to his friend, Kostiantyn Masyk, who was then acting prime minister. Army helicopters flew overhead. The streets were ominously empty. When McTaggart walked in, Masyk, sweating, was on the phone with Boris Yeltsin in Moscow. He motioned McTaggart to sit down. For three hours, McTaggart sat there, asking his interpreter to translate, while Masyk negotiated the coup on dedicated telephone lines.[37] Hearing he was number ten on the coup plotters' hit list, Masyk perspired yet more. "The situation," McTaggart later reported to the Greenpeace board, "was very serious. It was very tense."[38]

After the coup ended that night, McTaggart went to Tykhyy's apartment to share a bottle of vodka. That was one of McTaggart's last trips to Ukraine. For him, indeed, the Chernobyl project turned out to be career suicide. A few months later, McTaggart was pushed out as chairman of Greenpeace International. He took an early retirement and retreated to his house in Italy. Tykhyy stayed on only a few more months before he too was asked to leave.[39]

After the collapse of the Soviet Union, the crowd of Western scientists and charities thinned out. Many nonprofits moved on to other disasters. Despite the failures to get programs off the ground and the evident meddling by the KGB, Greenpeace remained in Ukraine. And that made the difference. International agencies that set up offices, staffed them with locals, and stayed through the rough years of transition transformed into more sensitive and effective organizations.

The International Red Cross was another organization that stuck around. The Red Cross donated to Gomel Province six mobile vans with diagnostic labs to test food, water, and bodies. They hired locals who had experience taking measurements. Aleksandr Komov, the sanitation inspector in Ukraine who sent seven tons of milk to Kyiv to prove people in his province were consuming radioactive food, went to work for the Red Cross in Belarus.[40] Initially, the program was intended to assure villagers that they were living in safe levels of radioactivity, but as they measured, Red Cross investigators themselves got worried, noticing that a fair portion of produce villagers ate was contaminated above permissible levels. "These figures," they wrote in 1993, "confirm the acute need for continuous monitoring of foodstuff."[41]

Between 1991 and 1993, the Red Cross group examined 70,000 people, many of whom had not seen a physician or radiologist since the accident. In just a few days, without the help of sensitive ultrasound equipment, doctors discovered 62 children in Gomel with thyroid cancers. In Rivne, Ukraine, they screened 400 adults and children, 360 of whom had illnesses that required immediate medical treatment. They told people on the spot the results of the tests. The doctors wrote that for residents to be trusted with information was empowering: "This cannot be overestimated."[42] As the Red Cross workers gained experience and spent time on the ground, they grasped that problems were far more serious than they had surmised in their first quick trips.

In Ukraine, Greenpeace also adapted after the Chernobyl Children's Program closed. The lab technician Anne Pellerin continued to collaborate with Ukrainian doctors by running tests on Greenpeace equipment imported from abroad. Her lab work won the respect of Ukrainian colleagues, which opened doors.[43] Greenpeace staffers who remained in Kyiv after 1991 were largely Ukrainian or Westerners who knew Russian. No longer working through interpreters, they made alliances with Ukrainian scientists, who invited them to their clinics. In one critical meeting, Ukrainian doctors laid out to Greenpeace staff the growing numbers of pediatric cancers they were seeing. At first, they said, they encountered mostly lymphomas, but then increasingly they found children had cancers of the kidneys, urinary tract, and bladder. Placing their patients in whole-body counters, technicians detected radiation from children's livers and intestines. They reasoned the cancers came about as children's digestive tracts processed radioactive nuclides.[44]

Greenpeace employees came to respect the work of local doctors, which they had first overlooked. They visited more labs and clinics and collected studies on Chernobyl-exposed people. Greenpeace became an important clearinghouse to collate and translate work of local scientists not known abroad. Working with Vasily Nesterenko in Belarus and the Russian scientist Alexey Yablokov, Greenpeace helped to publish a compendium of this work, which became the chief challenge to the UN assessments, repeated every few years, that Chernobyl caused no health problems, other than several thousand "easily treatable" children's thyroid cancers.[45] The experience of Greenpeace and other nonprofits navigating through Soviet Ukraine into a post-Soviet world taught something about the art of collaboration.

# PART VII

—————— // ——————

# SURVIVAL ARTISTS

# The Pietà

The scene plays over and over in a timeless loop. The camera alights around a hospital, taking in the crowded ward, shabby curtains, broken pipes, and tired nurses in starched white, working in what the narrator calls "primitive conditions." The photographers' pitiless stare follows the contours of disorder from architecture to human biology. The lens focuses on the twisted limbs of a toddler, pans to the flopping head of a hydrocephalic baby, and centers on a beautiful boy flat on a sheet, listless and leukemic. A girl turns from the videographer to the wall to imagine herself somewhere else. Mothers next to the little cots knead their hands helplessly and hold back tears.[1] This is filmmaking at its most brutal.

It must have taken courage to allow the photographers to turn the camera on your child. The mothers refer to their children's infirmity as a third party in the room. Some children, the narrator says, are dying. The mothers hold them and rock them. The camera frames the figure of mother and child in a modern, dystopic Pietà. The women present their children's bodies to the cameras in hopes of salvation, if not for their offspring, then for humanity, so this nightmare would end.

The image of the Chernobyl child is one of the most lasting cultural artifacts of the disaster. After Soviet medicine failed them, after foreign experts turned their backs, survivors staged their children's bodies as sites of pollution and disease in a desperate, last-ditch effort to be seen as worthy of care. Trying to win the attention of audiences abroad, they presented a *tableau vivant* of bodies in pain.[2]

Documentaries such as *Chernobyl Heart* and *Children of Chernobyl* appealed to audiences to donate to Chernobyl children's funds. The more needy and helpless post-Soviet medicine appeared, the more money the charities hoped to generate. The strategy had unfortunate consequences. Emphasizing helplessness compounded the assumption of Western superiority and former Soviet citizens' humiliation. In a vicious circle, the more Ukrainians and Belarusians made a case for aid, the more they appeared to be grasping, inferior, and devious.

By presenting the nation in need, the filmmakers walked right into the buzz saw of Westerners' arguments about failed Soviet medicine and the alleged graft and incompetency of the socialist system. Critics charged that Ukrainians and Belarusians pushed their children in front of cameras to rattle the cup for international aid. They claimed they used any sick child as a lure to snag handouts. As Chernobyl children's programs multiplied, survivors' alleged addiction to welfare became the chief problem, not radiation and the public health disaster. I find that to be an amazing misrepresentation. As long as anyone could remember, farmers of the Chernobyl territories had met most every need with their own labor—plowing, pumping, sawing, hauling, sowing, weeding, canning, milking, and healing. The accident took away villagers' economic independence and turned them into supplicants. Yet, when they requested shipments of clean food, they were depicted as beggars asking for giveaways. When they complained of health problems, they were "radio-phobic." If they kept farming to feed themselves, critics called them "nuclear fatalists" who refused to protect their families from danger.[3]

For many foreign consultants, the antidote was clear. They dished up medical metaphors in place of medical aid. "Shock therapy" would fix both the economy and the psychological problems that allegedly plagued the lost and passive Soviet people. Following this train of thought, IAEA officials insisted that relocating people from Chernobyl contaminated lands was not a question of health but of economics.[4] They hired a consultant, Simon French, a British decision-making expert, to work with high-placed leaders to brainstorm on how to once and for all conclude the Chernobyl event. In Minsk, French pulled out an easel and gave a lesson on capitalist-style cost-benefit analysis. To the Belarusians' astonishment, everything in actuarial science could be assigned an absolute value—disease, risk, safety, even human lives.[5] French explained that at each level of exposure, models could predict a numerical outcome. At a lifetime dose of 70 mSv, which Ukrainian and Belarusian leaders chose as the alternative to the Moscow-designated 350 mSv lifetime dose, he computed using charts from the A-bomb survivor Life Span Study that that they would save 240 people from getting cancer. If they elected a yet lower threshold, the nation would be spared 600 cancers.

But, French reminded them, if leaders lowered the permissible dose, then more people would have to move, hundreds of thousands more.[6] They also had to factor in the cost of relocation and of stress-induced health problems caused by uprooting people. In a flurry of calculations, French ran the numbers weighing protection against risk. With each lowered sievert and every beneficial step toward safety, the price tag rose. French calculated that resettling people at the dose set by Ukrainian and Belarusian officials would mount to over forty billion rubles. Where would they get that money? The state coffers were empty. Raising taxes was unpopular. It was a grim trade-off. The threat of bankruptcy, not socialist ideals, determined the future. The idea that life could be reduced to revenue coming in and expenses going out was so incendiary that Belarusian leaders asked that the meeting be kept secret.[7]

Industrial accidents are usually too expensive to clean up. No one at the meeting recalled that Soviet physicists, like promoters of nuclear energy elsewhere, had guaranteed that the Chernobyl nuclear power plant would be safe. After the exploding reactor released radioactive contaminants, the message and the promises to the public changed. It was too late to talk, as Ilyin did, about returning the landscape to its "preaccident form" where farmers could safely farm, where children could safely play in the grass.[8] The cost of return to the original state, French instructed, overran the benefit of achieving preaccident levels of safety. French's lesson for Soviet leaders was that risk—in this case in the form of man-made radioactivity—is inevitable and natural and needs to be brokered like anything else through the medium of capital.[9]

At first horrified by the Westerners' cold-blooded calculations, in subsequent years post-Soviet leaders succumbed to the rationale of cost-benefit analysis. New capitalist leaders (who were mostly former communist leaders) learned that dealing with Chernobyl was expensive and litigiously risky.[10] In independent Russia, Ukraine, and Belarus, subsidies for clean food and medical care dried up.[11] People slated for relocation in accordance with the 70 mSv guidelines adopted by Ukraine and Belarus received few subsidies to move. Nuclear power reactors earmarked for decommission, including the Chernobyl plant, continued to operate. Hospitals demanded hard currency for treatments that required foreign equipment or supplies. Few people had hard currency. Bus service slowed and then stopped in the countryside. Municipalities turned off streetlights and reduced heating. In shops, there finally appeared clean, packaged food, but few people could afford it.

Hunger followed, not famine like in the 1930s but a gnawing, low-level malnourishment. Blow by blow, new capitalist leaders hacked away at the socialist welfare state. They carved faster when they realized how much they and their family members could profit from privatizing Soviet industries. International consultants from the World Bank and the International Monetary Fund recommended more austerity measures to qualify for loans. The new leaders readily complied because foreign loans also generated personal fortunes. Dividing the wealth among themselves, the elected leaders voided the social contract.[12] The bitter understanding materialized quickly that capitalism would deliver no economic miracle, no Chernobyl cleanup, but crushing poverty instead.[13] Growing angry as their world constricted, citizens were handed a petty nationalism and a mean-spirited xenophobia as a consolation prize for their disappointments.

It follows that the 1990s were grim years for Chernobyl research. Inflation wiped out salaries and left labs in desperate need for just about everything. Most Russian scientists turned to other topics.[14] Chernobyl retreated from the media. Ukrainians and Belarusians kept research agendas alive but struggled alone.[15] Demodernization and economic crisis were bad for health. Life expectancy spiraled downward and with it fertility rates.[16] People fed themselves from garden plots.[17] Young people went abroad. Rural areas depopulated. On this landscape it became hard to distinguish the impacts of the economic disaster from that of nuclear catastrophe.[18]

Chernobyl became something that occurred once a year on the anniversary when leaders lay wreaths on monuments. Politicians learned not to deny or dismiss Chernobyl.[19] They showed sympathy, said all the right things, and continued on their course of transforming political power into personal wealth.

The UN helped its client states by repackaging the nuclear catastrophe. UN officials arranged for glossy brochures and articles in "upmarket" publications on science and medicine.[20] They designed programs aimed at transforming Chernobyl "victims" into responsible citizens.[21] Cut off from subsidies for clean food and medicine, residents were told they needed to learn how to "restore sustainable development" on Chernobyl farmland.[22] New manuals instructed farmers how to sort hay and filter milk of radioactivity using equipment they had no money to purchase. Overworked mothers were taught painstaking new recipes to prepare radiation-free food from contaminated produce.[23] Doctors underwent training in the latest medical technologies that few hospitals could afford. Villagers who had no means to evaluate levels of radioactivity were told to choose themselves whether to stay on contaminated land or go. Relocation became voluntary and largely self-financed.[24] As the safety network retreated, state and international agencies shifted the burden of managing the postaccident risk society onto the shoulders of exposed residents, the people with the fewest resources to manage it.[25]

In Belarus, President Aleksandr Lukashenko grew more hostile toward people who would not let the Chernobyl problem dissolve. Doctor Yuri Bandazhevsky, rector of the Gomel Medical School, fell in that category. A pathologist specializing in environmental factors in postnatal development, Bandazhevsky worked with his wife, Galina Bandazhevskaya.[26] The pair investigated how cesium-137 incorporated into the heart changed its function. They correlated children's electrocardiogram (ECG) patterns with incorporated doses and found a connection between cesium-137 in children's bodies and heart disease.[27] Several dozen of Bandazhevsky's students defended dissertations on research topics related to health problems and Chernobyl exposures. The evidence they published of a continuing public health disaster worked at cross-purposes to Lukashenko's message that his state had successfully rendered the accident into a unifying national memory and nothing more.

In 1998, Bandazhevsky wrote a report to the Belarusian government describing the lifeboat ethic of Lukashenko's administration: every man for himself as the state rafted away from its commitments to public health. Bandazhevsky complained that seventeen billion rubles earmarked for Chernobyl research had disappeared or had been spent on research in areas already well known.[28] Bandazhevsky's protest coincided with plans to build the first Belarusian nuclear power reactor. Most Belarusians opposed the plan, and Bandazhevsky's critical remarks enflamed antinuclear sentiment.[29]

A few months after he published his report, Bandazhevsky was arrested under a new antiterrorism law. He was held in solitary confinement and charged with treason. His sentence, if found guilty, was execution. Interrogators drugged and beat Bandazhevsky to force him to sign a confession. They needed a confession because they had no evidence of treason. Bandazhevsky refused to confess. Trying another tack, prosecutors brought the professor to a cell holding two men charged with murder. The inmates were the size of professional football players. Bandazhevsky recounted how he stood at the threshold and looked in: "I was filthy, beaten, embarrassed at how I smelled and looked. And I was so scared. And then up walked one of the men and he said, 'Do not fear Professor Bandazhevsky, we will take care of you.'" The prisoner made Bandazhevsky tea and gave him a cookie. With that act of kindness, Bandazhevsky found the courage to keep up his resistance.[30]

Finally, with no confession and no other evidence, the prosecutor downgraded Bandazhevsky's charge to bribery. The only evidence the prosecutor had was the denunciation of a colleague who later died under mysterious circumstances. Bandazhevsky was convicted and sentenced to eight years. With Bandazhevsky in jail, Chernobyl research projects slowed, and the Belarusian Ministry of Energy went ahead with plans to build the nuclear power plant in one of the few remaining portions of Belarusian territory that was not contaminated in 1986. The plant is scheduled to open in 2019.

A number of charities and the European parliament waged a human rights campaign and won Bandazhevsky's release in 2006. The doctor was given a European freedom passport, one of only twenty-five such documents ever issued. With it, he had the right to live in any European Union country. The doctor was also granted an apartment in southern France and a pension. Bandazhevsky could have stayed in France the rest of his life, living as a minor celebrity, but he didn't do that.

"It took a year in France to return to myself," Bandazhevsky said. "Then I wanted to get back to work." Bandazhevsky won a grant from the European parliament and found collaborators in a hospital in the Ukrainian town of Ivankiv, just outside the Chernobyl Zone of Alienation. "I didn't think," he reflected, "you could approach this topic from a distance."[31]

In 2015, I stopped in Ivankiv to visit Bandazhevsky at his clinic in a crumbling hospital complex of cement block structures scattered among fruit trees, flower beds, and buzzing insects. Bandazhevsky did not visibly wear his six years of incarceration. He was handsome in a tired, rumpled way. With a small budget, his group of three doctors is engaged in a simple and colossal mission. Each year a little yellow school bus brings children from the region's eighty villages to the clinic. The team looks at thirty children a day, each child annually. They run the children through a battery of exams—heart, kidneys, blood—and take an internal body count of cesium-137. They use modern, automated equipment so that every test is identical. Bandazhevsky's group has found that although ambient levels of radioactivity have decreased over the years, the rate of health problems has continued to rise among children born to exposed parents. The team also tracks health statistics among adults. They have found, as one would expect given long latency periods, that exposures earlier in life manifest later in cancers, heart disease, and other illnesses that cut patients' lives short by ten to fifteen years.[32] On a small scale, Bandazhevsky and his team are doing what dozens of powerful international scientists recommended for decades—a long-term study of Chernobyl health effects that tracks an exposed population and their offspring over many years to ask open-ended questions about human health when exposed to chronic low doses of radioactivity.[33]

Bandazhevsky's publications are in Ukrainian. They are rarely peer reviewed. That doesn't mean he is off target. Recent research by teams of international scientists tracks with Bandazhevsky's group and with the findings of Ukrainian and Belarusian scientists of the 1980s and 1990s. In a long-term study in Narodychi, an international team of scientists showed that children had low counts of leukocytes, platelets, and hemoglobin, which led to immune and blood system disorders and amounted to a "sick child syndrome."[34] Another study connected incorporated levels of cesium-137 in children with chronic digestive track disorders.[35] A Swedish group recorded an association between children's cesium-137 body burden at very low doses and impaired lung function.[36] Twelve years after the accident, breast cancer rates in contaminated areas of Ukraine and Belarus increased 80 percent and 120 percent, respectively.[37]

Belarusian researchers discovered a significant increase in cases of leukemia among children in the three years after the accident. American epidemiologists determined that prolonged exposures even at very low doses increased the risk of leukemia.[38] Children exposed to Chernobyl radiation, especially in utero, had lower IQ scores than children in a control group, researchers found, because of damage to neurological systems.[39] A large Russian study recorded a statistically significant increase in birth defects of the nervous system, while the number of congenital malformations in Belarus doubled between 1985 and 2004.[40] Even studies of dental health showed a connection to fallout. An American group learned that children in contaminated areas had more cavities than children randomly selected in clean areas because their endocrine glands secreted less saliva and that radically changed oral microflora.[41]

Work on other exposed groups corroborates these small-scale Chernobyl studies. In 2016, a massive investigation of 308,297 nuclear workers in France, the United States, and the United Kingdom discovered a statistically significant association between radiation doses and leukemia, circulatory and respiratory diseases, and illnesses of the digestive tract.[42] Western researchers are discovering, like Soviet scientists before them, that radioactive decay at low doses changes the way cells behave in subtle and life-changing ways.[43]

At the end of my visit, Bandazhevsky asked for a lift. To save on rent, he lives in a village outside Ivankiv. He has no car and normally takes the bus. As we drove, Bandazhevsky talked about finances, how hard it has been to underwrite his clinic. I heard that from every researcher engaged in Chernobyl research. Research grants, even among scientists who publish voluminously, are rare and skimpy. That too has been a long-term trend. Despite dozens of pledges over three decades to study the long-term effects of the Chernobyl disaster on a large scale, that funding has never materialized.

In the popular imaginary, Bandazhevsky and his patients do not exist as part of the Chernobyl event. Tourists flock to the Chernobyl Zone of Alienation to look at abandoned buildings. Gamers visit the Zone virtually to shoot mutants. Televison shows and photo essays loop images of the depopulated region around and around. Disaster tourists rarely stop at the inhabited towns and villages surrounding the Zone. Yet that is where the real drama is taking place.

# Bare Life

There has been a trend in the last thirty years. The constriction of social welfare in countries around the world keeps pace with the planet's state of ecological stress. As more and more people live in environments riddled with toxins, consequences of damage from exposures to those poisons have increasingly been privatized. In this neoliberal climate, charity in the Chernobyl territories became the most legitimate form of aid. Charity, seen as free of the compulsion of state welfare programs, was good because it supposedly gave people choices. Children, considered to be innocent of the poor socialist habits of their parents, were deemed the worthiest of care. Chernobyl children's charities hosted children for summer holidays in foreign countries to restore their health. The programs were plagued with corruption and did not come close to what was needed, but they helped.[1]

A summer abroad saved Irina Vlasenko's life. Irina was three years old when the accident occurred. She lived in Minsk, where her father was a doctor. Unfortunately, in April 1986, she was visiting her grandmother who lived in Gomel. She stayed the month of May. Her parents and grandmother were oblivious to the high levels of radioactive iodine that settled on the city. Six years later, Irina was on one of those wonderful visits for "Chernobyl Children" to Italy not because she was sick or in need, but because she was lucky. Spaces on these trips were often taken up by children like Irina with well-connected, professional-class parents. Irina's Italian host parents noticed she had a lump on her throat.

The girl seemed tired and had trouble sleeping at night. They brought her to a doctor, who felt a thyroid nodule, did a biopsy, and diagnosed cancer. Irina was sped home for an operation in Minsk.

IAEA pamphlets describe Chernobyl thyroid cancers as "easily treatable." Irina's was not. Her cancer metastasized. She soon had to have another operation and another round of radiation treatment. Her health worsened. Doctors in Minsk got in touch with Dr. Christoph Reiners, a German endocrinologist, who ran a clinic in Germany where he treated Chernobyl children for thyroid cancer.[2] Irina flew to Germany with other children who required care. She recovered. When she was sixteen, Irina's cancer returned and she went back to Reiners's hospital for more treatments and returned over the years for checkups. By the time Irina was twenty, she'd had a lot of experience in medical tourism.

I met up with Irina in Minsk on a weekday morning. Her two-year-old daughter, Lyra, was in tow. On that mild spring day, the girl was tightly girded against a stray draft in a fuzzy, pink hoodie with Big Bird on her chest. As she jumped over puddles, Lyra giggled. Irina, chastising her daughter for wet shoes, smiled proudly despite herself.

Doctors in Minsk told Irina she could not have a child. Reiners encouraged her to try. "I decided I wanted a child," Irina said. "I always hope for the best." Conceiving was difficult. Irina was hospitalized twice during her pregnancy. At birth, the newborn could not breathe, asphyxiated, and had to be revived. As an infant, Lyra was jaundiced, suffered from pneumonia, and had a congenital malformation, an opening in her heart, known locally as a "Chernobyl heart."

"It all worked out fine," Irina concluded her story, watching her daughter play with a plastic cup at the deli counter.

The Minsk doctors told Irina to have no more children. I asked her if she was going to follow that advice. Irina laughed in her quiet way and assured me she would do no such thing.[3]

Researching this story, I kept a lookout for people like Irina who adapted to the Chernobyl calamity, not by denying the consequence of exposure to radiation but by facing it head on. In the late 1990s, the physicist Vasily Nesterenko created Belrad, a foundation that helped residents in contaminated regions of Belarus identify the scope of their communities' danger. Of all categories of Chernobyl-exposed, those living on contaminated ground today have the most health problems and highest mortality rates—more than cleanup workers and more than evacuees, the categories that have received the most attention.[4] Belrad produced radiation monitoring devices and collaborated with village schools to measure food and bodies. After studying physics and chemistry, children fanned out to surrounding fields and forests with devices to map local spots of radiation. The idea was that children who grasped the parameters of contamination could help their parents navigate their way toward a healthy family diet.

With my research assistant Katia Kryvichanina, we visited a village in southern Belarus that collaborated with Belrad. In Valavsk, we met with two teachers, Irina Levkovskaya and her daughter Olga, and the school principal, Nikolai Kachan. Kachan reminded me of the huntsman from Slavic fairy tales. He had a bushy mustache, twinkling blue eyes, and a good laugh ready for any emergency. He showed me a chart recording whole-body scans that Belrad employees made in 2003 and 2015. Every one of the hundred names on the list, in both years, had radioactive cesium-137 in their bodies. Twenty becquerels of cesium-137 per kilogram (Bq/kg) of body weight was thought to be safe, but in 2003, only a handful of children passed that test. Most had two to ten times more. The good news is that the 2015 chart showed an improvement as Belrad staff taught families to be more careful about the food they ate. Half the kids had relatively safe levels. Kachan pointed to one person who scored higher than everyone else—504 Bq/kg. His eyes glittered. "Look at the name," he said.

"That's you," I said.

"Yes, that's me!"

I stared at him in wonder, waiting for an explanation. Kachan had a degree in physics and collaborated with Belrad because he took radioactive fallout seriously. How did he get that dose?

Kachan said that the night before the count he had gone to a friend's barbecue. The meat was delicious, but Kachan found out when he called the next day that it was wild boar. He shook his head merrily. "I would have never eaten it had I known it was wild." Kachan got the rest of the boar meat from his friend and gave it to the Belrad staff. They tested it and told him to take it to the forest, dig a very deep hole, and drop it in.[5]

Irina Levkovskaya showed me how she taught children to test the radioactivity of their food. Kachan gave her blueberries he had picked that morning. She weighed them, placed them in a radiometer, and calculated the number of becquerels per kilogram. She was delighted to find Kachan's berries were the cleanest she had recorded that summer—only 350 Bq/kg, which was 200 Bq/kg above the conservative Belarusian limit. She asked Kachan where he found them because she wanted to pick berries there too. Katia asked to buy the berries. Kachan gave them to her, and Katia immediately started eating them by the handful. Everyone was happy to have unearthed the cleanest berries of the summer. Everyone was happy but me. I wondered why they were thrilled to eat fruit that was more than twice the permissible level.

It took me a while to figure it out. In the markets, sellers display a little certificate giving the cesium-137 count of their produce. The exact same number is written on every certificate, which leads people to suspect that the merchants pay a bribe and no one actually monitors the berries. If you measure the berries yourself, you know what you are eating. Trusting your own calculations relieves the worry that you might be consuming berries as hot as Kachan's wild boar.

I asked if the accident had brought them any good in addition to all the misery.

"Yes, I would say so," Levkovskaya replied. "It opened our world. People came here from all over the globe who we would have never met. It made the planet a more intimate place."

With all the visiting teams of doctors, I asked if they ever feel like white mice. Kachan took a moment to answer. "I have felt their care and sympathy. And yes, if sometimes they take our blood samples and use it for their research that is good too. Something like this will happen again and they will, thanks perhaps to our bodies, have more information to deal with the next tragedy."

For thirty years, information about Chernobyl consequences was carefully controlled—first by Soviet censors, and then by others. KGB agents infiltrated international organizations and possibly stole computer files. International scientists' unwitting impulses marginalized evidence of a cancer epidemic among children, sidelined opposing scientists, and looped news releases with premature and incomplete evaluations of Chernobyl consequences.[6] Organizations engaged at a corporate level in secrecy, censorship, and propaganda produced a second kind of contagion. Skepticism and a cynical understanding of truth spread outward in ways that distorted politics and social life. The teachers of Valavsk restored my navigational bearings. They showed that one path to emerge from a society-threatening cynicism is to arm yourself with skills and alliances to produce knowledge you can trust.

# CONCLUSION

# Berry Picking
# into the Future

From the mid-1990s, over a million people soldiered on in contaminated territories, unnoticed in a land few wished to visit. Maybe it was better for residents to be cut loose from the unrealized hope of aid—the miracle cures, technological fixes, and wisdom of domestic and foreign experts alike. They were left to make their own world, if only a lean-to of a world, on the remnants of the swampy periphery of the Great Pripyat Marshes.

In summer 2016, Olha Martynyuk and I had a few meetings in Rivne Province of northern Ukraine. We noticed young people and women with wooden baskets filled with berries. On the forest roads, they walked, biked, hopped in and out of vans, hundreds of people. Every few miles along the road women sat under umbrellas with scales and stacks of shallow plastic crates. They bought berries from the pickers as they emerged from the forest. We learned that locals harvest blueberries in July, cranberries in August, and mushrooms in the fall. Since 2014, tens of thousands of tons of Polesian forest produce go for processing to Poland, where the goods enter the European market.[1] Polesians have foraged in the marshy forests for centuries, but the industrial-sized scale of this activity was novel.

I wondered about that trade. Thirty years and three hundred kilometers from the Chernobyl accident, I figured the berries were probably fine. But then again, the Pripyat Marshes had near-perfect conditions for cycling radioactive fallout from soils into plants. Biologists first predicted the ecological half-life of cesium would be fifteen years. Now, for reasons they do not fully understand, researchers from the U.S. Department of Energy estimate that the period for half of cesium-137 to disappear from Chernobyl forests will be between 180 and 320 years.[2] Since the laws of physics and the rate of nuclear decay of cesium-137 did not change, this meant that the berries' measured radioactivity was primarily determined by where they had been picked. If berries came from a still-existing "hot spot," then berries being sold today could be nearly as radioactive now as they were a couple of decades ago.

I also knew of a troubling study about birth defects in this same region. Wladimir Wertelecki, a researcher at the University of Southern Alabama, started tracking births in Rivne Province in 1998. Wertelecki was born in the city of Rivne when it was part of Poland before Soviet forces occupied it in 1939. His family fled during the war. As a child, he and his mother walked across Europe to reunite with his father, a Polish officer, in Switzerland. Raised in Zurich and Buenos Aires, Wertelecki did his medical residency at Harvard. He speaks many languages, each one with a slight accent.[3] Working with minimal financing, Wertelecki, his collaborator Lyubov Yevtushok, and their team have been giving medical exams to pregnant women and following them through the birth of their infants. The researchers discovered that in the six Polesian regions of Rivne Province, certain birth defects, such as microcephaly, conjoined twins, and neural-tube disorders occur six times more frequently than the European norm. They found Polesians living on the edge of the Pripyat Marshes continue to have elevated levels of cesium-137 in their bodies. Alcohol was not the source of birth defects because Polesians drank less frequently than the norm.[4] "We did not prove with this study that radiation causes birth defects," Wertelecki carefully elaborated. "We just have a concurrence, not proof, of cause and effect."[5]

Given the riddled history of the Pripyat Marshes, it is hard to say exactly what caused those birth defects. They could have been triggered by Chernobyl radiation, by Chernobyl radiation and preaccident radioactive isotopes circulating in the swamp from global fallout, by secret bomb tests in the marsh bombing range, or by a combination of radiation, nitrates, and pesticides added to crops before the accident to improve harvests and afterward as part of the cleanup effort. I have argued that Chernobyl is not an accident but rather an acceleration on a time line of exposures that sped up in the second half of the twentieth century. Seeing Chernobyl as an acceleration helps to grasp how it is connected to many other events; pollutants accumulate and change the environments in which they roost, one mutated cell building on another, one set of decisions commanding yet more actions to create a terrain of such toxic complexity that humans have a hard time grasping it, much less fixing it. Maybe that is why scientists and funding agencies have thrown up their hands in confusion and walked away.

Olha suggested we go undercover berry picking to find out more about the new mass trade in forest produce. An elderly farmer sold us a pair of handmade boxes from wood he harvested from the forest. The farmer's wife had just finished whitewashing the walls of their cottage, a traditional Ukrainian one-room hut. She dug the white clay plaster from the riverbank. I didn't see much in their cottage purchased at a store. The couple burned wood, kept chickens and goats, and put the ashes and manure on their vegetable garden. They ate everything they produced. They had never heard of a manual issued by any party—the Soviet government, a UN agency, or a charity—that advised against doing these things.

Taking the baskets, we armed ourselves with bug spray and set off into the woods. We tried to catch up to a group of pickers who pedaled bicycles. They sped down a primitive country road. Steering the car through water-filled potholes, the path rising and falling in waves of mud, I plowed into a deep crater that buried the front wheels in muck. The pickers, bouncing like rabbits, disappeared around a bend. It took the rest of the day to haul the car out and have village mechanics drive it onto a rickety birchwood ramp to fix it.

The next day, already smarter, we caught up with a group of pickers on foot. They were teens, siblings and cousins from a large Pentecostal family. The pickers spoke to us only briefly before hustling back to work. They carried scoops made from tin cans to quickly strip berries from bushes (which is faster than plucking berries by hand but more damaging to the plants). I watched them plunging through the berry bushes, sun rays refracting through the canopy and jeweling the purple fruit. The teens moved through the woods like a pack of juvenile bears, foraging with quick, efficient motions: stoop, scoop, step, the work silent except for the pinball roll of berries hitting plastic buckets.

The berry bushes tumble everywhere, low green shrubs, spreading wild for miles and miles under the cathedral of the Polesian forest. Anyone can pick the berries. The forests are democratic. And unlike expensive dairy farming, which was made costlier by the need for extra filters and chemicals to strain out radionuclides, berry picking requires little investment. Mostly women and children do the work. If they pick fast, they can earn $25 a day, which is good money in an economy where the monthly wage of a schoolteacher is $80.

Making an industry out of combing the woods for forest products is one of those creative adaptations Polesians have devised in order to persist in the contaminated spaces left to them. Nonprofits tried for years with little success to revive the local economy. With little outside help, Polesians have taken a place in the global economy by setting up a rural network to deliver freshly picked wild berries and mushrooms to wholesalers bound for Europe.

A group of pickers, sweaty and bug-bitten after a day in the forest, saw the irony of their new trade. Learning I am American, they joked, "We send you in the West our organic, wild berries, and get back berry-flavored drinks." Another chimed in, "That's right. We ship off good pine lumber and you send us faux-wood particleboard." The pickers were pointing to the traditional colonial exchange of raw materials for higher-priced manufactured goods. True enough, but there was more to this trade than classic economics.

After gathering our fill, we showed up at a berry wholesaler and noticed a radiation monitor who was stationed to meet buyers at the loading dock. The situation was tense. As the monitor waved a wand over each box of berries, measuring their gamma ray emission, she set aside about half of the boxes. Sellers argued with her over the gamma levels. I asked the monitor, a young townswoman, how many berries come up radioactive.

"All the berries from Polesia are radioactive," she replied, "but some are really radioactive. We've had berries measure over three thousand!" She could not describe what units she was referring to, microsieverts or microrems; the buyer only knew which numbers were bad. "The needle has to be between 10 and 15," she said, vaguely pointing to her wand, "and then I place it in this machine." She gestured toward a small gamma spectrometer. "If the readout is more than 450, then the berries are over the permissible level."

We stood around watching the sales transactions. The buyer placed berries above permissible levels aside but still purchased them, just at lower prices. The wholesaler said that the radioactive berries were used for natural dyes. The pickers claimed the hot berries were mixed with cooler berries until the assortment came in under the permissible level. The berries could then legally be sold to Poland to enter the European Union market, even if some individual berries measured nearly three times higher than the permissible level. Such mixing is legal as long as the overall average falls within the generous American and European Union limit of 1,250 Bq/kg. (The European Community established 600 Bq/kg for food in 1986 after the Chernobyl accident as a temporary emergency level and failed to lower it afterward. In 2016 soon after our visit, without public discussion, the European Union doubled the permissible level to 1,250 Bq/kg.)[6]

Since 1986, Polesians have consumed radioactive food from their forests. With this new trade, they pass Chernobyl-contaminated produce to European markets for wealthier consumers abroad. From Europe, the berries keep going. A nuclear security specialist told me that at the U.S.-Canada border, they stopped a truck with a "radiating mass" in its trailer. Alarmed about a possible dirty bomb, they looked inside. They found to their relief only berries from Ukraine. Because the fruit fell within the permissible norm, the guards let the truck pass into the United States.[7]

Normally poorer populations consume the most toxic by-products of the industrial world. The migration westward of Polesian berries upsets the global caste system. It also overturns the idea of national boundaries as discrete entities. In the twenty-first century, populist leaders keep up a drumbeat about the "black horde of migrants crossing the border."[8] The hardening of national borders occurs despite a dawning understanding of the planetary scale of even the smallest local actors, such as blueberries. Goods and their contaminants flow around the world passing from the membrane of one nation into that of another. Fungi, soils, and pesticides skate across oceans on the trade winds. Jellyfish slip aboard container ships to be ejected in foreign harbors. "Invasive" plants, insects, and viruses hop from continent to continent. And radioactive isotopes migrate around the globe. Despite populists' fantasy that borders can be sealed, events out there make it home.

Americans, Canadians, and Europeans may wake to a breakfast of Chernobyl blueberries. Some may randomly draw from the global marketplace berries measuring at 3,000 Bq/kg. That would be one unfortunate consequence of a policy promoted by UN agencies that recommended people remain living on contaminated ground.

Soviet leaders found that no amount of chemicals, bulldozers, scraping, or spraying cleaned radioactive isotopes from the landscape. But the berries are really good at this work. Berry bushes and mushrooms as well as the digestive tracts of milk cows and goats efficiently extract man-made isotopes from the environment and serve them in neat packages to people who desire them. In the simple exchange of money for berries is the ghost of a future transaction that makes more sense than sending Chernobyl fission products near and far in consumer markets. Rather than having radioactive berries become contamination-spreading agents, why not deploy them as allies in the cleanup?[9]

The people who argue that plants and animals in the Chernobyl Zone are thriving are wrong, but they are accurate in asserting that nature can help correct man-made disasters. That doesn't mean humans can step away and let nature do its work. Contaminated spaces require curation. As I learned in the Red Forest, trees are good repositories of radioactive isotopes—until they burn and send up radioactivity all over again. Creating systems so that places like the Red Forest do not go up in flames is the work left to humans in the twenty-first century. A more sustainable business model than exporting contaminated berries abroad would be to sell berry-picking tours to the thousands of end-of-world tourists who annually visit the Chernobyl Zone. Polesian pickers would serve as their guides and, rather than sell the berries and mushrooms, they would bury their harvests properly as radioactive waste.

To shift the terms of exchange in this way is to acknowledge that the child pickers with lips stained blue are actually nuclear waste workers. Despite decades of denials, there is no getting around that fact.

The global circulation of radioactive Chernobyl blueberries prompts a question: How did it happen that Chernobyl's reverberating explosions, which greatly altered planetary history, have so far done so little to revise human history? Over thirty years later, there is still no organization on the national or international level sophisticated enough to deal with a major nuclear emergency. The Chernobyl disaster illustrates how expansively states and international agencies failed the people they were tasked with protecting.

That is not for lack of trying. Soviet leaders responded to the Chernobyl crisis by calling up the tools of industrialized consumer society. They seeded clouds and scrubbed buildings with chemicals, built new houses with vinyl and asbestos, wrapped food in plastic packaging, and stored it in the first freezers many villagers had ever seen. They dumped tons of extra fertilizers and pesticides on crops, paved dirt roads, sent gas lines to villages, and plugged in new medical diagnostic equipment. Soviets fought admirably a battle against the invisible enemy that could not be named because censors banned discussion of Chernobyl fallout. Unfortunately, the platoons of professionals were defeated by the radionuclides that swarmed everywhere, reappeared after cleaning, and stubbornly entered the food chain.

Doctors found that radioactive contaminants infiltrated the bodies of their patients, who grew sicker each year. Gradually, sanitation officials understood they had a public health disaster on their hands. For three years, from 1986 to 1989, Soviet physicians had to sit on this information, telling no one but their bosses. Finally, in the spring of 1989, censors lifted the ban on Chernobyl topics. Residents made alliances with doctors and radiation monitors. They organized, petitioned, broke laws, defied their bosses, and carried on when dismissed as ignorant provincials in order to get the world to understand the new precarious life they led.

In many ways they succeeded. We know what we do about the Chernobyl accident because of the everyday heroes who refused to accept the assurances of the Soviet survival manuals. Natalia Lozytska, a physicist disguised as a housecleaner, discerned with few resources the dangers of the disaster and worked to warn Soviet leaders, international experts, and fellow citizens. Pavel Chekrenev and Aleksandr Komov did not stop until they forced their superiors to protect people who were drinking radioactive water and milk. Valentina Drozd, Olga Degtyariova, and Keith Baverstock risked their jobs to tell the world about the childhood cancer epidemic. Vasily Nesterenko and Yuri Bandazhevsky lost their jobs and nearly their lives because they would not be silenced. These people do not show up in the usual Chernobyl accounts of firefighters, masculine heroism, and official mendacity, but their story matters most in the future. Thousands of people made similar principled decisions. They lined up flank after flank to form a collective, a gathering force that strained to perceive and understand the chaos of millions of curies of radioactive nuclides let loose on earth.

For Soviet leaders, crowds on the streets were more threatening than radioactivity. They called in foreign experts for an "independent assessment." Scientists in the West used the A-bomb survivor Life Span Study to abstract the noticeable health damage from the accident into a series of risk assessments and promises of further study. In the Cold War decades, physicists became experts in measuring radioactive energy in ecological systems, but by 1990 they claimed to know little about how radioactivity, once in the body, affected biological processes. Five years after the accident, Soviet scientists working in contaminated areas were forced to recognize a far more sophisticated understanding of those processes—that plutonium, for example, lodged in a lung had a great deal more biological effect than the same amount of plutonium placed near a person's foot. This knowledge had been known for decades to researchers in the United States and USSR working on classified studies of victims of bomb fallout and nuclear weapons production.

Nonetheless, consultants from UN agencies dismissed the findings of scientists in Ukraine and Belarus. Using comparatively simple and generalizing computational models, UN experts stated that radioactivity at Chernobyl levels would cause no major damage to human health except for the risk of a small number of future cancers. They reiterated this statement for many years despite clear signs of a significant and alarming childhood cancer epidemic.

Today scientists repeat that we know little about the effects of low doses of radiation on human health. That claim is partly true because of the suppression of the record of catastrophic damage in the Chernobyl territories. And that statement is partly untrue. The first Western historian to work in the Ministry of Health archives, I found overwhelming evidence to confirm what was plainly visible to local observers: that soon after April 1986 healthy people in Chernobyl territories, especially children, fell ill. In subsequent years, rates of chronic disease increased. People suffered not just from cancers but also from diseases of the blood-forming system, digestive tract, and endocrine, reproductive, circulatory, and nervous systems.

Unfortunately, just as researchers in Ukraine and Belarus were able to pursue their research openly, the Soviet behemoth collapsed and Soviet science with it. Many commentators applauded the end of the Soviet state, convinced that political change would free citizens, liberate the truth, and bring relief to Chernobyl sufferers. But after the USSR disappeared, residents were left naked, bared to the raw forces of radioactivity, political disorder, and economic stress. In 1998, American medical anthropologist Sarah Phillips shadowed Ukrainian doctors working on contract for an international study. The doctors' mandate was only to draw blood, not to treat health problems. They tracked teens who had been children when the accident occurred. "The clinics had no electricity, running water, or doctors, and the pharmacies had no medicine," Phillips remembered. "It was striking how sick everyone was. All these kids looked ten years older than they were."[10]

Given the volume of evidence, the persistence over three decades of assertions that Chernobyl health problems are minimal is remarkable. The "worst nuclear disaster in human history," international experts continue to proclaim, amounted to only 54 deaths and 6,000 cases of "easily treatable" thyroid cancer. That was a risk, they insisted, the world could live with.[11]

Post-Chernobyl analyses are riddled with many understated quantities—from the number of hospitalizations to the estimated average dose of radiation people received, but the reporting of 54 Chernobyl fatalities stretches credulity the furthest. Unfortunately, the Russian, Ukrainian, and Belarusian states have no public tally of Chernobyl-related fatalities to update the count. The Ukrainian state pays compensation to 35,000 people whose spouses died from Chernobyl-related health problems.[12] This number only reckons the deaths of people old enough to marry. It does not include the mortality of young people, infants, or people who did not have records to qualify for compensation. The figure is only for Ukraine, not Russia or Belarus, where 70 percent of Chernobyl fallout landed. Off the record, a scientist at the Kyiv All-Union Center for Radiation Medicine put the number of fatalities at 150,000 in Ukraine alone. An official at the Chernobyl plant gave the same number. That range of 35,000 to 150,000 Chernobyl fatalities—not 54—is the minimum.

Underestimating Chernobyl damage has left humans unprepared for the next disaster. Chernobyl reporting focused on the "it could never happen here" argument emphasizes the particularly inept, corrupt, and secretive qualities of the Soviet state.[13] A disaster on the level of Chernobyl, commentators argued, could never occur in an open, democratic society where the energy business is run by private enterprise. Yet, when a tsunami crashed into the Fukushima Daiichi Nuclear Power Plant in 2011, Japanese businesspeople and political leaders responded in ways eerily similar to Soviet leaders. They grossly understated the magnitude of the disaster (a meltdown of three reactors), sent in firefighters unprotected from the high fields of radioactivity, and intentionally withheld from the public information about levels of radioactivity and health directives. They did not issue prophylactic iodine to children, yet raised the acceptable level of radiation exposure in schools from 1 to 20 mSv a year, which is the international standard for nuclear workers. In subsequent months, public health officials were reluctant to monitor food and dismissed parents' concerns about both their children's' health problems and the recorded increase in pediatric thyroid nodules and cancers. In 2011, Japan relied on nuclear power for 30 percent of its energy needs. As in the USSR, Japanese leaders dissembled and glossed over the disaster to privilege production and national pride over health and safety. "Did we learn nothing," anthropologist Sarah Phillips asks, "from Chernobyl?"[14]

Minimizing both the number of deaths so far and the ongoing health consequences of the Chernobyl disaster provided cover for nuclear powers to dodge lawsuits and uncomfortable investigations in the 1990s when, with the end of the Cold War, the record of four decades of reckless bomb production emerged from top-secret classification.

The declassified history of nuclear weapons testing has showed how little care or responsibility leaders took for damages caused by the detonation of the equivalent of 29,000 Hiroshima-sized bombs.[15] Between 1945 and 1998, bomb designers shot off bombs with abandon—bombs named after presidents; bombs named after scientists, wives, and uncles; bombs given the names of native tribes on whose land the explosions scattered billions of curies of fallout. The ill-advised detonation of nuclear weapons in Nevada delivered to milk-drinking Americans across the U.S. continent an average collective dose of radioactive iodine similar to that of people living in Chernobyl contaminated areas.[16] Alaskans, Welsh, and Scandinavians got hit with Soviet bomb tests at Novaya Zemlya. Australians and Pacific Islanders were strafed with fallout from French, British, and American bombs that blasted Pacific islands.[17] Indians exploded underground atomic bombs near the border of Pakistan. Pakistanis detonated their weapons upwind from India. Chinese and Soviet bomb designers together polluted the Eurasian interior. The tremendous quantity of radioactive fallout swirled in the atmosphere and came down in rain and snow. The more precipitation in a place, the more the radioactivity cascaded to earth. The period of nuclear testing qualifies as the most unhinged, suicidal chapter in human history. In the name of "peace" and "deterrence," military leaders waged global nuclear war.

The damage of a half century of nuclear bomb testing appears to have left a perceptible footprint on human health. Rates of thyroid cancer in the United States tripled between 1974 and 2013, and better detection did not account for all the increases. As global fallout sifted down, mostly in the Northern Hemisphere, thyroid cancer rates grew exponentially.[18] In Europe and North America, childhood leukemia, which used to be a medical rarity, increased in incidence year by year after 1950.[19]

Australia, hit by fallout from British and French tests, has the highest incidence of childhood cancer worldwide.[20] An analysis of 42,000 men showed sperm counts among men in North America, Europe, Australia, and New Zealand dropped 52 percent between 1973 and 2011.[21] These statistics show a correlation, not a causal link. They do, however, invite a lot of questions.

Thirty years after the Chernobyl accident, we are still short on answers and long on uncertainties. Ignorance about low-dose exposures is, I have argued, partly deliberate. Before 1986, Soviet and international experts knew about the connection between childhood thyroid cancer and radioactivity, but they suppressed and refuted evidence about the epidemic surrounding the smoking Chernobyl reactor because they had much larger radioactive skeletons in the closet from nuclear bomb tests. Thyroid cancer among children is the medical canary in the mine. Declassified Soviet health records demonstrate that thyroid cancer was just one outcome and that radioactive nuclides lodged in organs caused a wide range of illness among people in the Chernobyl territories. The Soviet medical records suggest it is time to ask a new set of questions that is, finally, useful to people exposed over their lifetimes to chronic doses of man-made radiation from medical procedures, nuclear power reactors and their accidents, and atomic bombs and their fallout. Few people on earth have escaped those exposures.

//

# ACKNOWLEDGMENTS

A complex, transnational event like Chernobyl calls up a complicated, transnational, and interdisciplinary research agenda that I could not do alone. Olha Martynyuk and Katia Kryvichanina, my research assistants, labored on this project as if it were their own. This book profits from their intelligence, keen insights, and willingness to hunt down even the faintest clues. Alane Mason at Norton pressed me to elucidate my inconsistencies. Tim Mousseau and Anders Møller generously invited me along as they worked in the Chernobyl Zone. Natalia Baranovska shared with me her wealth of knowledge and her Rolodex. Nadezhda Shevchenko became a local guide, warm and generous. Marjoleine Kars, Harry Bernas, Paul Josephson, Tatiana Kasperski, Tomasz Gałązka, and Ian Fairlie kept an eye out for sources and read multiple versions of the manuscript. They saved me many times over.

Several foundations generously funded this project and made it possible. The Andrew Carnegie Foundation, the American Council of Learned Societies (ACLS), the American Academy in Berlin, and the European University Institute (EUI) funded leave and research. The statements made and the views expressed in this book are solely the responsibility of the author. Many thanks to Rachel Brubaker, Jessica Berman, Eva Dominguez, and Scott Casper at University of Maryland, Baltimore County (UMBC) for grant support. My gratitude to Lynne Viola, Lewis Siegelbaum, John McNeill, Serguei Oushakine, and the anonymous reviewers for help in evaluating this project. Thanks to

Greta Essig at Carnegie; Pauline Yu at ACLS; and Pieter Judson, Regina Grafe, Laura Downs, Benedetto Zaccaria, Dieter Schlenker, and Anna Coda at the EUI. At the American Academy a long list helped in every way: René Ahlborn, Carmen Artmann, John Eltringham, Lutz Finkel, Johana Gallup, Thomas Heller, Reinold Kegel, Yolande Korb, R. Jay Magill, Carol Scherer, Gabriele Schlickum, and Michael Steinberg.

Along my travels many people helped as local guides in archives and in the world of science. In Belarus, they are Uladzimir Volodzin, Valentina Drozd, Alexei Nesterenko, Irina Vlasenko, Andrei Stepanov, Nikolai Kachan, Irina Levkovskaya and Olga Levkovskaya. In Ukraine, they are Lyubov Yevtushok, Aleksandr Komov, Wladimir Wertelecki, Serhy Yekelchyk, Aleksandr Kupny, Evgenii Abdulovich, Natalia Lozytska, Larysa Lavrenchuk, Maria Panova, Mykhailo Zakharash, Olha Bobyliova, Mykola Popelukha, and Tamara Viktorovna Kot. Toshi Higuchi, Horoko Takahasi, Bo Jacobs, Norma Field, and Dennis Riches served as my guides in Japan. Mike Faye, Susan Hyser, Sarah Phillips, Sudaba Lezgiveya, Susan Lindee, Nadezhda Kutepova, Lucas Hixson, Susan Lindee, Alan Flowers, Sonya Schmid, and Daniel Miller provided valuable information and analysis.

Special thanks to the organizers of workshops and conferences where material for this book was developed: Anna Tsing, Nils Bubant, and Andrew Mathews in Santa Cruze and Sintra, Portugal; Anna Storm in Oslo; Adriane Lentz-Smith at Duke University; James Cameron in Saõ Paulo, Brazil; Astrid Kirchhof and Jan-Henrik Meyer in Berlin; Melanie Arndt in Regensburg; Klaus Gestwa and Schamma Schahadat in Tübingen; Andrew Tompkins in Sheffield; Jacob Doherty at University of Pennsylvania; Paul Sabin at Yale; Dan Healey in Oxford; Victor Yakovenko in College Park; Tomasz Bilczewski in Krakow; Emily Elliott and Lewis Siegelbaum in East Lansing; Michael Lewis in Salisbury; Fred Corney and Hiroshi Katamura at William and Mary; and Choi Chatterjee and Ali Igmen in Los Angeles.

I am extremely grateful to the people who worked to put this book together. Sarah Lazin, Julia Conrad, and Margaret Shultz at Lazin Books, and at Norton and Penguin Alane Mason, Laura Stickney, Holly Hunter, Ashley Patrick, Donna Mulder, Rachel Salzman, Camille

Bond, Joe School, Dassi Zeidel, and Jessica Friedman all made major contributions.

Portions of several chapters of this book appeared in print previously and are reprinted with permission. Special thanks to editors Margaret Harris, Corey Powell, Anna Tsing, and R. Jay Magill. The articles are "Where Fruit Flies Fear to Tread," *Berlin Journal*, no. 30 (Fall 2016), published by the American Academy in Berlin; "Chernobyl's Black Hole," *Physics World* (April 2017) (© 2017 IOP Publishing); "The Harvests of Chernobyl," *Aeon Magazine* (November 2016); and "Photography in the Radioactive Depths," in *Arts of Living on a Damaged Planet: Stories from the Anthropocene*, edited by Anna Lowenhaupt Tsing, Nils Bubandt, Elaine Gan, and Heather Anne Swanson.

Friends and family have made this book possible in ways I cannot begin to list but mostly for making my life so rich and joyful. Special thanks to Marjoleine Kars and Sasha Bamford-Brown. Kama Garrison, Dave Bamford, Michelle Feige, Leila Corcoran, Tim Ahmann, Sally Hunsberger, Warren Cohen, Maggie Paxson, and Charles King provided meals, watered my garden, and looked after a certain little dog for long periods when I was away. My parents, Sally and William Brown, and siblings Liz and John Marston, Aaron and Denise Brown, and Julie and Kurt Hofmeister prop me up always in every way. Lisa Hardmeyer, Bruce Gray, Leslie Rugaber, and Prentis Hale give me support from afar. Tracy Edmonds is a source of inspiration for her courage and endurance. I wish she did not have to live a part of this history.

# NOTE ON TRANSLITERATION
# AND TRANSLATION

In the text and notes of this book, I have used a modified Library of Congress system to transliterate Russian and Ukrainian names of titles, places, and persons. To make it easier for English readers, I have omitted the non-English diacritics in Ukrainian transliteration. In cases of people who published or are known in English texts with a given spelling of their last name, I have kept that commonly recognized spelling. I have translated the title of archival documents from Ukrainian, Russian, and Belarusian archives for English-language readers. I did not translate book titles.

# LIST OF ARCHIVES AND INTERVIEWS

## Archive Acronyms and Abbreviations

Derzhavnyi arkhiv Chernihivskoi oblasti (DAChO)

Derzhavnyi arkhiv Zhytomyr'skoi oblasti (DAZhO)

European Union Archive (EUA), Florence, Italy

Glavnoie upravlenie iustistii Mogilevskogo oblispolkoma Uchrezhdenie "Zonal'nyi" gosudarstvennyi arkhiv v g. Kricheve (Arkhiv Kricheva), Krichev, Belarus

Greenpeace Archive (GPA), Amsterdam, Netherlands

Gosudarstvennyi arkhiv Gomel'skoi oblasti (GAGO), Gomel, Belarus

Gosudarstvennyi arkhiv Mogilevskoi oblasti (GAMO), Mogilev, Belarus

Gosudarstvennyi arkhiv obshchestvennykh ob'edinenii Mogilevskoi oblasti (GAOOMO), Mogilev, Belarus

Gosudarstvennyi Arkhiv Rossiyskoi Federatsii (GARF), Moscow, Russia

Haluzevyi derzhavnyi arkhiv Sluzhby bezpeky Ukrainy (SBU), Kyiv, Ukraine

Hoover Institution Archive (HIA), Palo Alto, CA

International Atomic Energy Agency (IAEA) Archive, Vienna, Austria

National Archives and Records Administration (NARA), College Park, MD

National Cancer Institute (NCI), Bethesda, MD

Natsional'naia akademiia nauk Belarusi (BAS), Minsk, Belarus

Natsionalnyi arkhiv respubliki Belarus (NARB), Minsk, Belarus

Nuclear Testing Archive (NTA), Las Vegas, NV

Rossiyskii gosudarstvennyi arkhiv ekonomiki (RGAE), Moscow, Russia

Tom Foulds Collection, University of Washington Special Collections (TFC UWSC), Seattle, WA

Tsentralnyi derzhavnyi arkhiv hromadskykh ob'yednan' Ukrainy (TsDAHO), Kyiv, Ukraine

Tsentralnyi derzhavnyi arkhiv vyshchykh orhaniv vlady Ukrainy (TsDAVO), Kyiv, Ukraine

Tsentralnyi derzhavnyi kinofotofonoarkhiv Ukrainy (TsDKFFA), Kyiv, Ukraine

United Nations Educational, Scientific, and Cultural Organization (UNESCO), Paris, France

United Nations Scientific Committee for the Effects of Atomic Radiation (UNSCEAR) Archive, Vienna, Austria

Uniformed Services University Archives (USU), Bethesda, MD

United Nations Archive (UN NY), New York, NY

World Health Organization Archive (WHO), Geneva, Switzerland

## Author Interviews

Telephone interview with Lynn Anspaugh, January 12, 2016

Interview by Olha Martynyuk with Anatoly Artemenko, December 22, 2017, Kyiv, Ukraine

Interview with Yuri Bandazhevsky, July 20, 2015, Ivankiv, Ukraine

Interview with Olha Oleksandrivna Bobyliova, June 6, 2017, Kyiv

Telephone interview with Dr. Michael Carome, Public Citizen Foundation, September 5, 2017

Interview with Nina Aleksandrovna Chekreneva, July 10, 2016, Zhytomyr, Ukraine

Telephone interview with Mona Dreicer, April 13, 2018

Interview with Valentina Drozd, February 12, May 13 (telephone), and June 15 and 23, 2016, Minsk

Telephone interview with Lars-Erik De Geer, December 18, 2017

Telephone interview with Robert Gale, October 29, 2017

Interview with José Goldemberg, Saõ Paulo, Brazil

Interview with Abel Gonzalez, June 3, 2016, Vienna, Austria

Interview with Dmitri Grodzinsky, July 14, 2015, Kyiv, Ukraine

Interview with Tamara Haiduk, July 7, 2016, Chernihiv, Ukraine

Telephone interview with Lucas Hixson, October 12, 2016

Interview with Nikolai Kachan, July 28, 2016, Valavsk, Belarus

Telephone interview with Alex Klementiev, August 6, 2017

Interview with Aleksandr Komov, August 2, 2016, Rivne, Ukraine

Interview with Tamara Kot, July 7, 2016, Chernihiv, Ukraine

Interview with Aleksandr Kupny, June 14, 2014, Slavutych, Ukraine

Telephone interview with Iryna Labunska, February 27, 2018

Interview with Irina Levkovskaya, July 28, 2016, Valavsk, Belarus

Interview by Olha Martynyuk with Boris Leskov, November 3, 2017, Kyiv, Ukraine

Telephone interview with Fred Mettler, January 7, 2016

Interview with Anders Møller, November 4, 2014, Paris, France

Telephone interview with Clare Moisey, February 19, 2018

Interview with Alexey Nesterenko, July 22, 2016, Minsk, Belarus

Interview with Maria Nogina, July 7, 2016, Chernihiv, Ukraine

Telephone interview with Sarah Phillips, May 18, 2018
Interview with Andrei Pleskonos, June 1, 2017, Kyiv, Ukraine
Interview with Oleksandr Popovych, June 1, 2017, Kyiv, Ukraine
Telephone interview with Jacques Repussard, February 28, 2018
Telephone interview with Olga Savran, January 17, 2018
Interview with Heorhii Shkliarevsky, June 1, 2017, Kyiv, Ukraine
Interview with Volodymyr Tykhyy, June 5, 2017, Kyiv, Ukraine
Interview with Irina Vlasenko, April 13, 2016, Minsk, Belarus
Interview with Wladimir Wertelecki, May 5, 2016, Washington, DC
Interview with Alexey Yablokov, June 5, 2015, St. Petersburg, Russia
Interview with Alla Yaroshinskaya, May 27, 2016, Moscow, Russia
Interview with Mykhailo Zakharash, July 1, 2016, Kyiv, Ukraine

# NOTES

## Introduction: The Survivor's Manual

1 "Instructions for Local Communities," August 25, 1986, Tsentralnyi derzhavnyi arkhiv vyshchykh orhaniv vlady (TsDAVO) 342/17/4390: 41–51.

2 *The Chernobyl Catastrophe: Consequences on Human Health* (Amsterdam: Greenpeace International, 2007), 1–15; and D. Kinley III, ed., *The Chernobyl Forum: Chernobyl's Legacy, Health, Environmental and Socio-Economic Impacts* (Vienna: IAEA, 2006).

3 Elisabeth Cardis, quoted in Mark Peplow, "Special Report: Counting the Dead," *Nature* 440 (2006): 982–83.

4 Telegram Gale to Beninson, June 27, 1986, and Giovanni Silini, "Concerning Proposed Draft for Long-Term Chernobyl Studies," Correspondence Files, August 1986, UN Scientific Committee for the Effects of Atomic Radiation (UNSCEAR) Archive.

5 Ihor Kostin, *Chernobyl: Confessions of a Reporter* (New York: Umbrage, 2006), 76–80.

6 "Informatsia MOZ URSR dlia Rady Ministriv Respubliky," April 30, 1986, in N. P. Baranovska, *Chornobyl'sk'ka trahedia: Narysy z istorii* (Kyiv: Instytut istorii Ukrainy NAN Ukrainy, 2011), 86.

7 Author interview with Olha Oleksandrivna Bobyliova, June 6, 2017, Kyiv.

8 "Internal Report," June 2, 1986, Haluzevyi derzhavnyi arkhiv Sluzhby bezpeky Ukrainy (SBU) Archive 16/1/1238, 148–53.

9 Aleksandr Zakharov, "Vospominania barnaul'skikh likvidatorov," http://milanist88.livejournal.com/12328.html; "To the Central Committee of the Communist Party of Belarus," August 6, 1986; and "On a Collective Petition," August 29, 1986, NARB 4R/154/393: 7, 3–4.

10 "Memo," May 4, 1986, TsDAVO 27/22/7701, 35.

11  "Situation Report," May 5, 1986, in Baranovska, *Chernobyl'sk'ka trahedia*, 75–
    76; and "Operative Group Information," May 11, 1986, Tsentralnyi derzhavnyi
    arkhiv vyshchykh orhaniv vlady Ukrainy (TsDAVO) 27/22/7703: 13.

12  Richard Wilson, "A Visit to Chernobyl," *Science* 236, no. 4809 (1987): 1636–40;
    and Alexander Sich, "Truth Was an Early Casualty," *Bulletin of the Atomic Sci-
    entists* (May/June 1996): 36.

13  Fernald Preserve, "Touring the Fernald Preserve PLEASE BE CAUTIOUS," U.S.
    Department of Energy.

## Liquidators at Hospital No. 6

1  "Professor Angelina Guskova: na lezvii atomnogo mecha," March 29, 2015,
   *Ozersk 74*, http://www.ozersk74.ru/news/politic/239887.

2  Christopher Sellers, "The Cold War over the Worker's Body: Cross-National
   Clashes over Maximum Allowable Concentrations in the Post–World War II
   Era," in *Toxicants, Health and Regulation since 1945*, ed. Soraya Boudia and
   Nathalie Jas (London: Pickering and Chatto, 2013), 24–45.

3  Hiroshi Ichikawa, "Radiation Studies and Soviet Scientists in the Second Half
   of the 1950s," *Historia Scientiarum* 25, no. 1 (2015): 86; and A. K. Gus'kova
   and G. D. Baisogolov, *Luchevaia bolezn' cheloveka* (Moscow: Meditsina, 1971),
   42, as cited in Adriana Petryna, *Life Exposed: Biological Citizens after Chernobyl*
   (Princeton, NJ: Princeton University Press, 2013), 119–20.

4  Andrei Ivanovich Vorobiev, "Gematologicheskii nauchnyi tsentr," accessed September 5, 2017, http://blood.ru/about/vse-rukovoditeli-tsentra/vorobjov-a-i
   .html.

5  For one case in 1970, see "Statement by the Executive Commission of the Acad-
   emy of Science of UkrSSR," February 9, 1970, Tsentralnyi derzhavnyi arkhiv
   hromadskykh ob'yednan' Ukrainy (TsDAGO) 1/25/365: 11–16. For overview, see
   Grigori Medvedev, *The Truth about Chernobyl: An Exciting Minute-by-Minute
   Account by a Leading Soviet Nuclear Physicist of the World's Largest Nuclear
   Disaster and Coverup* (New York: Basic Books, 1991), 5.

6  "Prichiny Chernobyl'skoi avarii," 1990, Rossiyskii gosudarstvennyi arkhiv
   ekonomiki (RGAE) 4372/67/9743: 86–99.

7  Sonja D. Schmid, *Producing Power: The Pre-Chernobyl History of the Soviet
   Nuclear Industry* (Cambridge, MA: MIT Press, 2015), 128–30.

8  "Informational Communiqué," April 28, 1986, SBU 16/1/1238, 53–57, 71–75;
   "Evidence from Yuri Tregub, Shift Foreman at Block N4," http://igpr.ru/library/
   dejstvija_personala_chast_1.

9  Vivienne Parry, "How I Survived Chernobyl," *The Guardian*, August 24, 2004,
   accessed August 24, 2017, https://www.theguardian.com/world/2004/aug/24/
   russia.health.

10  "Avariia na ChAES," Fireman.club, https://fireman.club/statyi-polzovateley/
    avariya-na-chaes-pervye-geroi-chernobylya/, August 24, 2017; and Serhii

Plokhy, *Chernobyl: The History of a Nuclear Catastrophe* (New York: Basic Books, 2018), 87–101.

11  "From the Report by Medical Sanitary Unit No. 1," no earlier than January 1991, Nataliia Baranovska, ed., *Chornobyl'—problemy zdorov"ia naselennia: zbirnyk dokumentiv i materialiv u dvokh chastynakh*, vol. 2 (Kyiv: Instytut istorii Ukrainy, 1995), 111–15.

12  Andrei Vorobiev, *Do i posle Chernobylia: vzgliad vracha* (Moscow: New Diamond, 1996), 58–60.

13  Karl Z. Morgan, "Reducing Medical Exposure to Ionizing Radiation," *American Industrial Hygiene Association Journal* (May 1975): 361–62.

14  "AFRRI Briefing for Sue Bailey, Deputy Assistant Secretary of Defense for Clinical Services," October 28, 1994, Uniformed Services University (USU) Archives.

15  Svetlana I. Vashchenko, "Liuberchane v Chernobyle," *Mestnye novosti iz pervykh ruk*, accessed April 26, 2017, http://lubgazeta.ru/articles/292856.

16  Miquel Macià, Anna Lucas Calduch, and Enric Casanovas López, "Radiobiology of the Acute Radiation Syndrome," *Reports of Practical Oncology and Radiotherapy* 16 (August 2011): 123–30.

17  Timothy Jorgensen, *Strange Glow: The Story of Radiation* (Princeton, NJ: Princeton University Press, 2016), 154.

18  Steve Weinberg, "Armand Hammer's Unique Diplomacy," *Bulletin of the Atomic Scientists* 42, no. 7 (September 8, 1986): 50–52.

19  Vashchenko, "Liuberchane v Chernobyle."

20  Richard Champlin, "With the Chernobyl Victims," *Los Angeles Times*, July 6, 1986.

21  "From the Minutes of Governmental Commission's Meeting No. 14," May 14, 1986; Baranovska, *Chornobyl'—problemy zdorov"ia*, 95–100.

22  Robert Peter Gale, "Two Chernobyl Doctors Were the First Humans to Get GM-CSF," Cancer Letters, May 29, 2015, 1-3.

23  Paul Jacobs, "UCLA Researcher Gets Reprimand for Marrow Transplant," *Los Angeles Times*, December 14, 1985.

24  Author telephone interview with Robert Gale, October 29, 2017.

25  Author telephone interview with Dr. Michael Carome, Public Citizen Foundation, September 5, 2017, and email correspondence, December 7, 2017.

26  "Informational Communiqué," June 5, 1986, SBU 16/1/1238: 178–79.

27  "From Protocol No. 7," May 6, 1986, Baranovska, *Chornobyl'—problemy zdorov"ia*, 58.

28  "Internal Report," SBU 16/1/1238: 118–24; and "Internal Report," July 18, 1986, SBU 16/1/1238: 289–92.

29  "Informational Communiqué," May 23–25, 1986, SBU 16/1/1238: 147–52.

30  "NBC and ABC Admit Reactor Film Was a Hoax," *New York Times*, May 15, 1986.

31  Alex S. Jones, "Press Sifts Through a Mound of Fact and Rumor," *New York Times*, May 1, 1986.

32  *Z arkhiviv VUChK-HPU-NKVD-KHB: Spetsvypusk* 16, no. 1 (2001): 79–80.

33 Stuart Diamond, "Long-Term Chernobyl Fallout: Comparison to Bombs Altered," *New York Times*, November 4, 1986, C3.

34 As cited in Alexander Sich, "Truth Was an Early Casualty," *Bulletin of the Atomic Scientists* (May/June 1996): 39; Stuart Diamond, "Chernobyl's Toll in Future at Issue," *New York Times*, August 29, 1986.

35 Baranovska, *Chornobyl'—problemy zdorov"ia*, 25–26. On bare feet—author interview with Dmitri Grodzinsky, July 14, 2015, Kyiv.

36 Vladimir Gubarev, *Strasti po Chernobyliu* (Kyiv: Algoritm; Arii, 2011), 35–40; and "Protokol No. 23," May 23, 1986, Baranovska, *Chornobyl'—problemy zdorov"ia*, 157–58.

37 Robert Peter Gale, "Two Chernobyl Doctors Were the First Humans to Get GM-CSF," *Cancer Letters*, May 29, 2015, 1–3.

38 Harry Nelson, "Ordinary Americans Also Chip in for Chernobyl," *Los Angeles Times*, May 18, 1986.

39 Robert Peter Gale and Alexander Baranov, "If the Unlikely Becomes Likely: Medical Response to Nuclear Accidents," *Bulletin of the Atomic Scientists* 67, no. 2 (March 2011): 10; and Gale and Baranov, "Bone Marrow Transplantation after the Chernobyl Nuclear Accident," *New England Journal of Medicine* 321, no. 4 (1989): 205–12.

40 "Gorbachev Meets Dr. Gale," *New York Times*, May 16, 1986; and William J. Eaton, "More Chernobyl Deaths," *Los Angeles Times*, May 16, 1986.

41 Vorobiev, *Do i posle Chernobylia*, 91–93, 112–15. On censorship, see "Ob ogranicheniiakh dlia pechati," July 31, 1986, Nesterenko Papers, Natsional'naia akademiia nauk Belarusi (BAS).

42 William J. Eaton, "More Chernobyl Deaths," *Los Angeles Times*, May 16, 1986.

43 On the paucity of expertise in radiation medicine in the West, see "Request for Waiver of Department Regulations," NCI Thyroid/Iodine 131 Assessments Committee, June 5, 1990, National Cancer Institute (NCI), RG 43 FY 03 Box 5, part 1.

44 Gale and Baranov, "Bone Marrow Transplantation"; and N. Parmentier and J. C. Nenot, "Radiation Damage Aspects of the Chernobyl Accident," *Atmospheric Environment* 23, no. 4 (1989): 771–75.

45 Robert Gillette, "Soviets Disparage Transplants for Chernobyl," *Los Angeles Times*, August 28, 1986; and "Information," November 11, 1986, Baranovska, *Chornobyl'—problemy zdorov"ia*, 46–48.

46 Harry Nelson, "Chernobyl 1 Year Later," *Los Angeles Times*, April 25, 1987.

47 William Sweet, "Chernobyl's Stressful After-Effects," *IEEE Spectrum*, November 1, 1999.

48 In an interview, Gale noted that he did not treat Soviet patients, Soviet doctors did. He assumed that their regulatory authorities approved of the treatment. A year later, he participated in testing the same drug on victims of a nuclear accident in Brazil. Author telephone interview with Robert Gale, October 30, 2017.

49 Vorobiev, *Do i posle Chernobylia*, 112–15.

50 William J. Eaton, "Gale-Soviet Atom Victim Study OKd," *Los Angeles Times*, June 7, 1986; and Robert Peter Gale, "Chernobyl: Answers Slipping Away," *Bulletin of the Atomic Scientists* 46, no. 7 (September 1990): 19.

51  Tony Perry, "Doctor Testifies That San Onofre Leaks Caused Leukemia," *Los Angeles Times*, January 14, 1994.

52  Robert Peter Gale and Eric Lax, "Fukushima Radiation Proves Less Deadly Than Feared," March 10, 2013, Bloomberg, https://www.bloomberg.com/view/articles/2013-03-10/fukushima-radiation-proves-less-deadly-than-feared.

### Evacuees

1  "On the Radiation Situation," May 3, 1986, Baranovska, *Chornobyl'—problemy zdorov"ia*, vol. 1, 35–38.

2  Of 55,000 people bused out of Pripyat on April 27, 42,000 went to the Poliske region and 13,000 to the Ivankiv region. These regions recorded higher levels of radiation than in Pripyat (2–4 mR/hr on the morning of April 27), "Informational Communiqué," April 28, 1986, SBU 16/1/1238: 73–74; "Memo about Evacation of the Population," July 15, 1986, TsDAVO 342/17/4391: 93–96; and "On Medical-Sanitary Provisions," May 7, 1986, in Baranovska, *Chornobyl'—problemy zdorov"ia*, vol. 1, 70–73, 7, 10, 13, 28, and vol. 2, 12.

3  "Protocol No. 10," May 10, 1986, in Baranovska, *Chornobyl'—problemy zdorov"ia*, vol. 1, 75–77.

4  Phil Taubman, "Soviet Challenges U.S. Milk Warning," *New York Times*, May 28, 1986; "Radiation in Soviet Veal Reported," *New York Times*, June 1, 1986; and "Protocol No. 7," May 7, 1986, Baranovska, *Chornobyl'— problemy zdorov"ia*, vol. 1, 67–68.

5  "Government Commission's Meeting," May 20, 1986; "From the Protocol," May 13, 1986; and "Protocol No. 25," May 26, 1986, Baranovska, *Chornobyl'—problemy zdorov"ia*, vol. 1, 25–26, 131–32.

6  "Protocol No. 30," June 3, 1986, Baranovska, *Chornobyl'—problemy zdorov"ia*, vol. 1, 202–4.

7  "Protocol No. 30," 21–23.

8  "Government Commission's Meeting," May 20, 1986, Baranovska, *Chornobyl'—problemy zdorov"ia*, 131–32.

9  Vorobiev, *Do i posle Chernobylia*, 6.

10  "On the Measures Taken," April 30, 1986, in Baranovska, *Chornobyl'—problemy zdorov"ia*, 16–18.

11  "Report on Health Examination of the Evacuated from 30-km Zone Children," 1986, TsDAVO 342/17/4391: 41–44.

12  "On Medical-Sanitary Provisions," May 7, 1986, in Baranovska, *Chornobyl'—problemy zdorov"ia*.

13  "On Medical-Sanitary Provisions in Gomel and Mogilev Provinces," May 28, 1987, RGANI 89/56/6: 199–204, and "On Progress in Implementing the Decree," May 16, 1987, RGANI 89/56/7: 210–14, Hoover Institution Archive (HIA). "Regions of 30-km Zone," 1986, and "Results of Screening," 1986, TsDAVO 342/17/4391: 73–79; "Protocol No. 12," May 12, 1986, and "Information," May 22, 1986, in Baranovska, *Chornobyl'—problemy zdorov"ia*, vol. 1, 81, 151–53; and

"Memo on Employment," August 1, 1989, Natsionalnyi arkhiv respubliki Belarus (NARB) 46/14/1264: 11–14.

14 For children's symptoms, see E. Stepanova et al., "Effekty vozdeistviia posledstvii Chernobyl'skoi avarii na detskii organizm," *Pediatriia* 12 (1991): 8–13.

15 "From Protocol No. 14," May 14, 1986, Baranovska, *Chornobyl'—problemy zdorov"ia*, 95–100.

16 "Medical Aspects of the Accident," June 16, 1987, RGANI 89/53/75; and A. K. Guskova and Iu. G. Grigor'ev, "Conclusions," November 16, 1986, RGANI 89/53/55, HIA.

17 Margaret Peacock, *Innocent Weapons: The Soviet and American Politics of Childhood in the Cold War* (Chapel Hill: University of North Carolina Press, 2014).

18 "Valentina Satsura to TsK BSSR," May 25, 1986, NARB 7/10/530: 287a–b.

19 Kyiv officials calculated that each month residents in contaminated areas were taking in 300 mSv to the thyroid. Baranovska, *Chornobyl'—problemy zdorov"ia*, vol. 1, 61–62.

20 Rudy Abramson, "Worst May Be Yet to Come," *Los Angeles Times*, May 4, 1986.

21 "Tasks for Sanitary-Epidemiology Service," September 23–26, 1986, TsDAVO 342/17/4355: 155–74.

22 "Memo from the Ministry," January 15, 1987, Baranovska, *Chornobyl'—problemy zdorov"ia*, vol. 2, 69–73.

23 See "To the First Secretary N. N. Sliun'kov," April 30, 1986, no. 588, Nesterenko Papers reprinted in *Rodnik*, no. 5–6 (1990): 56–58; "On Some Urgent Measures," May 5, 1986, Baranovska, *Chornobyl'—problemy zdorov"ia*, vol. 1, 46–47.

24 "Memo on Clinical Examination of Children," December 29, 1986, TsDAVO, 342/17/4391: 144–46a.

25 "Regulations," October 9, 1986, Nesterenko Papers, Akademia navuk Belarus' (BAS).

26 Author telephone interview with Maria Kuziakina, September 28, 2017.

27 "Memo," no earlier than October 1986, TsDAVO, 342/17/4391: 108–10.

28 "Methodical Recommendations," May 20, 1986, TsDAVO, 342/17/4390: 5–8.

29 "Ministry of Health Decree," May 18, 1986, and "Emergency-gram," May 20, 1986, Baranovska, *Chornobyl'—problemy zdorov"ia*, vol. 1, 125–26, 134. On complaints that local doctors did not have access to dose information, see "Operative Council," August 8, 1988, TsDAVO 324/17/4886, 15–17.

30 Vorobiev, *Do i posle Chernobylia*, 52.

31 Vorobiev, *Do i posle Chernobylia*, 52.

32 "Methodological Recommendations," June 4, 1986, TsDAVO, 342/17/4390: 17.

33 "Memorandum of Telephone Conversation between General Groves and Oak Ridge Hospital, 9:00 a.m., August 25, 1945," National Security Archive, accessed May 1, 2018, https://nsarchive.gwu.edu.

34 Stephen I. Schwartz, *Atomic Audit: The Costs and Consequences of U.S. Nuclear Weapons since 1940* (Washington, DC: Brookings, 1998).

35 Janet Farrell Brodie, "Radiation Secrecy and Censorship after Hiroshima and Nagasaki," *Journal of Social History* 48, no. 4 (Summer 2015): 842–964.

36 James V. Neel and William J. Schull, *The Criteria of Radiation Employed in the Study* (Washington, DC: National Academies Press, 1991).

37 "The Atom at Work," *Time* 65, no. 10 (March 7, 1955); and William C. Maloney, "Leukemia in Survivors of Atomic Bombing," *New England Journal of Medicine* 253, no. 3 (July 21, 1955): 89.

38 Kotaro Ozasa et al., "Japanese Legacy Cohorts: The Life Span Study Atomic Bomb Survivor Cohort and Survivors' Offspring," *Journal of Epidemiology* 28, no. 4 (April 5, 2018): 162–69.

39 Silini to Beebe, July 25, 1986, Correspondence Files, 1986; and Silini to Ilyin, February 5, 1987, Correspondence Files, 1987, UNSCEAR Archive.

40 Daniel L. Collins, "Nuclear Accidents in the Former Soviet Union: Kyshtym, Chelyabinsk and Chernobyl," 1991, Defense National Institute, Uniformed Services University (USU) DNA/AFRRI 4020, AD A 254 669.

41 "Design Institution—UkSSR Ministry of Health," TsDAVO 342/17/4390, no later than June 1986, 13–20; and Vorobiev, *Do i posle Chernobylia*, 14, 45–49.

42 A. K. Gus'kova, E. I. Chazov, and L. A. Ilyin, *Opasnost' iadernoi voiny* (Moscow: Novosti, 1982), 92–94.

43 Collins, "Nuclear Accidents in the Former Soviet Union."

## Rainmakers

1 "Iu. I. Izrael to Central Committee of KPSS," April 27, 1986, in V. I. Adamushko et al., eds., *Chernobyl': 20 let spustia* (Minsk: Natsional'nyi archiv Respubliki Belarus, 2006), 27–29.

2 "Central Aerological Observatory, Department of Cloud Physics and Modification," accessed September 17, 2017, http://www.cao-rhms.ru/OFAV/hist_of_dep/hist_of_dep_AktVoz.html.

3 Oleg Makarov, "Bitva s oblakami: razgon oblakov," *Populiarnaia mekhanika*, April 21, 2009.

4 "Playing with the Weather," BBC documentary, 2007.

5 "Progress Report," March–September 1994, Chernobyl Studies Project, Working Group 7.0, DOE, UCRL-ID-110062-94-6, Attachment E.

6 "Information," 1986, NARB 1088/1/989: 64.

7 Igor Elkov, "Chernobyl'skii tsiklon," *Rossiiskaia gazeta*, April 21, 2006.

8 Interview by Olha Martynyuk of Boris Leskov, November 3, 2017, Kyiv.

9 Albert A. Chernikov, "Works on Precipitation Modification in the Area of ChAES," *Moscow-Chernobyliu*, vol. 1 (Moscow: Voeenizdat, 1998), 479–83; and "Letter to the People's Deputy of UkSSR A. A. Dron'," April 28, 1990, TsDAVO 324/17/5328: 98–104.

10 The work of "not allowing precipitation" took place in Kyiv, Zhytomyr, Chernihiv, and Cherkasy Provinces. "Summarizing Memo," August 1, 1986,

TsDAVO 27/22/7701, 316–31; and "Memo No. 8095," October 9, 1986, SBU 68/483: 23.

11  "On Reorganization and Changes," May 19 and June 13, 1986, NARB 1088/1/986: 214–18.

12  "Dlia uskoreniia likvidatsii," *Tribuna energetika*, no. 13 (November 15, 1989): 1.

13  "On the Process of Liquidation," June 10, 1986, Gosudarstvennyi arkhiv obshchestvennykh ob'edinenii Mogilevskoi oblasti (GAOOMO), 9/181. "Directive of the Board of District Committee," July 4, 1986, Glavnoie upravlenie iustistii Mogilevskogo oblispolkoma Uchrezhdenie "Zonal'nyi" gosudarstvennyi arkhiv v g. Kricheve (Arkhiv Kricheva) 3/4/1503: 80–83. For measurements, see "No. 1353," September 19, 1986, Nesterenko Papers, BAS.

14  Denis Martinovich, "Nam dolbili, kak diatly: 'Tol'ko by ne bylo paniki," Tut.by, accessed June 22, 2018, http://news.tut.by/society/493766.html. On delayed reporting to Belarusian leadership, see "Memo-report," 1989, NARB 7/10/1938: 40–55.

15  "Statement," October 9, 1986, Nesterenko Papers, BAS.

16  "No. 899," June 23, 1986, Nesterenko Papers, BAS.

17  The first Belarusian document on the accident is "From Protocol No. 8," April 28, 1986. Local civil defense officials first detected radiation at 8 p.m. on April 27. "Protocol No. 5," June 19, 1986, in Adamushko, *Chernobyl'*, 29–30, 42–45.

18  For first orders, "Progress on Implementation," July 2, 1986, GAOOMO 9/184/55: 1–5, and "Progress in Liquidation," June 10, 1986, NARB 411/366, 157–58. For recognition that they were late, see "On Reorganization and Changes," May 19 and June 13, 1986, NARB 1088/1/986: 214–18. On lack of expertise in Belarus, see "Protocol No. 8," August 27, 1987, NARB 1088/1/1002: 59–61; "Memo of Environmental Protection, BSSR," April 13, 1989, NARB 83/1/767: 116–18.

19  "Statement on Republican Center of Radiational Measurements," October 9, 1986, Nesterenko Papers, BAS. For retrospective confirmation of doses, see A. I. Vorobiev, Ministry of Health, to B. E. Shcherbina, June 12, 1989, Gosudarstvennyi archiv Gomel'skoi oblasti (GAGO) 1174/8/2215, 24–27. On measurements, "Minutes of Meeting of Interdepartmental Commission," February 25, 1993, NARB 507/1/39: 1–5.

20  "No. 621, Nesterenko to Sliun'kov, N. N.," May 7, 1986, Nesterenko Papers, BAS.

21  They recorded 54 mSv/hr in Bragin and 3.2 mSv/hr in Chernev. Neither town was evacuated. "No. 899," June 23, 1986, Nesterenko Papers, BAS.

22  "No. 609, Data," June 21, 1986; "No. 900," June 21, 1986; "Meeting Minutes," September 15, 1986; and "No. 1353," September 19, 1986; "No. 1504/ File No. 54," October 27, 1986; and "No. 1354 from File ss.09.86/ no. 54," September 22, 1986, Nesterenko Papers, BAS.

23  "Residents of Bragin, Khoiniki and Narovlia Regions," May 11, 1986, NARB 4P/156/238: 104.

24  "No. 588," Nesterenko to N. N. Sliun'kov, April 30, 1986; and "To the Deputy Head of Council of Ministers of BSSR Petrov," May 29, 1886, Nesterenko Papers, BAS.

25  Vorobiev, *Do i posle Chernobylia*, 143.

26  Martinovich, "Nam dolbili, kak diatly."

## Operators

1  "Radiation Shield for Kyiv," October 1, 1986, Tsentralnyi derzhavnyi kinofoto-fonoarkhiv Ukrainy im. H. S. Pshenychnoho [TsDKFFA of Ukraine] No. 10219, film dossier.

2  Alla Yaroshinskaya, *Chernobyl: Crime without Punishment* (New Brunswick, NJ: Transaction Publishers, 2011), *Chernobyl: The Big Lie* (Moscow: Vremya, 2011), and *Chernobyl: The Forbidden Truth* (Lincoln, NE: Bison Books, 1995).

3  Author interview with Alla Yaroshinskaya, May 27, 2016, Moscow, and telephone interview, March 30, 2017.

4  Evgenii Chernykh, "Egor Ligachev: 'Stranno, konechno, chto Gorbachev ne s'ezdil v Chernobyl," *Komsomol'skaia pravda*, April 29, 2011.

5  "KPSS Central Committee Politburo Meeting," July 3, 1986, classified, single draft copy, Yaroshinskaya, personal collection.

6  "Boris Evdokimovich Shcherbina, 1919–1990," Gorod T, accessed June 21, 2018, https://gorod-t.info/people/obshchestvo-upravlenie/shcherbina-boris-evdokimovich/.

7  "On the Reactions of Foreign Reporters," July 8, 1987, SBU 16/1/1250: 193–94; and "Informational Communiqué," July 30, 1987, SBU 16/1256: 43–44.

8  "Interministerial Memo," July 1986, SBU 68/483: 15; and Anatolii S. Diatlov, *Chernobyl': Kak eto bylo* (Moscow: Nauchtekhlitizdat, 2003), chap. 4. For orders to restart reactor No. 3 by August, see "Directive of the Central Committee of KPSS," May 22, 1986, Baranovska, *Chernobyl'skaia trahedia*, 156.

9  "Lutsenko to SM USSR," December 2, 1988, Gosudarstvennyi Arkhiv Rossiyskoi Federatsii (GARF) 8009/51/4340: 166–71.

10  "KPSS Central Committee Politburo Meeting."

11  "Medical Aspects of the Chernobyl Accident," Conference Proceedings, Kyiv, May 11–13, 1988 (Vienna: IAEA, 1989), 49.

12  "KPSS Central Committee Politburo Meeting," 12–13.

13  "Informational Communiqué," March 17, 1987, SBU 16/1249: 48–51.

14  For a discussion of RBMK technical problems, see Sonja D. Schmid, *Producing Power: The Pre-Chernobyl History of the Soviet Nuclear Industry* (Cambridge, MA: MIT Press, 2015), 128–30.

15  "Causes of Chernobyl Accident, Facts and Fiction," 1990, RGAE 4372/67/9743: 86–99. On the RBMK's positive void coefficient, see Paul R. Josephson, *Totalitarian Science and Technology: Control of Nature* (Atlantic Highlands, NJ: Humanities Press, 1996), 308.

16  Schmid, *Producing Power*, 125.

17  "KPSS Central Committee Politburo Meeting."

## Ukrainians

1 For her latest history, see Natalia Baranovska, *Ispytanie Chernobylem* (Kyiv: Iustinian, 2016).

2 A. I. Avramenko to A. P. Kartysh, June 14, 1990, TsDAVO, 342/17/5220: 27.

3 "On Archival Programs," July 1, 1991, and Prof. Dr. Hans Booms, "Report about the Unesco-Mission to the UkSSR," August 29, 1991, UNESCO Archive, CII/PGI/MONT/4.

4 Author interview, June 13, 2014, Kyiv.

5 "Memo," 1987, TsDAVO 342/17/4391: 147–50.

6 Author interview with Natalia Baranovska and Irina, June 13, 2014, Kyiv.

7 "Sanitary-Hygiene Guidelines," August 4, 1986, Baranovska, *Chornobyl'—problemy zdorov"ia*, vol. 1, 18–20.

8 "Transcripts, May 3" and "Protocol No. 5," May 5, 1986, Baranovska, *Chornobyl'—problemy zdorov"ia*, vol. 1, 27–30, 52–56.

9 "Situation Report," May 5, 1986, *Z arkhiviv VUChK-HPU-NKVD-KGB* (Kyiv, 2001): 77–78; and "On Construction," October 28, 1991, TsDAVO 342/17/5357: 120.

10 "On a Question for Clarification," July 7, 1988, SBU 16/1262: 324–25.

11 "On Problematic Issues," December 6, 1988, SBU 16/1266: 248–50.

12 "Memo: On the Military," July 30, 1986, SBU 68/483: 13.

13 "On Shortcomings," August 1986, SBU 68/483: 16.

14 "From Protocol No. 15," May 15, 1986, Baranovska, *Chornobyl'—problemy zdorov"ia*, vol. 1, 104.

15 "Situation Report"; and V. K. Savchenko, *The Ecology of the Chernobyl Tragedy: Scientific Outlines of an International Programme of Collaborative Research* (Paris: Parthenon Publishing, 1995), 14, 95.

16 "Protocol," March 5, 1992, NARB 507/1/12: 8–9.

17 "Ministry Decree," May 15, 1986; "Objectives," September 1986, TsDAVO 342/17/4355: 155–74.

18 "Memo," 1987, TsDAVO 342/17/4391: 147–50; and "Objectives."

19 "Materials for Government Information," September 29, 1989, TsDAVO 342/17/5089: 159–61; and "Operative Meeting," August 8, 1988, TsDAVO 324/17/4886: 15–17.

20 Joseph Mangano, "Three Mile Island: Health Study Meltdown," *Bulletin of Atomic Scientists* (October 2004): 31–35; and Natasha Zaretsky, *Radiation Nation: Three Mile Island and the Political Transformation of the 1970s* (New York: Columbia University Press, 2018).

21 "Ministry Memo," January 15, 1987, Baranovska, *Chornobyl'—problemy zdorov"ia*, 69–73.

22 Author interview, Oleksandr Popovych, Kyiv, June 1, 2017.

23 "Minutes No. 17," May 17, 1986; and for the Polesia region, "Explanatory Note,"

June 14, 1986, in Baranovska, *Chornobyl'—problemy zdorov"ia*, vol. 1, 120, 217–18.

24 "Proposals from Science Department," May 4, 1986, Baranovska, *Chornobyl'—problemy zdorov"ia*, 109–10.

25 "On Some Urgent Measures," May 5, 1986, Baranovska, *Chornobyl'— problemy zdorov"ia*, vol. 1, 46–47.

26 "Methological Recommendations," May 20, 1986, TsDAVO, 342/17/4390: 5–8.

27 "Children Examination Results," 1986, TsDAVO 342/17/4391: 73–74. "From the Transcripts," May 14, 1986, "From the Protocol," May 13, 1986, Baranovska, *Chornobyl'—problemy zdorov"ia*, vol. 1, 101–2.

28 "Proposals from the Science Department."

29 "Transcripts of Meeting No. 7," May 11, 1986; "From the Protocol," May 13, 1986, Baranovska, *Chornobyl'—problemy zdorov"ia*, 46–48, 91–92.

30 "Information of the Operative Group of the State Committee of Industrial Agriculture," May 8, 1986, TsDAVO 27/22/7703: 10.

31 Elgė Rindzevičiūtė, *The Power of Systems: How Policy Sciences Opened Up the Cold War World* (Ithaca, NY: Cornell University Press, 2016), 187.

32 "Transcripts No. 7 of the Politburo TsK KPU Operative Group," May 11, 1986, in V. A. Smolii, *Chornobyl': Dokumenty operatyvnoi hrupy TsK KPU (1986–1988)* (Kyiv: Institut istoryi, 2017), 98.

33 A. K. Gus'kova, E. I. Chazov, and L. A. Il'yn, *Opasnost' iadernoi voiny* (Moscow: Novosti, 1982), 72.

34 "From the Protocol," May 13, 1986; and author interview with O. A. Bobyliova, June 6, 2017, Kyiv.

35 "On the Radiological Situation," May 3, 1986, "On Medical-Sanitation Provisions," May 7, 1986, and "Protocol No. 25," May 26, 1986, in Baranovska, *Chornobyl'—problemy zdorov"ia*, vol. 1, 38–39, 70–71, 166–68.

36 "From Protocol No. 7," May 6, 1986, Baranovska, *Chornobyl'—problemy zdorov"ia*, vol. 1, 58.

37 A. K. Gus'kova and Iu. G. Grigor'iev, "Conclusion," November 16, 1986, RGANI 89/53/55, HIA.

38 Valentyna Shevchenko, "Urodzhena mudrist'," in *Volodymyr Scherbyts'kyi: spohady suchasnykiv*, ed. E. F. Vozianov et al. (Kyiv: "In Yure," 2003), 47–48.

39 Author interview with Olha Oleksandrivna Bobyliova, June 6, 2017, Kyiv.

40 Memo, John Willis to Doug Mulhall and David McTaggart, August 14, 1990, Greenpeace Archive (GPA), 1625; and Bryon MacWilliams, "Climate Change: Crunch Time for Kyoto," *Nature* 431, no. 7004 (September 2, 2004): 12–13. An investigation of the Ukrainian Communist Party singled out Izrael and Ilyin for condemnation. "Minutes to the 28th Conference of the Communist Party of Ukraine," December 14, 1990, TsDAHO 1/2/1065: 112–43.

41 Monitors measured streets in Kyiv from 8–26 μSv/hr. "On the Radiological Situation."

42 "Report on Pediatric Screening," 1986, TsDAVO 342/17/4391: 41–44.

43 "30-km Zone Areas in Kyiv Province," 1986, TsDAVO 342/17/4391: 75–79.

44 "Meeting Minutes No. 1," May 3, 1986, in *Chornobyl': Dokumenty operatyvnoi hrupy TsK KPU*, 32–33. Three weeks later, Scherbytsky gave the order: "Information," May 22, 1986, Baranovska, *Chornobyl'—problemy zdorov"ia*, vol. 1, 151–53.

45 "Proposals from the Science Department," May 4, 1986, in Baranovska, *Chornobyl'—problemy zdorov'ia*, vol. 1, 46–48; and "Explanatory Note," 1986, TsDAVO 342/17/4390: 59–61.

46 "Science Department Memo," June 6, 1986, in Baranovska, *Chornobyl'— problemy zdorov"ia*, vol. 1, 206–8. Children from rural areas went at the end of May: "Memo," no earlier than May 20, 1986, TsDAVO, 342/17/4391: 14–19.

47 "From Protocol No. 14," May 14, 1986, Baranovska, *Chornobyl'— problemy zdorov"ia*, vol. 1, 93–100.

48 "From Protocol No. 13," May 13, 1986, and "From Protocol No. 14," May 14, 1986, in Baranovska, *Chornobyl'—problemy zdorov"ia*, vol. 1, 93–100; and Shevchenko, "Urodzhena mudrist'," 47–48.

49 "From the Transcripts," May 14, 1986, in Baranovska, *Chornobyl'—problemy zdorov"ia*, vol. 1, 101–2.

50 "Residents of Gomel to Gromyko," May 26, 1986, NARB 4R/154/362: 102–3ab; and "Protocol No. 12," May 12, 1986, in Baranovska, *Chornobyl'—problemy zdorov"ia*, vol. 1, 83–89.

51 "Informational Memo," 1989, NARB 7/10/1938: 40–55.

52 Levels were from 20 to 350 mSv/hr. "From Protocol No. 17," May 22, 1986, Baranovska, *Chornobyl'—problemy zdorov"ia*, vol. 1, 145–47; and "People to A. S. Kamai," May 11, 1986, NARB 4R/156/238: 104.

53 N. I. Rosha to the Commission, January 10, 1989, NARB 10/7/1851: 35–36. For thyroid dose reconstructions, see V. T. Khrushch and Iu. I. Gavrilin, "Verifikatsiia dozimetricheskikh dannykh i rekonstruktsiia individual'nykh doz oblucheniia shchitovidnoi zhelezy dlia zhitelei g. Minska" (Moscow: Institut biofiziki, 1991).

54 Baranovska, *Chornobyl'—problemy zdorov"ia*, vol. 1, 218–19. The use of Chernobyl children for farmwork continued in subsequent years. "On the Question," April 20, 1989, TsDAVO 342/17/5089: 10–11.

55 "Regions of 30-km Zone in Kyiv Oblast," 1986, and "Screening Results," 1986; "Radiometric Data," 1986, TsDAVO 342/17/4391, 73–74, 75–96; and "Government Commission's Meeting," May 20, 1986, and "Information," November 11, 1986, in Baranovska, *Chornobyl'—problemy zdorov"ia*, vol. 1, 131–34, 46–48.

56 "Screening Results"; "On the Procedure of Medical Examination," May 29, 1986, TsDAVO, 342/17/4390: 2–4; and "Memo," 1986, TsDAVO 342/17/4391, 34–40.

57 "On the Results of Medical Examination of Pregnant Women and Children," no earlier than July 1986, "Memo on Clinical Examination of Children," December

29, 1986, and "Report on Examination of Children," 1986, TsDAVO 342/17/4391: 77–78, 144–46a, 41–44.

58 "Memo on Medical-Sanitary Provisions for Pregnant Women," no earlier than October 1986, TsDAVO, 342/17/4391: 108–10.

59 Baranovska, *Chornobyl'—problemy zdorov"ia*, vol. 1, 224–25.

60 "Temporary Methodological Recommendations," May 23, 1986, TsDAVO, 342/17/4390: 9–11.

61 "Report on the Examination of Children," 1986, TsDAVO 342/17/4391: 41–44.

62 Andrei Vorobiev, *Do i posle Chernobylia*, 136.

63 Internal measurements from 0.002–0.01 mSv/hr continued from several weeks up to two months. "Estimation of Incorporated Radioactive Elements in Bowels," June 20, 1986, TsDAVO 342/17/4390: 32–33.

64 "Estimation of Incorporated Radioactive Elements in Bowels"; "Methodological Recommendations," TsDAVO 342/17/4390, June 4, 1986, 13–20; and Vorobiev, *Do i posle Chernobylia*, 85.

65 "Decree," May 12, 1986, and "Protocol No. 12," May 12, 1986, in Baranovska, *Chornobyl'—problemy zdorov"ia*, vol. 1, 90–91.

66 "Memo," no earlier than May 20, 1986, and "Memo on Medical-Sanitary Examination of Pregnant Women and Children," no earlier than October 1986, TsDAVO, 342/17/4391: 14–19, 108–10.

67 "On the Management of Clinic Examinations," June 5, 1986, NARB, 46/14/1261: 43–46.

68 Jorgensen, *Strange Glow*, 230–31.

69 "Memo," no earlier than May 20, 1986, TsDAVO, 342/17/4391: 14–19.

## Physicists and Physicians

1 "Protocol No. 21," May 21, 1986, Baranovska, *Chornobyl'—problemy zdorov"ia*, vol. 1, 142–43; and "Transcripts of the Operative Group," May 15, 1986, in V. A. Smolii, ed., *Chornobyl: Dokumenty operatyvnoi hrupy* (Kyiv: Instytut istoriï Ukraïny, 2017), 129–33.

2 "From Protocol No. 16," May 20, 1986; "From Protocol No. 17," May 22, 1986, Baranovska, *Chornobyl'—problemy zdorov"ia*, vol. 1, 130–31, 145–47; "Commission Meeting Minutes," June 10, 1987, NARB 7/10/1524: 44–46; "To the Head of USSR Council of Ministers," July 17, 1986 (Drugie berega: Moscow, 1992), Yaroshinskaya, *Sovershenno sekretn*; 426–27.

3 ~200 kBq per m²; "Temporary Instructions," May 30, 1986, TsDAVO 342/17/4340: 50–75.

4 Vorobiev, *Do i posle Chernobylia*, 87–91.

5 One curie is $3.7 \times 10^{10}$ becquerels.

6 Jorgensen, *Strange Glow*, 62.

7 "Technical Memo," September 26, 1990, GAMO 7/5/3999: 1–9.

8  "Transcripts of the Meeting," July 28, 1986, in Baranovska, *Chornobyl'—problemy zdorov"ia*, vol. 2, 9–18.

9  Author telephone interview with Lynn Anspaugh, January 12, 2016.

10 Ellen Leopold, *Under the Radar: Cancer and the Cold War* (New Brunswick, NJ: Rutgers University Press, 2009), 231, 142–44.

11 Soraya Boudia, "Managing Scientific and Political Uncertainty," in *Powerless Science: Science and Politics in a Toxic World*, ed. Soraya Boudia and Nathalie Jas (New York: Berghahn, 2014), 95–112; and Scott Frickel, "Not Here and Everywhere: The Non-Production of Scientific Knowledge," in *Routledge Handbook of Science, Technology and Society* (New York: Routledge, 2014), 263–76.

12 Vorobiev, *Do i posle Chernobylia*, 12–15.

13 Vorobiev, *Do i posle Chernobylia*, 12–13, 151; and "Nakaz Ministerstva," May 28, 1986, Baranovska, *Chornobyl'—problemy zdorov"ia*, 186–87.

14 "Progress Report," March–September 1994, Chernobyl Studies Project, Working Group 7.0, DOE, UCRL-ID-110062-94-6, attachment H; and Awa A. Akio et al., "Biodosimetry: Chromosome Aberration in Lymphocytes and Electron Paramagnetic Resonance in Tooth Enamel from Atomic Bomb Survivors," *World Health Statistical Quarterly* 46 (1996): 67–71.

15 I. A. Gusev et al., "Monitoring of Internal Exposure," presented at "Medical Aspects of the Chernobyl Accident," *Conference Proceedings*, Kyiv, May 11–13, 1988 (Vienna, IAEA: 1989), 201.

16 Vorobiev, *Do i posle Chernobylia*, 27, 117.

17 Email correspondence with Lucas Hixson, October 12, 2016.

18 "Protocol No. 12," May 12, 1986; and "Protocol No. 23," May 23, 1986, Baranovska, *Chornobyl'—problemy zdorov"ia*, vol. 1, 83–86, 157–59.

19 "On the Installation of the WBS [whole-body counter]," September 9, 1986, NARB, 46/14/1261: 71–72.

20 "Council of Ministers of USSR, Protocol No. 29," June 23, 1986, as reproduced in "Chernobyl'skaia katastrofa," *Sil'nye novosti*, Gomel.today, April 26, 2011.

21 "From Protocol No. 12," May 12, 1986, in Baranovska, *Chornobyl'—problemy zdorov"ia*, vol. 1, 80–81.

22 "Transcripts of the Meeting No. 38," August 13, 1986, in *Chornobyl': Dokumenty operatyvnoi hrupy TsK KPU*, 411–33.

23 Vorobiev, *Do i posle Chernobylia*, 27–28.

24 Email correspondence with Andrei Vorobiev, October 8, 2017. For the order classifying doses on June 27, 1986, see "Chernobyl'skaia katastrofa," *Sil'nye novosti*, Gomel.today, April 26, 2011.

25 William J. F. Standring, Mark Dowdall, and Per Strang, "Overview of Dose Assessment Developments and the Health of Riverside Residents Close to the 'Mayak' PA Facilities, Russia," *International Journal of Environmental Research and Public Health* 6, no. 1 (2009): 174–99.

26 L. A. Buldakov et al., "Theory and Practice of Establishing Radiation Standards before and after the Chernobyl Accident," presented at "Medical Aspects of the

Chernobyl Accident," *Conference Proceedings*, Kyiv, May 11–13, 1988 (Vienna: IAEA, 1989), 83–84.

27  "Protocol No. 15," May 15, 1986, in Baranovska, *Chornobyl'—problemy zdorov"ia*, vol. 2, 104–5.

28  "A. I. Vorobiev to B. E. Shcherbina," June 12, 1989, GAGO 1174/8/2215: 24–27.

29  They later cut the number to thirty-nine villages. L. A. Il'in and K. I. Gordeev, "Explanatory Note," no earlier than August 1, 1986, TsDAVO, 342/17/4390, 59–61; and Iu. A. Israel, "On the Assessment of Radiological Situation," May 21, 1986, in Yaroshinskaya, *Bol'shaia lozh'*, 14.

30  An international group of scientists ruled that direct measurements are more accurate; "Progress Report," March–September 1994, Chernobyl Studies Project, Working Group 7.0, DOE, UCRL-ID-110062-94-6, attachment H.

31  "From the Meeting Transcripts," July 28, 1986, and "Memo," 1986, TsDAVO 342/17/4391: 16, 14–19; and "From the Transcripts," May 20, 1986, Baranovska, *Chornobyl'—problemy zdorov"ia*, vol. 1, 135–36.

32  "From the Transcripts," July 28, 1986, Baranovska, *Chornobyl'—problemy zdorov"ia*, 9–22.

33  "From the Transcripts," July 28, 1986, 9–22.

34  "Ministry Decree," May 15, 1986, Baranovska, *Chornobyl'—problemy zdorov"ia*, vol. 1, 110–12.

35  "Resolution No. 304 of the Governmental Commission," November 13, 1986, TsDAVO 342/17/4348: 37.

36  "V. V. Malashevsky to Iu. Spizhenko," October 12, 1989, TsDAVO 342/17/5091: 95–96. See also "Proposal for a Work Plan," July 3, 1989, NARB 10/7/1851: 108.

37  "Information on the Radiological Situation," no earlier than May 31, 1989, TsDAVO 342/17/5092: 44–48.

38  "Protocol of the Commission Meeting," January 20, 1987, NARB 10/7/1524: 12–17; "Protocol," October 6 and November 30, 1987, NARB 7/10/1524: 44–48, 70–75. For discussions on more "re-evacuations," see "E. I. Sezhenko to Council of Ministers UkSSR," January 5, 1988, NARB 7/10/1525: 106; "A. Grishagina to V. G. Evtukh," September 5, 1988, NARB 7/10/1523: 56; and "Letter from VRIO [acting authority], No. 1/571," July 3, 1989, NARB 10/7/1851: 106.

39  "TsK KPB, No. 4972," April 7, 1987, and "TsK KRSS No. 148225," April 20, 1987, NARB 4r/156/438: 60–66, 69–75.

40  "Meeting Protocol," December 26, 1990, January 16, 1992, and December 9, 1992, NARB 507/1/12: 1–4, 27–29.

41  "No. 934," June 30, 1986; "No. 1354 from ss.09.86/File No. 54," September 22, 1986, and "No. 1433, A. L. Grishaginu," October 11, 1986, Nesterenko Papers, BAS; "List of Populated Locales," August 22, 1986, TsDAHO 1/25/3: 121–22.

42  "Decree No. 25c," August 29, 1986, TsDAVO 27/22/7703: 20; "Information about the Situation," June 5, 1987, NARB 4R/156/393: 45–60; "Protocol," January 20, 1987, NARB 7/10/1524: 12–16. On Ukrainian evacuations, see "On Resettling People from Several Villages in Narodychi Region," February 2, 1989, "On a Draft

Directive," February 22, 1989, and "On Question No. 3," July 24, 1989, TsDAVO 342/17/5089: 6, 8–9, 124; and V. N. Sych to E. P. Tikhonenkov, May 24, 1991, GAGO 1174/8/2445: 45–53. On protest of schoolchildren working in radioactive fields and living in areas with more than 40 ci/km, see "Obrashchenie," August 29, 1990, and "V. Voinov to V. F. Kebich," November 5, 1990, NARB 46/14/1322: 54, 202.

43  "On Amendments," March 2, 1987, RGANI 89/56/1; "On Additional Included Villages," May 23, 1987, RGANI 89/56/5: 196, HIA; and "On Management of Aid," May 30, 1988, NARB 7/10/1523: 21–23.

44  "Proposal," July 30, 1986, Yaroshinskaya, *Sovershenno sekretno*, 418–19; "List of Populated Locales," August 22, 1986, in V. I. Adamushko et al., eds., *Chernobyl': 20 let spustia* (Minsk: NARB, 2006); and "On Radioactive Pollution of BSSR Territory," 1, 1986, NARB 7/10/439: 54.

45  As quoted in "Chernobyl'skaia katastrofa," *Sil'nye novosti*, accessed May 17, 2016, https://gomel.today/rus/article/society/4594/.

46  "Report," April 14, 1987, 55/1: 347 Nesterenko Papers, BAS.

47  Ales Adamovich, *Imia sei zvezde Chernobyl'* (Minsk: Kovcheg, 2006), 91–94.

48  "To the Military Command of the Belarusian Military Region," June 19, 1987, NARB 4P, 156/393: 71–75.

49  "E. Sokolov to N. I. Ryzhkov," June 23, 1987, NARB 4R/156/393: 70.

50  "On the Radiological Situation," June 23, 1987, NARB 4R/156/393: 76–81; and Adamovich, *Imia sei zvezde Chernobyl'*, 99.

51  For measurements, see "Upon the Directive of UkSSR Council of Ministers," June 7, 1990, TsDAVO 324/17/5238: 115–18.

## Woolly Truths

1  "To the Office of the Minister from V. F. Larikov," August 27, 1991, TsDAVO 342/17/5357: 91–92.

2  "Memo of the State of Conditions for Sanitary-Hygiene," July 11, 1987, TsDAVO, 342/17/4672: 74–81.

3  "Review of the Living Question," February 4, 1987, Derzhavnyi arkhiv Chernihivs'koï oblasti (DAChO) 2347/4/1678: 11–13.

4  Author interview with Tamara Haiduk, July 7, 2016, Chernihiv.

5  "Towards Securing Safety of Movement," October 30, 1986, DAChO 2341/1/1651: 69.

6  "Certification of the State of Conditions for Sanitary-Hygiene," June 15–18, 1987, TsDAVO 342/17/4672: 82–85.

7  Author interview with Tamara Kot, July 7, 2016, Chernihiv.

8  Certificiation of the State of Conditions for Sanitary-Hygeine.

9  In 1986, they fulfilled the plan by 358 percent. "On Bonuses," June 1, 1986, DAChO 2341/1/1650: 1–4.

10  "To the Office of the Minister from V. F. Larikov," August 27, 1991, TsDAVO 342/17/5357: 91–92.

11  "On Regulation of Truck Shipments," July 16, 1986, DAChO 2341/1/1650: 57.

12  "To the Office of the Minister from V. F. Larikov." Sheep in the fields in the Ivankiv region in late May measured 3.2 mR/hr. "Transcripts of Meeting No. 15," May 20, 1986, in *Chornobyl': Dokumenty operatyvnoi hrupy TsK KPU*, Kyiv, 168–72.

13  Oral interview with Maria Nogina, main engineer, Chernihiv Wool Factory. "Thematics," November 24, 1988, DAChO 2347/4/1683: 133–34.

14  "Certification of the State of Conditions for Sanitary-Hygiene"; and "Belarusian SSR Ministry of Health," August 8, 1987, GARF P8009/51/3559: 22–23.

15  Author interview with Maria Nogina, July 7, 2016, Chernihiv. The first records of problems in the wool industry were noted in August. "Transcripts of Meeting No. 38," August 13, 1986, in *Chornobyl': Dokumenty operatyvnoi hrupy*, 411–33.

16  "Certification of the State of Conditions for Sanitary-Hygiene."

17  "V. K. Solomakha to A. A. Tkachenko," May 5, 1986, TsDAVO 27/22/7701: 20.

18  "Transcripts of Meeting No. 15," and "Information of the Operative Group of the State Committee of Industrial Agriculture," June 4, 1986, TsDAVO 27/22/7703: 37.

19  "On Dosimetric Control at the Factory," August 4, 1986, DAChO 2341/1/1650: 116.

20  "V. S. Iarnykh to the Sovet Ministers, Ukraine SSR," August 22, 1991, TsDAVO 324/17/5357: 99–100; and "To the Office of the Minister from V. F. Larikov."

21  At 0.18 mSv/hr, workers working twelve-hour shifts, six days a week for twenty weeks received a dose of 259 mSv, surpassing the annual permissible norm of 100 mSv.

22  "Temporary Methodological Recommendations for Diminishing the Level of Radioactivity of Wool Raw Material," April 9, 1987, RGAE 650/1/183: 2–8.

23  "On the Elimination of Shortcomings in the Protection of Labor," July 25, 1986, DAChO 2341/1/1650: 86–91.

24  "Planning Chart," November 8, 1986, DAChO 2341/1/1650: 126.

25  "On the Creation of a Radiological Laboratory," April 10, 1987, DAChO 2347/4/1667: 26.

26  More instructions followed: "On Providing Rules for Radiation Safety in Decontaminating Wool with Radioactive Substances," April 22, 1987, DAChO 2347/4/1667: 47.

27  "Certification of the State of Conditions for Sanitary-Hygiene" [appendix], 82–88.

28  "From M. S. Mukharsky to the Chief State Sanitary Doctor," July 1, 1986, TsDAVO 342/17/4340: 169.

29  "Certification of the State of Conditions for Sanitary-Hygiene."

30  "Certification of the State of Conditions for Sanitary-Hygiene," appendix, 84–85.

31  Author interview with employees, sorting shop, Chernihiv Wool Factory, July 6, 2016, Chernihiv.

32 "Technical Report," October 11, 1991, DAChO 9014/1/18: 279.

33 Lewis H. Siegelbaum, *Stakhanovism and the Politics of Productivity in the USSR, 1935–1941* (Cambridge: Cambridge University Press, 1988).

34 "Certification of the State of Conditions for Sanitary-Hygiene," June 15–18, 1987, TsDAVO 342/17/4672: 82–85.

35 "Protocol of the Meeting of the Professional Union Committee," December 16, 1986, DAChO 2347/4/1618: 106.

36 "Protocol," September 25, 1987, DAChO 2347/4/1678: 78–80; "Protocol," April 20, 1989, DAChO 2347/1/1705: 21–24; and "On Directions to the Group of Residents of Chernihiv," GARF 8009/51/ 4340: 59.

37 To A. Yablokov from E. B. Burlakov and V. I. Naidiach," no earlier than 1990, RGAE 4372/67/9743: 393–95. For other symptoms, see "To Respected Comrade K. I. Masyk," January 5, 1990, TsDAVO 342/17/5238: 5–7.

38 Vorobiev, *Do i posle Chernobylia*, 26.

39 Jorgensen, *Strange Glow*, 99.

40 Kate Moore, *The Radium Girls: The Dark Story of America's Shining Women* (New York: Simon and Schuster, 2016).

41 Eileen Welsome, *The Plutonium Files: America's Secret Medical Experiments in the Cold War* (New York: Dial Press, 1999), 50.

42 Hematologist Andrei Vorobiev lists these symptoms under the description of chronic radiation syndrome. Vorobiev, *Do i posle Chernobylia*, 85.

43 "Dear Deputies!" no later than June 1990, TsDAVO 342/17/5238: 21–22.

44 Iu. A. Izrael, "An Evaluation of Radiological Conditions," May 21, 1986, in Alla Yaroshinskaya, *Chernobyl': Bol'shaia lozh'* (Moscow: Vremia, 2011), 14.

45 "Khulap to A. A. Grakhovskii," July 18, 1986, GAGO 1174/8/1940, 31–36; "Decree No. Pr184-DSP," June 30, 1986, Derzhavnyi arkhiv Zhytomyr'skoi oblasti (DAZhO) 3756/1/1440: 231–34; and "On Establishing Higher Up to 25% Tariff Rates," January 2, 1990, GAGO 1174/8/2336: 152–53.

46 V. P. Platonov, E. F. Konoplia, "Information on the Major Results of Scientific Work Connected with the Liquidation of the Accident at the ChAES," April 21, 1989, RGAE 4372/67/9743: 490–571.

47 "On Norms for Primary Processing of Wool Contaminated with Radioactive Substances," August 18, 1986, DAChO 2347/1/1650: 135–36; and "On Disposal of Solid Radioactive Waste," August 18, 1986, DAChO 2347/1/1650: 137.

48 Author interview with Maria Nogina.

49 "On the Disposal of Unwashed Wool Contaminated with Radionuclides Higher Than 1 mR/hr," December 14, 1987, DAChO 2347/4/1669: 122; and "Protocol of the Professional Union Committee Meeting," December 16, 1986, DAChO 2347/4/1618: 106.

50 "Towards Securing Safety of Movement," October 30, 1986, DAChO 2341/1/ 1651: 69.

## Clean Hides, Dirty Water

1 "Deposition No. 247," May 9, 1986, DAZhO 1150/2/3017: 9–11.

2 "Telephonogram to the Ministry of Health, UkSSR, Comrade M. S. Mukharskiy," August 27, 1986, TsDAVO 324/17/4348: 15.

3 "Act of Sanitary Investigation," August 11, 1986, and "Directive from the Soviet Ministers," August 20, 1986, TsDAVO 324/17/4348: 14, 66.

4 "To M. S. Mukharskiy from P. I. Chekrenev," August 27, 1986, TsDAVO 342/17/4348: 15–16.

5 "Evaluation on Work on Processing Hides," August 22, 1986, TsDAVO 342/17/4348: 5–10.

6 The hides measured from 30 to 500 microroentgen/hr. "Act of Sanitary Investigation," and "On Processing Leather Raw Material," August 18, 1986, TsDAVO 324/17/4348: 26.

7 Amie Ferris-Rotman, "The Scattering of Ukraine's Jews," *The Atlantic*, September 21, 2014.

8 Author interview with Nina Aleksandrovna Chekreneva, July 10, 2016, Zhytomyr.

9 Only later did they monitor household possessions for radioactivity. "Disposition 172-r," March 19, 1990, GAMO 7/5/3964: 162–64.

10 "Telephonogram to the Ministry of Health."

11 "Directions of the Council of Ministers, UkSSR," September 1, 1986, TsDAVO 324/17/4348: 66; and "On Processing Leather Raw Material."

12 Author interview with Nina Aleksandrova Chekreneva, July 10, 2016, Zhytomyr; and "Foreword, P. I. Chekrenev," August 20, 1986, TsDAVO 324/17/4348: 13.


## Making Sausage of Disaster

1 "Transcripts of Meeting No. 1," May 3, 1986, in *Chornobyl': Dokumenty operatyvnoi hrupy TsK KPU*, 32–33.

2 "Proposal for Reprocessing Livestock of Kyiv Province," May 1986, TsDAVO 27/22/7701: 13; "Information of the Operative Group of the State Committee of Industrial Agriculture," May 4, 1986, TsDAVO 27/22/7703: 5; "Case 20, Various Information," May 22, 1986, TsDAVO 27/22/7701: 7–8: and "L. K. Filonenko to A. N. Tkachenko," May 1986, TsDAVO 2605/9/1601: 8; "Information on the Situation in the Territory of Gomel' and Mogilev Provinces," June 5, 1987, NARB 4R/156/393: 45–60.

3 The order for the "necessary slaughter" was issued on May 5. "Decree No. 186," December 29, 1986, DAZhO 5005/1/546: 109–11. Belarusian officials rounded up 50,900 head. "TsK KP BSSR "Certification on the Course of Liquidation," 1989, NARB 4r/156/627: 126–38.

4 "Operative Information of the Directory of Mechanization and Electrification," May 7, 1986, TsDAVO 27/22/7701: 51.

5   "V. K. Solomakha to A. A. Tkachenko," May 5, 1986, TsDAVO 27/22/7701: 20.

6   "Recommendations for the Use of Meat Raw Material," June 18, 1986, RGAE 650/1/556: 1–3; and "From M. S. Mukharsky to the Chief State Sanitary Doctor," July 1, 1986, TsDAVO 342/17/4340: 169.

7   "Recommendations for the Use of Meat Raw Material," June 18, 1986, RGAE 650/1/556: 1–3.

8   The threshold to be considered waste was $2.10 \times 10^{-6}$ ci/kg of beta activity. "On Classification of Meat as Waste," no later than June 15, 1986, TsDAVO 342/17/4370: 31–33.

9   "Certification on the Course of Instruction of Teachers," May 12, 1986, TsDAVO 27/22/7701: 19; and "Decree No. 186."

10  "Evaluation of the Berdychiv Regional Sanitation-Epidemiological Station," n.d. 1986; "Annual Assessment of Radiation Hygiene," DAZhO 3950/1/1296: 13–21, 36–40.

11  The new permissible level on the job was 50 microroentgen/hr. "On Classification of Meat as Waste." On worker injuries, see "Decree on Reducing Inadequacies," March 17, 1987, Arkhiv Kricheva 154/1/34: 105–6. On decontaminating a packing plant, see "Decree," July 28, 1986, Arkhiv Kricheva 154/1/33: 310.

12  "Certification," May 4, 1986, TsDAVO 27/22/7701: 35; and "General Union Sanitary-Hygiene Anti-Epidemic Regulations and Norms," 1986, TsDAVO 342/17/4370: 174–90.

13  "Decree on Bringing to Justice for Disciplinary Responsibility," September 22, 1986, Arkhiv Kricheva 154/1/33: 363; and "Decree on Reducing Inadequacies."

14  "Emergency-gramma No. 129 from the Ministry of Health SSSR," June 23, 1986, TsDAVO 342/17/4340: 162. See also "Contamination of Produce with Radioactive Substances," May 30, 1986, GARF, P8009/51/3559: 25–26.

15  "Dispatch No. 15," May 17, 1986, TsDAVO 27/22/7701: 217–18; "A. N. Tkachenko to E. V. Kachalovskyi," February 10, 1987, TsDAVO 27/22/7808: 76; and "Transcripts of Meeting No. 39," August 20, 1986, in *Chornobyl': Dokumenty operatyvnoi hrupy*, 435–46.

16  "Certification of the State of Special Provisions," August 15, 1987, SBU 16/1/1256: 81–84.

17  "Protocol No. 23," May 23, 1986, and "Protocol No. 26," May 27, 1986, Baranovska, *Chornobyl'—problemy zdorov"ia*, vol. 1, 157–58, 175–77.

18  "A. M. Kasianenko to A. N. Tkachenko," June 16, 1986, TsDAVO 342/17/4370: 40.

19  "On the Use of Meat with Elevated Levels of Radioactive Substances," January 9, 1987, TsDAVO 27/22/7808: 10–11.

20  "Information on Meat Production," April 17, 1987, NARB 4R/156/393: 16.

21  Vorobiev, *Do i posle Chernobylia*, 91–93, 112–15; and author interview with Dmitri Grodzinskii, Kyiv, July 14, 2015.

22 "On the Use of Meat with Elevated Levels of Radioactive Substances."

23 "Telegram from Khusainov to V. S. Murakhovsky," October 9, 1986; "M. N. Der-
gachev to SM BSSR, No. 13/341-233," November 25, 1986, NARB 7/10/475: 79,
103; "M. V. Kovalev to K. Z. Terekhov," September 15, 1987, NARB 7/10/475: 68;
and "E. F. Sukhorukov to Sovmin BSSR," May 30, 1988, NARB 7/10/1523: 20.

24 "Results of Detection of Strontium-90 and Cesium-137," December 8, 1986,
TsDAVO 27/22/7808: 16-41.

25 "Radiation in Soviet Veal Reported," *New York Times*, June 1, 1986.

26 "Information for Meat Production"; "Telegram from Khuseinov"; "M. N. Der-
gachev to SM BSSR, No. 13/341-233," November 25, 1986; "State Telegram,"
December 18, 1986, NARB 7/10/475: 79, 103, 110-12; and "Protocol No. 17/2-SP,"
April 12, 1988, NARB 7/10/1523: 11-21.

27 "On the Arrival at the Southwest Railroad of Meat with Elevated Levels of
Radioactivity," March 28, 1990, SBU 16/1/1284: 151.

28 "On the Exacerbation of the Situation at the Southwest Railroad," May 14, 1990,
SBU 16/1/1288: 47-48.

29 "On the Arrival at the Southwest Railroad of Meat with Elevated Levels of
Radioactivity," and "On the Situation Concerning the Refrigeration Section of
Meat Production," August 23, 1990, SBU 16/1/1284: 158-59.

30 "To A. N. Tkachenko from A. M. Kas'ianenko," June 25, 1986, TsDAVO
243/17/4370: 63; "Information of the Operative Group of the State Committee
of Industrial Agriculture," June 2, 1986, TsDAVO 27/22/7703: 35; "Transcripts
from Meeting No. 38," August 13, 1986, in *Chornobyl': Dokumenty operatyvnoi
hrupy*, 411-33.

31 "Explanatory Notes," August 12, 1986, TsDAVO 342/17/4370: 131-32.

32 "On Results of Investigation of Strawberries," May 6, 1986, TsDAVO
342/17/4370: 1, 7; "On Use of Goods," June 20, 1986, TsDAVO 243/17/5411: 69-70;
"On Purchase and Processing of Berries," June 25, 1986, TsDAVO 342/17/5411:
62; "I. M. Chaban from V. V. Vetchinin," May 10, 1986; "To A. N. Tkachenko
from A. M. Kas'ianenko," June 25, 1986; "On Results of Spectrological Investigation of Meat," July 10, 1986; "Temporary Directions," May 11, 1986; "General
Union Sanitary-Hygiene Anti-Epidemic Regulations and Norms," 1986, TsDAVO
243/17/4370: 3, 39, 63, 84, 179-90.

33 "From the Commission of the European Communities, Com (87), Minutes 894,"
November 4, 1987, PSP 133, European Union Archive (EUA); and "Wheat 'A La
Chernobyl' for the Third World," *IPS*, September 28, 1990.

34 V. K. Savchenko, *The Ecology of the Chernobyl Tragedy: Scientific Outlines of an
International Programme of Collaborative Research* (Paris: Parthenon Publishing, 1995), 30-31.

35 "On Measures to Liquidate the Consequences of the Accident," September 1986,
NARB 7/10/429: 29-32; "Additional Recommendations," June 30, 1986, Yaroshinskaya, *Sovershenno sekretno*, 386-87.

36 "On Measures to Liquidate the Consequences of the Accident" and "Reckoning of Need for Commercial Feed," 1988, NARB 7/10/1851: 53.

37 "M. Kovalev to N. I. Ryzhkov," October 16, 1986, NARB 7/10/439: 63; and "From 17 December 1986," December 17, 1986, GAOOMO 15/44/5: 150–52.

38 "E. E. Sokolova to N. I. Ryzhkov," November 6, 1987, NARB 4R/156/393: 143–44.

39 "On Measures to Liquidate the Consequences of the Accident."

40 "Directions for Decontaminating Milk," January 27, 1987; and "Temporary Instructions," October 21, 1986, RGAE 650/1/555: 1, 2–10.

41 "On the Use of Milk," May 17, 1986; "V. K. Solomakha to A. A. Tkachenko," no earlier than May 19, 1986, TsDAVO 27/22/7701: 224–25, 258; and "Temporary Instructions for Extinction of Directives." For milk collection within the 30 km zone, see "Protocol of Meeting of Council on Questions," December 26, 1991, NARB 507/1/12: 1.

42 "A. V. Romanenko to E. V. Kachalovsky," May 17, 1986, TsDAVO 2605/9/1601: 21–22; "Information of Operative Group of the State Committee of Industrial Agriculture," June 28, 1986, TsDAVO 27/22/7703: 119; and "On the Work of State Committee of Industrial Agriculture on Liquidation," January 25, 1987, TsDAVO 27/22/7808: 58–64.

43 "To I. M. Chaban from V. V. Vetchinina," May 10, 1986, TsDAVO 243/17/4370: 39; "Protocol No. 22," May 22, 1986, in Baranovska, *Chornobyl'—problemy zdorov"ia*, 148–50.

44 "Informational Communiqué," August 31, 1989, SBU 16/1/1279: 19–23.

45 "Information of Operative Group of the State Committee of Industrial Agriculture," June 5, 1986, TsDAVO 27/22/7703: 38.

46 "On the Procurement and Processing of Berries," June 26, 1986; "On Use of Berries in Children's Food," June 27, 1986; "Minsdrav UkSSR to C. A. Nanasiuk," June 26, 1986, TsDAVO 243/17/4370: 41–42, 73–74, 80–81; and "On Radiometric Control," September 17, 1986, TsDAVO 324/17/4348: 21.

47 "Commentary on the Active Norms of Radioactive Substances in Food Products," July 20, 1989, TsDAVO 342/17/5089: 88–90.

48 "Information on the State of Vegetables," no earlier than May 4, 1986; "On the Condition of Vegetables in Measures Taken," no earlier than May 5, 1986, TsDAVO 27/22/7701: 21–22, 31–32. "Information on the Condition of Shipments and Supply," May 9, 1986, TsDAVO 2605/9/1601: 92–93; and "The State of Radiological Control of Food Products," November 26, 1986, TsDAVO 342/17/4370: 207–10.

49 "Information of the Special Group of the State Committee of Industrial Agriculture," May 2, 1986, TsDAVO 27/22/7703: 2; and "On Inspection of Radiometric Control," June 1986, TsDAVO 342/17/4370: 27–30.

50 "On the State of Control of the Harvest Collection," May 15, 1986, TsDAVO 2605/9/1601: 184.

51 "Information on the Quality of Water in Resevoirs," May 4, 1986, TsDAVO 27/22/7701: 28–29.

52 "General Union Sanitary-Hygiene Anti-Epidemic Regulations and Norms," 1986, TsDAVO 342/17/4370: 174.

53 "To M. S. Mukharsky from V. I. Smoliar," August 4, 1986, TsDAVO 243/17/4370: 115.

54 "Certification on the Distribution of Evacuated Families from the Zone of the Accident," September 19, 1986, SBU 68/483: 20–21.

55 "On Creation of a Special Map No. 3 for Disposal of Biomass," October 1986, TSDAVO 342/17/4348: 31–32.

56 "Radiation Shield of Kyiv," October 1, 1986, Tsentralnyi derzhavnyi kinofoto-fonoarkhiv Ukrainy (TsDKFFA) No. 10219, film dossier.

57 "Information," May 15, 1986, TsDAVO 27/22/7701: 179–80.

## Farms into Factories

1 "On Results of Inspection," July 18, 1986, GAGO 1174/8/1940: 31.

2 "Information on Circumstances," June 5, 1987, NARB 4R/156/393: 45–60; "A. A. Grakhovskii to V. S. Murakhovsky," September 16, 1988, GAGO 1174/8/2113: 116; and "On Carrying Out of the Decree," May 26, 1987, RGANI 51/7, HIA, reel 1.1008.

3 "Commander of the Army of the Belarusian Military Region," June 19, 1987, NARB 4R/156/393: 71–75.

4 "Province Consumer Union to Gomel' Province Executive Committee," July 3, 1986, GAGO 1174/8/1940: 25.

5 "Temporary Recommendations for Allotment of Agro-Industrial Produce," May 30, 1986, TsDAVO 342/17/4340: 50–75; "General Union Sanitary-Hygiene Anti-Epidemic Regulations and Norms"; "On Directions for Decontaminating Milk," January 27, 1987; "Temporary Instructions for Decontaminating Milk," October 21, 1986, RGAE 650/1/555: 1, 2–10.

6 "Decree No. 102," June 27, 1986, DAZhO 219/1/404: 119; "Resolutions," August 27, 1986, NARB 1088/1/1002: 73; "On Reorganization and Changes," June 13, 1986, NARB 1088/1/986: 216–18; "Protocol No. 8," August 27, 1987, NARB 1088/1/1002: 59–61; and "Sovmin BSSR to Belorusian Consumer Cooperative," June 13, 1988, NARB 7/10/1524: 85-87.

7 "On Hermeticization of Cabins," November 8, 1986, NARB 7/10/467: 110; "Narodychi Regional Agri-Industrial Collective," July 7, 1986, DAZhO 219/1/404: 127; "Order for Narodychi Regional Agri-Industry," April 24, 1987, DAZhO 219/1/428: 70; "On the Supply of Tractors with Hermeticized Cabins," July 11, 1988, NARB 7/10/1523: 50–52; and "On Additional Measures," March 20, 1990, DAZhO 1150/2/3278: 17–20.

8 On prisoners, "Protocol of Meeting of the Council on Questions," October 13, 1992, NARB 507/1/12: 2–4. "On the Redistribution of Construction Points of

Special Refabrification," February 20, 1987, NARB 10/7/1524: 87; and "Certification Report," 1989, NARB 7/10/1938: 40–55.

9 "Tasks," June 1, 1988, DAZhO 1150/2/3164: 21; "On Measures to Liquidate the Accident," September 5, 1986, NARB 7/10/1851: 35–36; "Decree No. 107," May 22, 1987, DAZhO 5068/1/139: 156–59; and "Decree No. 132," June 3, 1988, DAZhO 5068/1/162: 7–9.

10 "Residents to A. S. Kamai and TsK KPSS," May 11, 1986, NARB 4R/156/238: 104; "On the Necessary Battle with Insect-Enemies," July 31, 1986, NARB 7/10/466: 259; "Information of the Operative Group of the State Committee of Industrial Agriculture," June 10, 1986, TsDAVO 27/22/7703: 42; "On Measures to Liquidate the Accident" and "On the Termination of Production of Disinfectants from DDT," June 14, 1988, TsDAVO 342/17/4900: 88–89.

11 "Draft Report on Toxic Chemical Contamination in Ukraine," September 3, 1990, GPA 999.

12 "Tasks of the Sanitary-Epidemiological Service of the Republic," September 23–26, 1986, TsDAVO 342/17/4355: 155–74; "On Work of State Committee of Industrial Agriculture USSR on Liquidation," January 25, 1987, TsDAVO 27/22/7808: 58–64; "List," May 5, 1987, DAZhO 76/36/42: 32–34; "Protocol," October 20, 1987, NARB 7/10/1524: 49–53; and "Clarification of Accounting," 1990, NARB 507/1/2: 278–83.

13 "Assessment of the Possibility of the Contamination of the River," September 15, 1986, and "On Effectivity of Water Reservoir Activities," September 30, 1986, NARB 7/10/467: 91, 99; "On the Circumstances and Pace of Investigation of the Accident at the Chernobyl NPP," May 5, 1986, *Z arkhiviv VUChK-HPU-NKVD-KHB*, 1 (Kyiv, 2001): 77–78; "On Unfortunate Circumstances," December 21, 1988, SBU 16/1/1266: 300–304; and "On Additional Construction," September 5, 1986, NARB 7/10/469: 130–31.

14 "Certification of Processes," September 8, 1986, SBU 68/483: 18.

15 "On Gasification of the Poliske Region," July 17, 1989, TsDAVO 324/17/5091: 50; and "L. Riabev on the Heating-Energy Complex," September 27, 1989, GAGO 1174/8/2215: 77.

16 "Instructions," November 21, 1986, NARB 7/10/467: 124.

17 "On Supplying the Population with Clean Food," June 16, 1988, NARB 7/10/1523: 42–43.

18 "Food and Consumer Goods in Khoiniki and Bragin Regions," August 28, 1986, GAGO 1174/8/1940: 28; "From the Gomel Sanitary-Epidemiological Service," May 16, 1986, GAGO 1174/8/1940: 19; and "On Party-Political Work," August 28, 1986, Arkhiv Kricheva 621/1/941: 51–52.

19 "Z. A. Khulap to V. K. Levchik," October 18, 1986, GAGO 1174/8/1940: 95–96; and A. Drozdov, "Chem zhivut Khoiniki?" *Sovetskaia belorussiia*, May 5, 1988, 3.

20 "V. A. Tsalko to A. A. Rakhnovskii," September 4, 1989, GAGO 1174/8/1940: 92–94; and "E. I. Sizenko to First Deputy Chair of State Committee of Industrial Agriculture," April 20, 1988, GAGO 1174/8/2113: 56–57.

21 "To M. S. Mukharsky from V. I. Smoliar," August 4, 1986, TsDAVO 243/17/4370: 115.

22 "Information of the Operative Group of the State Committee of Industrial Agriculture," August 29, 1986, TsDAVO 27/22/7703: 120; "On the Preparation of the School," August 29, 1986, Arkhiv Kricheva, 466/1/1120: 19–20; "Certification on the State of Children's Food," no later than June 24, 1992, TsDAVO 324/19/32: 25–28; and "On State Sanitation Oversight," 1989, NARB 46/14/1263: 98–114.

23 "An Especially Important Issue," May 2, 1989, TsDAVO 342/17/5089: 29–31.

24 Deborah Kay Fitzgerald, *Every Farm a Factory: The Industrial Ideal in American Agriculture* (New Haven, CT: Yale University Press, 2003); and Ted Genoways, *The Chain: Farm, Factory, and the Fate of Our Food* (New York: Harper, 2014).

25 For the KGB's acknowledgment of this fact, see "On Some Problems in the Liquidation of the Consequences of the Accident," December 6, 1988, *Z arkhiviv VUChK-HPU-NKVD-KHB*, 1 (Kyiv, 2001), 370–71. For reports of failed cleanup efforts, see "On Work," August 4, 1989, and "Certification," 1989, DAZhO 1150/2/3230: 1–16; "Resolutions," June 20, 1989, and "Dear Comrades," 1989, DAZhO 1/1/850: 225–29, 230–33; and "Certification," April 29, 1990, DAZhO 1150/2/3281: 52–56.

26 Eugene P. Odum, *Fundamentals of Ecology*, 2nd ed. (Philadelphia: Saunders, 1959), 481, quoted in Joel Hagen, *An Entangled Bank: The Origins of Ecosystem Ecology* (New Brunswick, NJ: Rutgers University Press, 1992), 116.

27 "On the Clarification of the Structure of the Planting," December 23, 1986, TsDAVO 27/22/7808: 12–13; "To the Commander of the Belorusian Military District," June 19, 1987, NARB 4R/156/393: 71–75; "Certification of Report on the Results of Recontamination Efforts," November 13, 1987, NARB 4R, 156/393: 145–49; "On the Relocation of Residents of Villages of Narodychi Region," February 2, 1989, TsDAVO 342/17/5089: 6.

28 "On Some Questions of the Liquidation of the Consequences of the Accident at the ChAES," February 22, 1991, SBU 16/1/1292: 143–45; and "First Draft, Report of Meetings in Kyiv," December 11, 1992, GPA 994.

29 "On Additional Included Villages," May 23, 1987, RGANI 89/56/5: 196, HIA; "No. 3-50/753," June 26, 1989, TsDAVO 342/17/5089: 70; and "On Circumstances Forming in the Narodychi Region of Zhytomyr Province," GARF 5446/150/1624: 13–18.

30 "On Additional Measures to Secure Safety," August 19, 1987, NARB 4R/156/393: 129–32.

31 "Gomel Province," no earlier than December 1, 1986, NARB 4R/154/392: 11–39; "A. A. Grakhovskii to V. S. Murakhovskii," September 16, 1988, GAGO 1174/8/2113: 116; and "Certification," June 19, 1986, NARB 7/10/530: 74–78.

32 "On Additional Allocation of Gomel and Mogilev Provinces with Feed," December 11, 1987, NARB 4R/156/393: 153.

33 "Data on Labs and Dosimetrical Posts for Objects," June 1987, NARB 4R/156/393: 35.

34  "On Agricultural Produce," June 1, 1987, NARB 4R/156/393: 61–65; and "On State Oversight," 1989, NARB 46/14/1263: 98–114.

35  "On Agricultural Produce."

36  That observation was borne out. See "On the Collective Letter from Residents of Rudnia Radovel's'ka," October 19, 1989, TsDAVO 342/17/5089: 183.

37  Nadezhda Koroleva interview with Angelina Gus'kova, "Serdtse v rukakh radiologa," *Atomnaia strategiia*, no. 14 (November 2004); author interview with Abel Gonzalez, June 3, 2016, Vienna.

38  For a range of requests to be relocated, see "Women from Chernihiv Province to M. S. Gorbachev," June 19, 1986, and "Residents of the Village Luhovyky, Kyiv Province," October 21, 1986, TsDAHO1/41/106: 148 ob, 191 ob; "Certification," November 20, 1987, *Z arkhiviv VUChK-HPU-NKVD-KHB*, 1 (Kyiv, 2001), 290–92; "V. I. Chazovu," n.d. 1988, GARF 8009/51/4340: 111–14; "Ministry of Health SSR," December 20, 1988, GARF 8009/51/4340: 103–4; "Notes from the Protocol No. 4 Meeting of the Soviet Labor Collective, Strelichevo Farm," October 4, 1989, GAGO 1174/8/2336, 142–43; "Protocol, Meeting of Workers, Cherykaw," June 1, 1989, GAGO 1174/8/2215: 71–72, 74; "To the Central Committee KPSS," March 23, 1989, GARF 8009/51/4340: 140–43. For a poll showing that 93 percent of respondents wished to move, see "Certification," 1992, NARB 507/1/20: 66–70.

39  Only in 1990 did they drop the quota. "On Measures to Speed the Implementation of the State Program for Liquidation," September 29, 1990, NARB 507/1/1: 28–30.

40  "On State Oversight," 1989, NARB 46/14/1263: 98–114; "Information on Milk Quality," April 1987, NARB 4R/156/393: 30; and "Annual Account," 1987, Arkhiv Kricheva 588/1/173: 29–30. A 1991 report prepared for Moscow by those responsible for safety in the State Committee of Industrial Agriculture gives different numbers that show cleaner milk from 1986 to 1990: "On Measures Taken in the Agri-Industrial Complex," March 18, 1991, GAGO 1174/8/2445: 62–68. For charges that the Ukrainian State Committee of Industrial Agriculture falsified radiation data, see "Minutes to the 28th Conference of the Communist Party of Ukraine," December 14, 1990, TsDAHO1/2/1065: 112–43.

41  "Explanatory Note," 1988, Arkhiv Kricheva 588/1/176: 22–23; "Certification," no later than March 1989, TsDAVO 324/17/5359: 51–65; and "Health Indicators of the Region," 1992, DAZhO 2959/2/1209: 1–219.

42  "On Order of SM UkSSR No. 5182/86," May 20, 1989, TsDAVO 342/17/5089: 67–69; and "A Group Deputy Question," June 9, 1989, TsDAVO 324/17/5089: 38–42.

43  "On Circumstances Forming in the Narodychi Region of Zhytomyr Province."

44  "On Radiological Conditions in Korosten'," October 19, 1989, TsDAVO 342/17/5089: 159–61.

45  "On Review of Petition," April 14, 1989, TsDAVO 342/17/5089: 24–26.

46 "On Building Up the Lab a Prep-Solution for Milk/Meat Produce," June 25, 1990, GAMO 11/5/1557: 104.

47 For a sample of documents ordering decontamination measures: "On Additional Measures," December 14, 1986, GAOOMO 40/50/7: 143–46; "Protocol of Commission Meeting," November 30, 1987, NARB 10/7/1524: 70–74; "Decree No. 161," December 8, 1987, DAZhO 219/1/428: 219; "On Additional Measures—Not for Publication," June 6, 1988, DAZhO 1150/2/3164: 15–19; "E. I. Sizhenko to First Deputy Chair of State Committee of Industrial Agriculture," April 20, 1988, GAGO 1174/8/2113: 56–57; "Protocol No. 17/2-SP," April 12, 1988, NARB 7/10/1523: 11–21; "Protocol of Commission Meeting," July 21, 1988, NARB 7/10/1523: 44–47; "On Creation of a Working Group," March 25, 1988, NARB 7/10/1525: 32–33; "On Problems in Leadership of Some Organizations," March 30, 1988, GAOOMO 9/187: 59; "Resolution No. 556," January 20, 1989, GAGO 1174/8/2215: 1–4; "Resolution of the Zhytomyr Party Committee," March 2, 1989, DAZhO 1150/2/3178: 10–14; "Certification," n.d. before June 1989, DAZhO 1150/2/3230: 7–11; "Resolution No. 5-29-8/1," May 15, 1989, GAMO 7/5/3833: 295–306; "On Measures to Improve the Food for Population in the Zone of 'Strict Control,'" September 14, 1989, TsDAVO 342/17/5089, 148–49, 156–58; "Decision No. 601 Carried Out for 1990," October 29, 1990, NARB 507/1/1: 89–95; "Vetkovskii Regional Committee to Chair of the Supreme Soviet, BSSR, N. I. Dementei," November 5, 1990, NARB 507/1/1: 142–43; "Belarus Consumer Cooperative to Deputy Chair of Soviet," November 10, 1990; "Project," Gomel, 1990, NARB 507/1/2: 245–46, 266–68; "Certification," no later than March 1989, TsDAVO 324/17/5359: 51–65; "Decision No. 580," November 16–17, 1989, GAGO 1174/8/2215: 88–91; "On Food for the Population of Controlled Territories," June 4, 1990, TsDAVO 324/17/5328: 85–86; "V. N. Sych to E. P. Tikhonenkov," May 24, 1991, GAGO 1174/8/2445: 45–53; and "Protocol No. 1," March 30, 1993, GAMO 7/5/4183: 1–10.

48 "Directive," May 23, 1989, GAMO 7/5/3856: 154–57. On unfulfilled orders to resettle, see "Radiological Circumstances," December 29, 1989, GAGO 1174/8/2336: 41–42; and Urii Bondar, ed., *20 let posle Chernobyl'skoi katastrofy: Sbornik nauchnykh trudov* (Minsk, 2006), 72. On partial closures of swampy land and pastures, see "Decision," September 19, 1989, GAMO 7/5/3837: 128–29. On full closures, see "Schematic," 1993, GAMO 7/5/4288: 15–20.

49 "On Medical-Sanitary Provisioning of Gomel and Mogilev Provinces," May 28, 1987, RGANI 89/56/6: 199–204, HIA; and "On Lowering Limits of Exposure of Population," March 16, 1987, NARB 4R/156/393: 93–96.

50 "To Commander of the Belarusian Military District."

51 "Resolution No. 556," January 20, 1989, GAGO 1174/8/2215: 1–4.

52 "On Question No. 3," July 24, 1989; and "On Citizens' Letters," n.d., TsDAVO 342/17/5089: 124, 129.

53 "Residents of Village Luhovyky, Kyiv Province," October 21, 1986, TsDAHO 1/41/106: 191 ob.

54 "To Efimovich from Voters of Kosiukovicheskii Region," 1989, GAGO 1174/8/2215: 75–76.

## The Swamp Dweller

1 David Blackbourn, *The Conquest of Nature: Water, Landscape, and the Making of Modern Germany* (New York: Norton, 2006), 239.

2 U.S. Army Aerial Maps, September 1943, RG 373 GX 16058 SK, National Archives Research Administration (NARA), College Park, MD.

3 Joice M. Mankivell and Sydney Loch, *The River of a Hundred Ways: Life in the War-Devastated Areas of Eastern Poland* (London: G. Allen and Unwin, 1924), 53–66.

4 On pesticides and fertilizers, see Owen Hoffman, "Interim Report: The Toxicological Program," December 21, 1992, GPA 1002.

5 Łukasz Łuczaj et al., "Wild Edible Plants of Belarus: From Rostafiński's Questionnaire of 1883 to the Present," *Journal of Ethnobiology and Ethnomedicine* 9, no. 1 (January 2013): 21–38; and Yuriy V. Movchan, "Environmental Conditions, Freshwater Fishes and Fishery Management in the Ukraine," *Aquatic Ecosystem Health and Management* 18, no. 2 (April 2015): 195–204.

6 Volodymyr Heorhiienko, *Utro Atomograda*. Documentary. Ukrtelefil'm, 1974.

7 "V. K. Solomakha to A. N. Tkachenko," May 5, 1986, TsDAVO 2605/9/1601, 20; and "To M. S. Mukharsky from P. I. Chekrenev," August 27, 1986, TsDAVO 342/17/4348, 15–16.

8 For this history, see my *A Biography of No Place: From Ethnic Borderland to Soviet Heartland* (Cambridge, MA: Harvard University Press, 2004).

9 On the increase of five to seven times the levels of radioactivity from forest fires, see "To Deputy Director of Mogilev Oblispolkoma, A. S. Semkin," October 30, 1990, GAMO 7/5/3990: 21–22; and "Fire Protection," December 12, 1990, and "Unsatisfactory Execution of Directives," April 1992, GAMO 7/5/4126: 82–83, 69–71.

10 Timothy A. Mousseau et al., "Highly Reduced Mass Loss Rates and Increased Litter Layer in Radioactively Contaminated Areas," *Oecologia*, March 4, 2014.

11 For evidence that animals in the field are more sensitive than in the lab, see Jacqueline Garnier-Laplace et al., "Radiological Dose Reconstruction for Birds Reconciles Outcomes of Fukushima with Knowledge of Dose-Effect Relationships," *Scientific Reports* 5 (November 16, 2015).

12 Anna Lowenhaupt Tsing, *The Mushroom at the End of the World: On the Possibility of Life in Capitalist Ruins* (Princeton, NJ: Princeton University Press, 2015).

13 T. Mousseau and A. Møller, "Reduced Abundance of Insects and Spiders Linked to Radiation at Chernobyl 20 Years After the Accident," *Biology Letters* (2009): 5.

14 Anders Pape Møller et al., "Ecosystems Effects 25 Years after Chernobyl: Pollinators, Fruit Set and Recruitment," *Oecologia* 170 (2012): 1155–65.

15  Møller et al., "Elevated Mortality among Birds in Chernobyl as Judged from Skewed Age and Sex Ratios," *PLOS ONE* 7, no. 4 (April 11, 2012).

16  Rachel Carson, *Silent Spring* (Boston: Houghton Mifflin, 1994).

17  Rachel Nuwer, "Forests Around Chernobyl Aren't Decaying Properly," March 14, 2014, Smithsonian.com, accessed June 7, 2016, https://www.smithsonianmag .com/science-nature/forests-around-chernobyl-arent-decaying-properly -180950075/.

18  Author interview with Anders Møller, November 4, 2014, Paris.

19  A. V. Stepanenko, "On a Conception of Residence on Territory Contaminated with Radionuclides," 1990, RGAE 4372/67/9743: 11–22; and V. P. Platonov and E. F. Konoplia, "Summary of Main Results of Scientific Work on Liquidation of Consequences of the Accident at the ChAES," April 21, 1989, RGAE 4372/67/9743: 490–571.

20  "To First Deputy, E. V. Kachalovsky," June 6, 1989; "On Conditions in the Narodychi Region, Zhytomyr Province," June 16, 1989, SBU 16/1/1275: 27–28; "For the period of 1987," no earlier than March 1989, "Circumstances Forming in Narodychi Region, Zhytomyr Province," GARF 5446/150/1624: 28–29, 13–18. For State Committee of Industrial Agriculture refutation of reported rise in birth defects among animals, see "On Question III," July 24, 1989, TsDAVO 342/17/5089: 124.

21  Nicholas A. Beresford and Davide Cobblestone, "Effects of Ionizing Radiation on Wildlife: What Knowledge Have We Gained Between the Chernobyl and Fukushima Accidents?" *Integrated Environmental Assessment and Management* 7 (2011): 371–73.

22  See the difference between earlier and later publications. Robert Baker with Jeff Toney, "Dean's Corner," *Translating the Endless Wonderment of Science*, March 16, 2011; Robert J. Baker et al., "Mitochondrial Control Region Variation in Bank Voles Is Not Related to Chernobyl Radiation Exposure," *Environmental Toxicology and Chemistry* 26, no. 2 (February 2007): 361–69; and Robert J. Baker et al., "Elevated Mitochondrial Genome Variation after 50 Generations of Radiation Exposure in a Wild Rodent," *Evolutionary Applications* 10, no. 8 (September 1, 2017): 784–91.

23  T. G. Deryabina, S. V. Kuchmel, L. L. Nagorskaya, T. G. Hinton, J. C. Beasley, A. Lerebours, and J. T. Smith, "Long-Term Census Data Reveal Abundant Wildlife Populations at Chernobyl," *Current Biology* 25, no. 19 (October 5, 2015): R824–26.

24  On the shortcomings of computational studies, see Leopold, *Under the Radar*, 142–44.

## The Great Chernobyl Acceleration

1  "On Measures," June 9, 1986, DAZhO 76/36/42: 50–53; "Several Problems," January 22, 1991, SBU 16/1/292: 59–63.

2  Author interview with Aleksandr Komov, August 2, 2016, Rivne, Ukraine.

3  Michael S. Pravikoff and Philippe Hubert, "Dating of Wines with Cesium-137," accessed July 23, 2018, https://arxiv.org/ftp/arxiv/papers/1807/1807.04340

.pdf. On releases, see Owen Hoffman et al., "A Perspective on Public Concerns about Exposure to Fallout from the Production and Testing of Nuclear Weapons," *Health Physics* (2002): 736–49.

4  A. N. Marei et al., *Global'nye vypadeniia Cs-137 i chelovek* (Moscow: Atomizdat, 1974), 3–13.

5  A. N. Marei et al., *Global'nye vypadeniia Cs-137 i chelovek*, 113.

6  *Federal Radiation Council Protective Action Guides: Hearings before the Subcommittee on Research, Development and Radiation of the Joint Committee on Atomic Energy, U.S. Congress, June 29–30, 1965* (Washington, DC: U.S. Government Printing Office, 1965), 217.

7  Marei, *Global'nye vypadeniia*, 141.

8  "Minutes to the 28th Conference of the Communist Party of Ukraine," December 14, 1990, TsDAHO1/2/1065: 112–43.

9  "Information about the Case of Irradiation," February 12, 1970; "On the Case of Irradiation," February 10, 1970; "AKT: Commission Presidium Academy of Science UkSSR," February 9, 1970; "To the Science Division, TsK KPU," August 10, 1970; and "Measures to Execute the Resolution," August 10, 1970, TsDAHO1/25/365: 3–8, 11–20, 49–50.

10  "Iadernyi vzryv v Khar'kovskoi oblasti," June 27, 2008, Prestupnosti.net, accessed June 21, 2018, https://news.pn/en/politics/1503.

11  V. I. Zhuchikhin, *Podzemnye iadernye vzryvy v mirnykh tseliakh* (Snezhinsk: RFIaTs, 2007), 23–43.

12  "Measures to Liquidate and Cover the Gas Fountain," May 11, 1972, TsDAHO 1/16/109: 51–57.

13  L. F. Chernogor, "Vzryvy na gazoprovodakh i avarii na gazovykh khranilishchakh– istochnik ekologicheskikh katastrof v Ukraine," *Ekologiia i resursy* 19 (2008): 56–72; and Olha Martynyuk interview with Leonid Chernogor, Kharkiv, October 18, 2017.

14  On radioactive contamination of Kharkhiv oblast, see A. M. Koz'mianenko, "On Possibility of Allocation of NPP," May 1, 1989, TsDAVO 342/17/5091: 4–6.

15  Chernogor, "Vzryvy."

16  Medvedev, *The Truth about Chernobyl*, 46; and "On Possible Lowering of Power at the Chernobyl NPP," February 4, 1986, *Z arkhiviv VUChK-HPU-NKVD-KHB*, 1 (Kyiv, 2001): 62–63.

17  "Notes of Report, UKDB URSR," March 12, 1981, SBU 65/1/5: 1–74; and "Informational Communiqué," April 8, 1986, SBU 16/1/1283: 9.

18  "Accident at NPP," September 13, 1982, *Z arkhiviv VUChK-HPU-NKVD-KHB*, 1 (Kyiv, 2001): 47; and Medvedev, *The Truth about Chernobyl*, 5.

19  Borys Kachura, *Volodymyr Scherbytsky: Spohady suchasnykiv* (Kyiv: Vidavnychyi dim In Yure, 2003), 104–6; and Valentyna Shevchenko, "Urodzhena mudrist'," 47–48.

20  N. K. Vakulenko, "Notes from a Report," March 12, 1981, SBU 65/1/5: 74.

21  Vorobiev, *Do i posle Chernobylia*, 119.

22  Kachura, *Volodymyr Scherbytsky*, 104–6; and V. K. Vrublev'sky, *Vladimir Shcherbytsky: Pravda i vymysly* (Kyiv: Dovira, 1993), 204.

23  "Informational Communiqué," September 16 and 30, October 17 and 28, December 27, 1986, SBU 16/1/1245:137–38, 164–65, 208, 233–34, 388–89.

24  "Address to Cherep," December 14, 1990, Baranovska, *Chernobyl'sk'ka trahedia*, 623–28.

25  For speculation that Chernobyl was one of the RBMK plants designated for military use, see David R. Marples, "Chernobyl: The Political Fallout Three Years Later," *EIR Science and Technology* 16, no. 20 (May 12, 1989): 24–31.

26  March 14, 1983, N. S. Neporozhnii, *Energetika strany glazami ministra: Dnevniki 1935–1985 gg* (Moscow: Energoatomizdat, 2000), accessed December 21, 2016, http://prozhito.org/person/460.

27  "Information on Circumstances," June 5, 1987, NARB 4R/156/393: 45–60.

28  "Execution of Directives of Council of Ministers UkSSR," July 10, 1989, TsDAVO 342/17/5089: 91–93.

29  "Letter of Chief State Sanitary Doctor," January 26, 1988, Baranovska, *Chornobyl'sk'ka trahedia*, 499–500.

30  "Inventory Material," October 21, 1988, TsDAVO 342/17/4886: 25–30; and "Medical Sanitary Provisioning of Population," March 1, 1989, TsDAVO 342/17/5092: 9–10.

31  "Letter to Minister of Health UkRSR," May 19, 1988, in Baranovska, *Chornobyl'sk'ka trahedia*, 527–28.

32  Author interview with Aleksandr Komov, August 2, 2016, Rivne, Ukraine.

33  Author interview with Nikolai Davidovich Tervonin, July 28, 2016, Ol'shany, Belarus.

34  B. V. Sorochinsky, "Molecular-Biological Nature of Morphological Abnormalities Induced by Chronic Irradiation in Coniferous Plants from the Chernobyl Exclusion Zone," *Cytology and Genetics* 37 (2003): 49–55.

35  Adamovich, *Imia sei zvezde Chernobyl'*, 73. Greenpeace recorded "rumors" of testing in the marshes. "VT to Science Unit," April 7, 1993, GPA 1002. On the importance of strategic nuclear weapons, see David Holloway, "Nuclear Weapons and the Cold War in Europe," in Mark Kramer and Vit Smetana, eds., *Imposing, Maintaining, and Tearing Open the Iron Curtain: The Cold War and East-Central Europe, 1945–1989* (Lanham, MD: Lexington Books, 2013), 440–43.

36  Author telephone interview with Valentina Drozd, February 12, 2016. For evidence of abnormally high rates of birth defects, see "Memo on Pediatric Service, Dobrush Region," 1988, NARB 46/14/1261: 134–36.

37  "Protocol Council Meeting on Questions," January 16, 1992, NARB 507/1/12: 2–4.

38  L. Smirennyi, "Predtecha Chernobylia," *Nauka i zhizn'*, no. 10 (2003).

39 Martin Bürgener, *Pripet-Polessïe: Das Bild Einer Polnischen Ostraum-Landschaft* (Gotha: Justus Perthes, 1939).

40 "Informational Communiqué," November 30, December 6, and December 14, 1987, SBU 16/1/1256: 273–74, 303–4; February 11, February 17, April 8, May 10, and October 22, 1988, SBU 16/1/1262: 61, 76, 151, 192–93; "Informational Communiqué," October 22, 1988, SBU 16/1/1266: 179; January 19, 1989, SBU 16/1/1273: 31-3; October 13, and October 24, 1989, SBU 16/1/1279: 112, 139; "Informational Communiqué," September 8, 1988, SBU 16/1/1266: 96–97; "Some Problems of Safety at NPP," August 6, 1990, 16/1/1288, 127–29; "Results of Safey Inspection at Rivne NPP with IAEA Methods," April 17 to May 18, 1987, RGAE 859/1/592: 118–27.

41 "Informational Communiqué," November 23, December 6, and December 14, 1987, SBU 16/1/1256: 273–74, 303–4, 307–8; February 11, February 17, April 8, and May 10, 1988, SBU 16/1/1262: 61, 76, 151, 192–93; January 19, 1989, SBU 16/1/1273: 31–33; October 13 and October 24, 1989, SBU 16/1/1279: 111–12, 139; "Informational Communiqué," October 22, 1988, SBU 16/1/1266: 179; "To V.A. Ivashko from N. Holushko on Problems with Raising Safety at ChAES," SBU 16/1/1279: 111–12, 139, 178–80; and "Some Problems with Liquidating Consequences of ChAES Accident," November 14, 1990, SBU 16/1/1289: 43–48. On accidents at Soviet-built plants in Bulgaria, see "Circumstances at the Bulgarian NPP," November 13, 1990, SBU 16/1/1289: 41–42.

42 John McNeill, *The Great Acceleration* (Cambridge, MA: Harvard University Press, 2016).

## The Housekeeper

1 Silini, "Concerning Proposed Draft for Long-Term Chernobyl Studies," Correspondence Files, 1986, UNSCEAR; and National Cancer Advisory Board convened on September 11–12, 1998; 1999 National Cancer Institute (NCI) Annual Report.

2 "Medical Aspects of the Chernobyl Accident," Kyiv, May 11–13, 1988 (Vienna: IAEA, 1989), 9–12.

3 "Medical Aspects of the Chernobyl Accident," 349.

4 "Medical Aspects of the Chernobyl Accident," 100, 137.

5 "Medical Aspects of the Chernobyl Accident," 331.

6 "Medical Aspects of the Chernobyl Accident," 55.

7 "Medical Aspects of the Chernobyl Accident," 14.

8 Author interview with Natalia Lozytska, July 16–17, 2016, Kyiv.

9 "To Rector of Kyiv University," 1986, Lozytska personal papers.

10 "Information on Radioactive Particles," n.d., Lozytska personal papers; and Lozits'ka, "Iak ziti dali pislia Chornobylia," *Ukrains'ki obrii*, no. 3 (1991): 2.

11 Author telephone interview with Lars-Erik De Geer, December 18, 2017.

12 Lars-Erik De Geer et al., "A Nuclear Jet at Chernobyl around 21:23:45 UTC on April 25, 1986," *Nuclear Technology* (2017).

13 Alasdair Wilkins, "Why a Nuclear Reactor Will Never Become a Bomb," March 17, 2011, accessed December 18, 2017, https://www.gizmodo.com.au/2011/03/why-a-nuclear-reactor-will-never-become-a-bomb/.

14 "Natalia Lozytska to Deputy Riabchenko," 1989, GARF 1007/1/212: 16–18.

15 "Lozytska to Skopenko," August 18, 1986; and "To M. S. Gorbachev," Lozytska personal collection. For recognition of the importance of hot particles twenty-five years later, see "Radioactive Particles in the Environment," IAEA-TECDOC-1663 (Vienna: IAEA, 2011).

16 "Examination of Pediatric Population," 1989, Lozytska personal papers.

17 "Natalia Lozytska to Deputy Riabchenko," 1989, GARF 1007/1/212: 16–18.

18 On industrial accidents, see "Informational Communiqué," March 30, April 13, April 21, 1987, SBU 16/1/1249: 93–96, 136–37, 164; "Informational Communiqué," May 4, May 18, May 22, June 11, June 30, 1987, SBU 16/1/1250: 20, 54–57, 72–75, 128–29, 170–71; and "Informational Communiqué," October 19 and December 10, 1987, 16/1/1262: 214, 305–6; January 19, February 11, February 12, May 26, and June 28, 1988: 37, 61, 67–68, 229, 311.

19 "Unsatisfactory Circumstances in Production," December 21, 1988, SBU 16/1/1266: 300–4.

20 "Informational Communiqué," April 1986, SBU 16/1113: 12.

21 In support of the idea that foreign experts should decide the impact of the disaster, see "Zelenyi svit u mikrofona," *Trudovaia vakhta*, no. 12(58) (March 10, 1989): 2.

22 B. Danielsson, "Poisoned Pacific: The Legacy of French Nuclear Testing," *Bulletin of the Atomic Scientists* 46, no. 2 (March 1990): 22.

23 L. R. Anspaugh, R. J. Catlin, and M. Goldman, "The Global Impact of the Chernobyl Reactor Accident," *Science* 242 (1988): 1513–19.

24 C. C. Lushbaugh, M.D., Oak Ridge Associated Universities, to John Kozlowich, Knolls Atomic Power Laboratory, June 18, 1980, Steve Wing Personal Files.

25 "Human Radiation Studies: Remembering the Early, Oral History of Pathologist Clarence Lushbaugh, M.D.," 19950401, accessed December 29, 2017, https://ehss.energy.gov/OHRE/roadmap/histories/0453/0453toc.html; and Harriet A. Washington, *Medical Apartheid: The Dark History of Medical Experimentation on Black Americans from Colonial Times to the Present* (New York: Doubleday, 2006), 235.

26 Lisa Martino-Taylor, *Behind the Fog: How the U.S. Cold War Radiological Weapons Program Exposed Innocent Americans* (New York: Routledge, 2018); Eileen Welsome, *The Plutonium Files: America's Secret Medical Experiments in the Cold War* (New York: Dial Press, 1999); and Kate Brown, *Plutopia* (New York: Oxford University Press, 2013).

27 "Human Radiation Studies: Remembering the Early, Oral History of Dr. John W.

Gofman, M.D., Ph.D.," 19941220, accessed January 4, 2018, https://ehss.energy .gov/ohre/roadmap/histories/0457/0457toc.html.

28 "Informational Communiqué," March 11, 1987, SBU 16/1/1249: 28–30.

29 "Director General's Statement to the Board of Governors," May 12, 1986, IAEA Box 15717.

30 TASS announcement quoted in Andrei Mikhailov, "Podlodka, napugavshaia Gorbacheva i Reagana," October 6, 2012, http://www.pravda.ru/society/ fashion/models/06-10-2012/1130459-k_219-0/; and "Soviet Nuclear Submarine Carrying Nuclear Weapons Sank North of Bermuda in 1986," posted October 7, 2016, National Security Archive, http://nsarchive.gwu.edu/NSAEBB/ NSAEBB562-Soviet-nuclear-submarine-sinks-off-U.S.-coast/.

31 Author interview with José Goldemberg, Saõ Paulo, Brazil, August 9, 2017; and William Long, "Brazil Deaths Bring Fallout of Fear," *New York Times*, November 8, 1987.

32 For a critique of the failure of European authorities to prepare for nuclear emergencies, see Christopher Auland, "Chernobyl Reactor Accident and Its Aftermath," Brussels, August 3, 1986, CA-0007, European Union Archive (EUA).

## KGB Suspicions

1 "Negative Processes among a Portion of Soviet Youth," April 16, 1987, SBU 16/1/1249: 147–50.

2 Olha Martynyuk interview with Anatoly Artemenko, December 22, 2017, Kyiv.

3 "Informational Communiqué," April 27 and 29, 1987, SBU 16/1/1249: 179–180; and "Interruption of Antisocial Behavior," December 29, 1987, SBU 16/1/1256: 321–22.

4 Anders Åslund, *How Ukraine Became a Market Economy and Democracy* (New York: Columbia University Press, 2009), 15–17; and "Informational Communiqué," March 21, 1987, SBU 16/1/1249: 68–72.

5 Olha Martynyuk interview with Oles' Shevchenko and Oleksandr Tkachuk, December 1, 2017, Kyiv.

6 "Pre-Meditated Creation of Antisocial Groups," August 4, 1987, SBU 16/1/1256: 61–62; "Preparation for a Meeting in Kyiv on Ecological Problems," November 14, 1998, SBU 16/1/1266: 225–27.

7 "Intensification of Anti-Soviet Slander Abroad," January 24, 1990, SBU 16/1/1284: 49–51; and Taras Kuzio and Andrew Wilson, *Ukraine: Perestroika to Independence* (New York: St. Martin's Press, 1994), 70–73.

8 Interview with Shevchenko and Tkachuk.

9 "On Prevention of Antisocial Activity," December 28, 1987, SBU 16/1/1256: 321–22; and "On Intentions to Organize Antisocial Community," August 4, 1987, SBU 16/1/1256: 61–62.

10 Alexander Statiev, *The Soviet Counterinsurgency in the Western Borderlands* (Cambridge: Cambridge University Press, 2010).

11 Oleksandr Shvets', "Teatr tinei," *Vechirniy Kyiv*, October 19, 1987; and "On Termination of the Meeting of Anarcho-Syndicalists," October 26, 1989, SBU 16/1/1279: 146–48.

12 "Informational Communiqué," September 2, 1986, SBU 16/1/1245: 100–102; "Informational Communiqué," April 10, 1987, "Internal Report," April 29, 1987, SBU 16/1/1249: 130–31, 187–90; and "Report on Subversive Attempts by the Antagonist," January 19, 1988, SBU 16/1/1262: 33–36. By 1988, KGB agents reviewed 686,000 cases of people repressed from 1930 to 1953. "Organization of Review Work," August 3, 1988, SBU 16/1/1266: 55–57.

13 "To Comrade V. A. Ivashko from N. Golushko," February 10, 1989, SBU 16/1/1273: 81–82.

14 "Notes of the Secretary-General's Meeting with the Permanent Representative of the Ukrainian Soviet Socialist Republic," March 11, 1988, SG Country File, S-1024-87-8, UNA.

15 Interview with Shevchenko and Tkachuk.

16 Vitalii Shevchenko, "Persha nekomunistychna demonstratsiia," in *Kyivs'ka vesna*, ed. Oles' Shevchenko (Kyiv: Oleny Telihy, 2005), 348–52; "V upravlinni vnutrishnikh sprav mis'kvykonkomy," *Prapor komunizmu*, April 28, 1988; and interview with Shevchenko.

17 Interview with Shevchenko and Tkachuk.

18 Jane Dawson, *Econationalism* (Raleigh, NC: Duke University Press, 1996).

19 "Seminar on Ecology," September 10, 1988, TsDKFFA, No. 11266, Film Dossier.

20 David R. Marples, *Ukraine under Perestroika: Ecology, Economics and the Workers' Revolt* (New York: St. Martin's, 1991), 91, 137.

21 For history of these movements, see Kuzio and Wilson, *Ukraine*, 77–80.

22 "Public Ecology Demonstration in the City of Kyiv," November 14, 1988, SBU 16/1/1266: 225–28.

23 "Creation of an Initiative Group of Social Movements," November 24, 1988, SBU 16/1/1266: 243–45; "Public Ecology Demonstration in the City of Kyiv," November 14, 1988, SBU 16/1/1266: 225–27.

24 Interview with Shevchenko and Tkachuk.

### Primary Evidence

1 "Porog," 1988, Rollan Sergienko, Kinostudia im. Oleksandra Dovzhenka.

2 For liquidator testimony, see "D. I. Moisa, Chief Clinic Doctor, VNTsRM, to M. I. Selikhov," March 1988; "M. I. Velikhov to A. P. Samokhvalov," September 8, 1988, TsDAVO 2605/8/17: 12, 24–25; "O. I. Shamov to V. V. Red'kin," November 29, 1988, GARF 8009/51/4340: 24; "Aleksandr P. Borshchevsky to Evgeny I. Chazov," October 30, 1988, GARF 8009/51/4340: 67–84; and "K. L. Gorshunov to S. M. Riabchenko," July 6, 1989, GARF 1007/1/212: 2–5.

3 "Head Department of Medical Problems," March 17, 1994, TsDAVO 324/19/261: 17–18. On rules for establishing a "connection," see "Decree, No. 57," October 3,

1990, TsDAVO 342/17/5220: 51; and Adriana Petryna, *Life Exposed: Biological Citizens after Chernobyl* (Princeton, NJ: Princeton University Press, 2002), 105–17.

4 "To E. M. Luk'ianova and L. M. Zhdanova from A. M. Serdiuk," 1988, TsDAVO 324/17/4886: 24–32; "Operative Meeting," August 8, 1988, TsDAVO 342/17/4886: 15–18, 19–22; "Medical-Sanitary Provision of Population," March 1, 1989, TsDAVO 342/17/5092: 9–10; and "Material for Report to the Government," September 29, 1989, TsDAVO 342/17/5089: 159–61.

5 E. I. Chazov, "Medical Aspects of the Accident," June 16, 1987, RGANI 89/53/75: 1–7, HIA; "Decisions of the Coordinating Meeting," Kyiv, May 22–24, 1989, TsDAVO 342/17/5090: 34–43; "On Question III," July 24, 1989, TsDAVO 342/17/5089: 124–28.

6 For quote, "Report," no earlier than March 11, 1990, TsDAVO 342/17/5240: 88–98; and "Notes on the Secretary General's Meeting with Mr. Anatoliy Zlenko, Foreign Minister of Ukrainian USSR," September 20, 1990, SG Country File, S-1024-87-8, UNA.

7 See, for example, the difference in reporting levels of radioactivity in Chernihiv Province between a 1989 internal document: "Deputy Minister of Health UkSSR, A. M. Kas'ianenko, 04-r-16 DSP," June 7, 1989, TsDAVO 342/17/5089: 68, and an appeal for external aid: "Material for UNESCO," UNESCO Archive, Paris, 1994, 9014/1/26: 27. "Assessment of State of Health of Population Living on Territory Contaminated with Radionuclides," no earlier than April 1991, TsDAVO 342/17/5358: 8–11.

8 Most famously, Alla Yaroshinskaya and Yuri Shcherbak were both journalists who became Chernobyl activists and politicians.

9 For discussions of the political uses of the Chernobyl disaster, see Petryna, *Life Exposed*, and Olga Kuchinskaya, *The Politics of Invisibility: Public Knowledge about Radiation Health Effects after Chernobyl* (Cambridge, MA: MIT Press, 2014).

10 "Results of Inspection of Financial Activity of the Polesian Reserve," no earlier than January 1992, NARB 507/1/6: 65–68.

11 Author telephone interview with Alex Klementiev, August 6, 2017.

12 Robert Proctor, *Cancer Wars: How Politics Shapes What We Know and Don't Know about Cancer* (New York: BasicBooks, 1995); Naomi Oreskes and Erik M. Conway, *Merchants of Doubt: How a Handful of Scientists Obscured the Truth on Issues from Tobacco Smoke to Global Warming* (New York: Bloomsbury Press, 2011).

13 "To E. M. Luk'ianova and L. M. Zhdanova from A. M. Serdiuk."

14 "Inspection of Medical Services," October 28, 1988, TsDAVO 342/17/4877: 14–30. For similar numbers, see "Director of Division of Medical-Prophylactic Aid," January 18, 1988, TsDAVO 342/17/4886: 4; and "Distribution of the Decreed Contingent," no earlier than December 1988, TsDAVO 342/17/5092: 120–26. In Belarus, see "Executive Committee of the Province Council of People's Deputies," January 5, 1990, and "To Tamara Vasikas'ko, chair, ispolkom," GAGO 1174/8/2336: 68–69.

15  "Report Academy of Science UkSSR," July 7, 1988, Baranovska, *Chornobyl'sk'ka trahedia*, 540–51; "Information on the Appearance of Sick Children," 1989, GARF 5446/150/1624: 30–35; "To E. M. Kuk'ianova and L. M. Zhdanova from A. M. Serdiuk," and "Certification on the State of Medical and Pharmaceutical Help," no earlier than January 1989, TsDAVO 342/17/5240: 21–31; "To Head of the Division," January 18, 1988, TsDAVO 342/17/4886: 4; and "Inspection of Organization of Medical Services," 1988, TsDAVO 342/17/4877: 14–30.

16  "Certification of State of Health of Population in Koriukovskii Region," April 5, 1990, TsDAVO 342/17/5238: 62–67.

17  "Circumstances Forming in Narodychi and Zhytomyr Provinces," 1989 GARF 5446/150/1624: 26–27.

18  The most common background rate for birth defects is 3–4 percent. "Inspection of Medical Services of the Population of Northern Regions of Rivne Province," February 9, 1989, and "Inspection of Quality of Medical Service," no earlier than April 1989, TsDAVO 342/17/5092: 1–8, 91–106; "Certification," no later than March 1989, TsDAVO 324/17/5359: 51–65; "Operative Meeting," August 8, 1988, TsDAVO 342/17/4886: 15–17, 25–30; and "Inspection of Medical Services," October 28, 1988, TsDAVO 342/17/4877: 14–30.

19  "Mortality of Children Age 0–14, 1989," TsDAVO 342/17/5241: 45–47.

20  "Memo," December 1990, TsDAVO 342/17/5240: 32–54; "Operative Meeting"; "To E. M. Kuk'ianova and L. M. Zhdanova from A. M. Serdiuk"; and "Medical-Sanitary Provisions for Population of Controlled Regions," January 1989, TsDAVO 342/17/5092: 27. For retrospective studies, see "Investigation of Activity of Institute of Epidemiology and Prophylactics," February 12, 1992, TsDAVO 324/19/34: 1–46; "Evaluation of Scientific-Research Work," December 24, 1992, TsDAVO 324/19/25: 1–25; "Review of Organization of Medical Services and Clinical Observation of Population of Kyiv Province," October 28, 1988, and "Review of Organization of Medical Services and Clinical Observation of Population of Zhytomyr Province," October 10, 1988, TsDAVO 342/17/4877: 1–13, 14–30; and "Report," no earlier than March 11, 1990, TsDAVO 342/17/5240: 88–98.

21  "Review of Organization of Medical Services," "Distribution of the Decreed Contingent," and "Medical-Sanitary Provisioning of the Population."

22  "Report of Academy of Science UkSSR," July 7, 1988; N. A. Loshchilov and B. S. Prister, "Informational Report," 1989, RGAE 4372/67/9743: 4–10; and "Answer: State Committee of Industrial Agriculture, Ukrainian Republic," February 15, 1989, Baranovska, *Chornobyl'sk'ka trahedia*, 540–51, 571.

### Declassifying Disaster

1  E. I Bomko, "From the Meeting," August 8, 1988, and "Informational Material," October 21, 1988, TsDAVO 342/17/4886: 198.

2 For the full list of censored topics, see "On Press Limitations Connected with the Accident at Reactor No. 4 of the Chernobyl NPP," July 31, 1986, Nesterenko Papers, BAS.

3 Author interview with Heorhii Shkliarevsky, June 1, 2017, Kyiv.

4 "Shcherbina to Ryzhkov," January 7, 1987, RGANI 89/53/55, HIA.

5 "Reactions of Workers to the Arrival of M. S. Gorbachev," February 28, 1989, SBU 16/1/1273: 91–93.

6 Leonid Kravchuk, "Shcherbytsky buv liudynoiu voliovoiu z syl'nym, zaharto-vanym kharakterom," in *Volodymyr Shcherbytsky: spohady suchasnykiv*, ed. E. F. Vozianov et al. (Kyiv: Vydavnichyi dim "In-Yure," 2003), 71–72.

7 "To Chair of the Council of Ministers, USSR, N. I. Ryzhkov," May 16, 1989, GARF 5446/150/1624: 1–3.

8 "Review of Collective Letter," April 14, 1989, TsDAVO 342/17/5089: 24–26.

9 "Olevs'k Regional Council to V. A. Masol," October 8, 1989, TsDAVO 342/17/5089: 178.

10 "Resolution of Collective Party-Economic Activity, Bykhovskii Region, Mogilev Province," June 26, 1990, NARB 46/14/1322: 15–18.

11 "Supreme Soviet of Belarus' SSR," September 26, 1990, NARB 46/14/1322: 176.

12 "Instruction to Group of Residents of Chernihiv," 1988, GARF 8009/51/4340: 59; "V. Sharavara to K. I. Masyk," March 10, 1990, TsDAVO 324/17/5283: 39–41; "To the Leader of Supreme Soviet of USSR, Ryzhkov, N. I.," May 16, 1989, GARF 5446/150/1624, 1–3; and "To the Deputies of Supreme Council," April 11, 1990, GARF 1007/1/212: 11–113.

13 Heorhiy Shkliarevsky, *Mi-kro-fon*, Kyiv, 1988.

14 "Collective Petition of the Participants of a Public Demonstration in City of Malyn," April 29, 1990, TsDAVO 342/17/5238.

15 "A. M. Serdiuk to V. M. Ponomarenko," May 8, 1990, TsDAVO 342/17/5220: 15; and "On the Assessment of Health Indicators in Malyn Region," July 7, 1990, TsDAVO 342/17/5240: 62–72.

16 "On Radiological Situation in Ivankiv Region of Kyiv Province," February 2, 1990, TsDAVO 342/17/5238: 10–11.

## The Superpower Self-Help Initiative

1 Petryna, *Life Exposed*, 43; and Baranovska, *Chernobyl'—problemy zdorov"ia*, 67, 144.

2 For a report of what Soviet scientists knew at the time, see Daniel L. Collins, "Nuclear Accidents in the Former Soviet Union: Kyshtym, Chelyabinsk and Chernobyl," 1991, Defense National Institute, DNA/AFRRI 4020, AD A 254 669.

3 Silini to Beebe, July 25, 1986, Correspondence Files, 1986, UNSCEAR Archive.

4 National Cancer Institute, Annual Report, 1999, accessed November 4, 2015, https://archive.org/stream/annualreport199173nati/annualreport199173nati_djvu.txt.

5  Gayle Greene, *The Woman Who Knew Too Much: Alice Stewart and the Secret of Radiation* (Ann Arbor: University of Michigan, 2001).

6  Gayle Greene, "Science with a Skew: The Nuclear Power Industry after Chernobyl and Fukushima," *Asia-Pacific Journal*, December 25, 2011; and David Richardson, Steve Wing, and Alice Stewart, "The Relevance of Occupational Epidemiology to Radiation Protection Standards," *New Solutions* 9, no. 2 (1999): 133–51.

7  "Testimony of Dr. Rosalie Bertell," U.S. Senate Committee on Veterans' Affairs, April 21, 1998; and Tom Foulds Collection, University of Washington Special Collections, Seattle, WA.

8  Robert Alvarez, "The Risks of Making Nuclear Weapons," in *Tortured Science: Health Studies, Ethics, and Nuclear Weapons in the United States*, ed. Steve Wing et al. (Amityville, NY: Baywood Publishing, 2012), 181–98.

9  Greene, "Science with a Skew."

10  C. C. Lushbaugh, MD, Oak Ridge Associated Universities, to John Kozlowich, Knolls Atomic Power Laboratory, June 18, 1980, Steve Wing Personal Files.

11  "Informational Communiqué," March 11, 1987, SBU 16/1/1249: 28–30; and "Travel Report, P. J. Waight," October 23–30, 1991, WHO E16-445-11: 6.

12  "Material for a Report to the Government," September 29, 1989, TsDAVO 342/17/5089: 159–61.

13  "G. I. Razumeeva to I. A. Liashkevich," August 23, 1989, TsDAVO 324/17/5091: 61–63; and *Nauka i suspils'tvo*, no. 9 (September 1989).

14  "To E. M. Luk'ianova and L. M. Zhdanova from A. M. Serdiuk"; "Outcomes of the Advancement of the All-Union Registry," Kyiv, May 22–24, 1989, TsDAVO 342/17/5090: 34–43; "On the Health of Children Exposed to Radioactive Substances," June 21, 1989, TsDAVO 324/17/5091: 35–36; and "Certification," July 7, 1990, TsDAVO 342/17/5240, 62–72.

15  "Materials for the Council of Soviet Ministry of Health," October 21, 1988, TsDAVO 342/17/4886, 25–30.

16  "Materials for the Governmental Report," September 29, 1989, "Debate on the Letter of Executive Committee," November 21, 1989, "On Question No. 3," July 24, 1989, TsDAVO 342/17/5089: 124–28, 159–61, 193–94.

17  V. G. Bebeshko, N. V. Bugaev, V. K. Ivanov, and B. A. Ledoshchuk, "Estimate of Future Scientific Research," 1989, TsDAVO 342/17/5090: 20–28.

18  O. O. Bobyliova, "Memo," no earlier than March 11, 1990, TsDAVO 342/17/5240: 88–98.

19  "Monitoring of Provision of Health Services to the Population," no earlier than December 1990, TsDAVO 342/17/5240: 32–54.

20  "On State of Medical Service for Residents of Kyiv Province," no earlier than January 1989, "Spravka," July 7, 1990, TsDAVO 342/17/5240, 21–31, 62–72.

21  "Inspection of Medical Services to Residents of Zhytomyr Oblast," 1988, TsDAVO 342/17/4877: 1–8; "On the State of Medical Services for Children in City of Chernihiv," no later than June 24, 1992, TsDAVO 324/19/32: 1–5; "Inspec-

tion of the Organization of Medical Services," February 9, 1989, TsDAVO 342/17/5092; Sergei G. Wamruk, no later than May 1991, NARB 507/1/7: 21; "On Medical Services," November 11, 1986, NARB 7/10/608: 71–73; "Information about Clinical Examination in Areas under [radiological] Control," August 7, 1989, NARB 46/14/1262: 45–54; "On the Amendment of the Long-Term Program," July 8, 1991, NARB 46/14/1373: 1–3; "On Medical Provisions for Children," November 21, 1990, NARB 507/1/1: 144; "List," 1989, NARB 507/1/2: 2–43b; "Evaluation of Incidence of Disease," April 18, 1989, NARB 46/14/1261: 110–12; "Letter from N. I. Rosh No. 06-11/21," January 10, 1989, NARB 7/10/1851: 35–36; "Memo," "Yearly Report," 1987, Arkhiv Kricheva 588/1/173: 25–27; "Memo," June 19, 1986, NARB 7/10/530: 74–78; and "Memo," no earlier than November 1993, GAMO 7/5/4156: 81–94. For doctors' appeal to leave the Zone, see "Announcement," September 19, 1989, NARB 46/14/1218: 79.

22  "Belarus SSR Ministry of Health to UkSSR Ministry of Health," August 30, 1990, TsDAVO, 342/17/5220, 47–48.

23  "V. G. Perederii to V. S. Kazakov," August 30, 1990, TsDAVO, 342/17/5220: 49.

24  "On Clinical Examination of Children Who Were Exposed to Radiation," August 31, 1988, TsDAVO 342/17/4886: 18–19.

25  "Operative Meeting," August 8, 1988, TsDAVO 342/17/4886: 15–18.

26  "Memo," February 9, 1989, TsDAVO 342/17/5092: 1–8. Another village discovered to have 584 ci/km was also overlooked. "On Resettling Residents of Pershotravneve," December 29, 1990, TsDAVO 342/17/5357: 20–21.

27  G. I. Razumeeva, "Medical-Sanitary Provisions for Population in Controlled Regions," January 1989, TsDAVO 342/17/5092: 26.

28  "Information Material for the Council of Soviet Ministry of Health," October 21, 1988, TsDAVO 342/17/4886: 25–30.

29  "On the State of Medical and Pharmaceutical Provisions," no earlier than January 1989, TsDAVO 342/17/5240: 21–31. For similar conditions in rural Belarus, see "On Medical Services of Rural Residents of BSSR," November 11, 1986, NARB 7/10/608: 71–72; "To the State Deputy of Supreme Council I. G. Chigrinov," July 6, 1989, NARB 46/14/1218: 187; and "To the Gomel Province Executive Committee," November 4, 1986, GAGO 1174/8/1940: 98.

30  "Memo," no earlier than March 11, 1990, TsDAVO 342/17/5240: 88–98.

## Belarusian Somnambulists

1  "Work Plan," October 17, 1990, NARB 507/1/1: 75–87.

2  "Directive of the Klimavichy Regional Administration," July 4, 1986, Arkhiv Kricheva, 3/4/1503: 80–83; "On the Progress of Liquidation," June 10, 1986, GAOOMO 9/181/66: 17ob; and "On the Progress of Decontamination," August 22, 1986, GAOOMO 15/44/5: 123–25.

3  "Enactment of State Sanitary Control," 1986, NARB 46/14/1263, 98–108; "On

Additional Questions," March 2, 1987, RGANI 89/56/1, HIA; and "On Improvements for the Long-Term Program," July 8, 1991, NARB 46/14/1373: 1-3.

4 "On the Work of the Medical Radiology Division of the Special Department of BSSR Ministry of Health," 1989, NARB 46/12/1264: 1-10; "Resolution No. 580," November 16-17, 1989, GAGO 1174/8/2215: 88-91; "On the Needs of Medical Radiology Research Institute," July 4, 1990, NARB 46/14/1322: 142; "Central Commission of KPB, Memo on Progress with Liquidation," 1989, NARB 4R/156/627: 126-38; and "V. A. Matiukhin to E. E. Sokolov," June 1, 1989, NARB 4R/156/627: 209-12.

5 "On Additional Measures," December 14, 1986, GAOOMO 40/50/7: 143-46; "Information," July 15, 1986, "Decree of the Chief Doctor of Cherykaw Region," April 7, 1987, Arkhiv Kricheva 588/1/166: 1-9, 90; "On the Assessment of Children's Health," April 18, 1989, "V. Ia. Latysheva, on Neurological Services," 1989, NARB 46/14/1261: 110-12, 152-62; and "On the Assessment of Endocrinology Services," March 23, 1988, NARB 46/12/1262: 83-85. For regulations about secrecy, see "On Providing Information," April 28, 1988, NARB 7/10/1525: 45. On equipment and failure to monitor, see "Protocol No. 8," August 27, 1987, NARB 1088/1/1002: 59-61; and "V. N. Sych to E. P. Tikhonenkov," May 24, 1991, GAGO 1174/8/2445: 45-53.

6 "Information about the Situation in Gomel and Mogilev Provinces," June 5, 1987, NARB 4R/156/393: 45-60; "On the Organization of Aid," May 30, 1988, NARB 7/10/1523: 21-23; "On the Work of BSSR Ministry of Health," 1989, NARB 46/14/1260: 1-15; and "List of Populated Settlements to Be Relocated," 1989, NARB 507/1/2: 2-43ob. On newly discovered contaminated territory in Brest Province, see "P. P. Shkapich to V. F. Kebich," July 3, 1990, and "Iu. M. Pokumeiko," September 9, 1990, NARB 46/14/1322: 4-6, 142.

7 "Survey of the Incidence of Disease among Children and Adults in Mogilev Province," March 13-25, 1989, NARB 46/14/1263: 1-15.

8 "On the Structure of Diseases of the Endocrine System," 1989, NARB 46/14/1261: 17-21.

9 "On the Structure of Diseases of the Endocrine System," 17-21.

10 "Incidence of Disease among Children Living on Territory Contaminated from 15-40 ki/km² 1983-1988," 1989, NARB 46/14/1261: 80-84.

11 Author telephone interview with Valentina Drozd, May 13, 2016. On assembly-line quality of medical exams in Belarus, see "To the Soviet Ministry of Health," 1988, GARF 8009/51/4340: 103-4.

12 "Residents of Komarin and Bragin Region to the Kremlin (Moscow)," October 18, 1989, and "Protocol of the Public Demonstration in City of Cherykaw," June 1, 1989, GAGO 1174/8/2215: 71-72, 74, 103-6.

13 "Ales Adamovich, Speech to the Plenum of the USSR Union of Writers," April 28, 1987, NARB 4R/156/436: 4-10; "E. P. Petriaev to E. E. Sokolov," March 27, 1987, NARB 4R/156/437: 6-7; and "E. E. Sokolov to N. I. Ryzhkov," June 23, 1987, NARB 4R/156/393: 70.

14  On nuclear solutions to modernization, see Paul Josephson, *Red Atom: Russia's Nuclear Power Program from Stalin to Today* (New York: W. H. Freeman, 1999).

15  "On Inspection of Letters," 1986, NARB LA 4R/157/86: 11–32.

16  "Appeal Claim to V. B. Nesterenko," January 6, 1988, NARB 4R/158/538: 153–58.

17  "Explanations to the Commission of the Minsk Regional Party Committee," September 27, 1988, NARB 4R/137/538: 212–18.

18  Budakovsky refers to this letter in "Transcripts," February 3, 1987, NARB LA 4R/157/86: 106.

19  "An Employee of BSSR Academy of Science to the Politburo TsK KPSS," August 5, 1986, "Citizens of Minsk, Patriots of Their Motherland, to M. S. Gorbachev," n.d., and "Citizens to M. S. Solomentsev," June 22, 1987, NARB 4R/156/441: 3–8, 152–55, 221–25.

20  "V. N. Ermashkevich to M. I. Demchuk," February 17, 1987, and "To M. S. Solomentsev," July 23, 1987, NARB 4R/156/441: 43–52, 197–200.

21  Wladimir Tchertkoff, *The Crime of Chernobyl: The Nuclear Gulag* (London: Glagoslav Publications, 2016), 110.

22  "Solutions of the Science-Technology Commission," no earlier than November 28, 1986, and "Decree of the Ministry of Medium Machines," November 20, 1986, NARB LA 4R/157/86: 35–39, 51–54.

23  "Party Control Commission," July 7, 1987, NARB 4R/156/441: 183–86.

24  "P. Budakovsky to N. N. Sliun'kov," January 7, 1987, NARB 4R/157/86: 57–60.

25  "Transcript," February 3, 1987, NARB 4R/157/86: 89–119.

26  "Memo on Inspection of Letters," January 29, 1987, NARB LA 4R/157/86: 76–84.

27  "I. Isakov to M. S. Solomentsev," July 27, 1987, and "To Comrade M. S. Gorbachev," November 16, 1987, NARB 4R/156/441: 178–82, 196–98.

28  "Nesterenko to President of BSSR Academy of Science," no later than January 27, 1988, NARB 4R/157/548: 50–51; "A. V. Stepanenko," April 12, 1987, and "M. Demchuk," December 29, 1987, NARB 4R/157/538: 96, 98–99.

29  Author interview with Alexey Nesterenko, July 22, 2016, Minsk.

30  "Memo," 1990, Arkhiv Kricheva 588/1/173: 8–23; "Directive No. 5-29-8/1," May 15, 1989, GAMO 7/5/3833: 295–306; and "Resolution," May 25, 1990, GAMO 7/5/8940: 25–34. Nesterenko first measured 80 ci/km and warned the minister of health about it. "No. 1353," September 19, 1986, Nesterenko Papers, BAS.

31  "Memo," February 21, 1990, Arkhiv Kricheva 588/1/181: 11–15.

32  "Memo," 1987, "Godovoi otchet," 1987, Arkhiv Kricheva 588/1/173: 25–27.

33  "Information," July 15, 1986, Arkhiv Kricheva 588/1/160: 1–9.

34  "Doses of Exposure of Cesium-134-137 among the Population," Arkhiv Kricheva 588/1/160: 7.

35  "Information," July 15, 1986, Arkhiv Kricheva 588/1/160: 1–9.

36  "Information," May 20, 1987, Arkhiv Kricheva 588/1/173: 1–12.

37  "Yearly Report," 1987, Arkhiv Kricheva 588/1/173: 29–30.

38  "Explanatory Note," 1988, Arkhiv Kricheva 588/1/176: 22–23; and "Memo on Radiation Protection," Arkhiv Kricheva 588/1/181: 48–49.

39 "Protocol from the Public Demonstration of Workers in City of Cherykaw," June 1, 1989, GAGO 1174/8/2215: 71-72.

40 "Memo," 1990, Arkhiv Kricheva 588/1/173: 8-9; and "State Statistical Report," 1991, Arkhiv Kricheva 588/1/182: 1-10.

41 "Information," 1988, Arkhiv Kricheva 588/1/176: 20.

42 "Information," May 20, 1987, and "Information," 1987, Arkhiv Kricheva 588/1/173: 1-12, 20-25.

43 "Information," 1988, Arkhiv Kricheva 588/1/176: 20.

44 "Clinical Examination,"1988, Arkhiv Kricheva 588/1/176: 19; and "Lab Tests Results for Newborns and Infants," NARB 46/14/1261: 106-9.

45 "Information on Obstetric and Gynecological Services in Cherykaw Region," 1988, Arkhiv Kricheva 588/1/176: 17-19.

46 "Conclusions," May 20, 1989, Arkhiv Kricheva 588/1/181: 23-30.

47 "On the Assessment of the Incidence of Disease among Children," April 18, 1989, NARB 46/14/1261: 110-12.

48 "Dynamics in Disease Incidence among Children," 1989, NARB 46/14/1161a: 23-24.

49 "Malignant Tumor Incidence in BSSR," 1989, NARB 46/14/1161a: 32.

50 "Cherykaw Region, Tables," 1988, Arkhiv Kricheva 588/1/176: 15, 32; and "Memo" 1990, Arkhiv Kricheva 588/1/173: 8-23.

51 "Demographic Data," 1989, Arkhiv Kricheva 588/1/176: 33; "Acute Leukemia among Citizens of Mogilev Province," 1989, NARB 46/14/1161a: 27-30; and "G. V. Tolochko, MZ BSSR," March 30, 1989, NARB 46/14/1264: 111-25.

52 "Child Population of Cherykaw Region," October 20, 1989, Arkhiv Kricheva 588/1/181: 29-30.

53 "State Statistical Report," 1991, Arkhiv Kricheva 588/1/182: 1-10.

54 "Disease Incidence among Children Who Live on a Territory with Pollution of 15-40 ki/km²," 1989, NARB 46/14/1261: 80-84.

55 "On the Medical Aspects of Liquidating the Consequences of the Accident," June 22, 1989, NARB 46/14/1264: 83-88; and "Analytical Memo on the Work of Neurological Services in Gomel Province," 1989, NARB 46/14/1261: 152-62.

56 "On the Work of the Ministry of Health, BSSR from 1986-1989," 1989, NARB 46/14/1260: 1-15.

57 "Memo on the Incidence of Hypothyroidism," June 1989, NARB 46/14/1264: 87-88.

58 "Evaluation of the Incidence of Disease among the Pediatric and Adult Population of Mogilev Province," March 13-15, 1989, NARB 46/14/1263: 1-11. See for similar arguments, "On the Work of BSSR Ministry of Health, 1986-1989."

59 "Lab Tests for Key Health Indicators among Newborns," 1988, NARB 46/14/1261: 106-9; and "Memo," March 17, 1989, NARB 46/14/1263: 18-24.

60 "To the Head of SM V. A. Masol," April 26, 1990, SBU 16/1/1284: 190-93; and "On the Evaluation of Medical Consequences of Chernobyl Accident," n.d., 1996, SBU 35/68: 1-12.

61 "On Problems Liquidating the Consequences of the Accident at ChAES," November 14, 1990, SBU 16/1/1289: 43–48; and author interview with Mykhailo Zakharash, July 1, 2016, Kyiv.

62 Author interview with Mykhailo Zakharash, July 1, 2016, Kyiv.

63 "On the Rate of Children Born with Health Defects," 1989, NARB 46/14/1262: 89–93.

64 "To the Director of of the Kazan' Synthetic Rubber Factory," July 4, 1990, NARB 46/14/1332: 19.

65 "Enacting State Sanitary Control," 1989, NARB 46/14/1263: 98–114; "On the Situation Emerging in Narodychi Region of Zhytomyr Province," 1989, and "V. Mar'yin to V. E. Shcherbina," June 19, 1989, GARF 5446/150/1624: 13–18, 51; "Technical Memo," September 26, 1990, GAMO 7/5/3999: 1–9; "On Enhancement of the Laboratories in the Meat and Dairy Industries," June 25, 1990, GAMO 11/5/1557: 104; "Information," March 16, 1992, GAMO 7/5/4126: 145–46; "Protocol No. 1," March 30, 1993, GAMO 7/5/4783: 1–10; and "Measurements of Permissible Content of Cesium-137," 1989, Arkhiv Kricheva 588/1/181: 5–7.

66 "A. I. Vorobiev, MZ USSR, to Academician B. E. Shcherbina," 1989, GAGO 1174/8/2215: 24–27; "On Results of Measurement of the Plutonium Body Burden among Residents of Gomel' Province in 1989," May 11, 1990, BAS 242/2/5: 1–5; "On the Central Allotment [of goods]," October 5, 1989, GAGO 1174/8/2215: 216; and "Conclusions on the Radiological Situation and Living Conditions," 1990, GAGO, 1174/8/2336: 43–45.

67 "S. N. Nalivko, Memo," January 9, 1990, NARB 46/14/1264: 49–66; "Memo on Assessment," 1989 NARB 46/14/1262: 1–15; "Information about the State of Clinical Examinations in the Regions under Control in 1989," August 1, 1989, NARB 46/14/1262: 45–54; and "On the Work of the Health Services of Brest Province in Liquidating Consequences of the Accident," August 1, 1989, NARB 46/14/1264: 11–14.

68 For general health: "On Pediatric Services in the Dobrush Region," 1988, "Disease Incidence among the Population," 1988; "Analysis," no earlier than March 1989, NARB 46/14/1161a: 1–4, 6, 15; "Health Indicators among the Population in the Areas under Control," March 10, 1989, NARB 46/14/1264: 32–44; "On the Evaluation of Primary Care Services in Krasnopole and Karma Regions," April 18, 1989, NARB 46/12/1262: 1–12; "On the State of the Auto-Immune System," May 31, 1989, NARB 46/14/1264: 127–33; "Information on the State of Clinical Examinations"; "On the Estimation of Disease Incidence among the Pediatric Population of Chechersk Region, Gomel Province," April 18, 1989, NARB 46/14/1261: 110–12; "On the Epidemiological Situation Related to Non-Specific Diseases of Respiratory Organs," 1989, "On Major Outcomes of Clinical Examinations in 1989," August 14, 1990, NARB 46/14/1264: 67–76, 135–39; "A. A. Romanovskii to Gomel Province Executive Committee," July 18, 1989, GAGO 1174/8/2215: 195–96; "To the Head of the Department of Health Care and Pre-

vention," January 18, 1988, TsDAVO 342/17/4886: 4; "On the Evaluation of Medical Services," 1988, TsDAVO 342/17/4877: 14–30; "To the Minister of Health Iu. P. Spizhenko," May 3, 1990, TsDAVO 342/17/5241: 1–5; "Materials for Release to the State," September 29, 1989, TsDAVO 342/17/5089: 159–61; "On the Situation Emerging in the Narodychi Region of Zhytomyr Province"; "On Work of the Commission of the Soviet Ministry of Health in Poliske Region, Kyiv Province," March 4, 1990, TsDAVO 342/17/5240: 57–59; and "Materials for Presentation to UNESCO," 9014/11/26: 27–37.

On endocrine disorders, including the appearance of tumors: "Evaluation of Endocrinology Services," March 23, 1988, "On Results of a Survey of Disease Incidence," March 23, 1989, NARB 46/12/1262: 83–85, 92–95; "Measures for a Survey of Hypothyroidism among Children in Komarin," November 28, 1988, GAGO 1174/8/2215: 118; "Memo," 1989, NARB 46/14/1264: 23–25; "The Gomel Province Executive Committee and Council of People's Deputies," January 5, 1990; and "Tamara Vasikas'ko, Chair of Executive Committee," GAGO 1174/8/2336: 68–69.

On birth defects and infant mortality: "Memo on Rates of Children with Birth Defects," March 23, 1989, "Memo about Analysis of Lethal Malformations among the Population of the City of Gomel," March 24, 1989, NARB 46/14/1262: 89–92, 121–26; "Still Births," 1988, NARB 46/14/1161a: 8–15; "Informational Material for the Council of the Soviet Ministry of Health," October 21, 1988, TsDAVO 342/17/4886: 25–30; and G. I. Razumeeva, "On Medical-Sanitary Provisions for Populations in Areas under Control," January 1989, TsDAVO 342/17/5092: 29–30.

## The Great Awakening

1   "Explanatory Note," 1989, TsDAVO 342/17/5089: 166–67.
2   "Resolution No. 566," January 20, 1989, GAGO 1174/8/2215: 1–5.
3   "Workers of Karma to Politburo TsK KPSS," 1989, GAGO 1174/8/2215: 205–7.
4   L. A. Ilyin et al., "Strategiia NKRZ po obosnovaniiu vremennykh predelov Doz godovogo oblucheniia naseleniia posle avarii na Chernobyl'skoi AES, Konseptsiia pozhiznennoi dozy," *Meditsinskaia radiologia* 8 (1989): 3–11.
5   "On the Work of BSSR Ministry of Health," 1989, NARB 46/14/1260: 1–15; and "On the Results of Debate over Problematic Issues," July 9, 1989, TsDAVO 342/17/5091: 41–43.
6   "Gomel Province," no earlier than December 1, 1986, NARB 4R/154/392: 11–39.
7   "On Trends in the Growth of Negative Events in the Republic," February 26, 1990, SBU 16/1/1284: 96–98.
8   "On Undesirable Conditions," January 16, 1990, SBU 16/1/1284: 24–25.
9   "E. P. Petriaev to E. E. Sokolov," March 27, 1987, NARB 4R/156/437: 6–7.
10  "Elaboration of Methods for Cytology Diagnostics," 1991, BAS 242/2/28: 12–15;

and "On the Results of Measuring Radioactive Body Burdens among Residents of Gomel and Mogilev Provinces," 1990, BAS 242/2/11: 1–27.

11 "On the Results of Screening for Plutonium Body Burdens among Residents of the Gomel Province in 1989," May 11, 1990, BAS 242/2/5: 1–5.

12 K. V. Moshchik et al., "Study of the Incidence of Non-Infectious Disease," vol. 1, 1988, BAS 242/2/1: 1–30.

13 V. P. Platonov and E. F. Konoplia, "Information on Major Findings of Scientific Work Related to Liquidating Consequences of ChAES Accident," April 21, 1989, RGAE 4372/67/9743: 490–571.

14 V. P. Platonov and E. F. Konoplia, "Information on Major Findings of Scientific Work Related to Liquidating Consequences of ChAES Accident"; and "Resolution of the First All-Union Convention on Radiobiology," Moscow, August 21–27, 1989, RGAE 4372/67/9743: 399–403.

15 For records of higher doses of radioactivity in food in cleaner areas, see "Technical Memo," September 26, 1990, GAMO 7/5/3999: 1–9.

16 "Evalution of External Pediatric Hematology, Ministry of Health, BSSR," April 26, 1990, NARB 46/14/1264: 96–101. Later reports of leukemia in Belarus strangely do not include the years 1986–1990, when one would expect the appearance of the disease. See Urii Bondar, ed., *20 let posle Chernobyl'skoi katastrofy, natsional'nyi doklad* (Minsk, 2006), 46–47.

17 Platonov, "Information on Major Findings"; and A. V. Stepanenko, "Toward a Concept of Living on Territory Contaminated with Radionuclides," January 27, 1990, RGAE 4372/67/9743: 11–22.

18 Platonov, "Information on Major Findings."

19 "Explanatory Note," 1989, TsDAVO 342/17/5089: 166–67.

20 "Resolution No. 566," April 27, 1989, TsDAVO 342/17/5089: 14–19.

21 "On Carrying Out Plan No. 566 of the State Commission," May 18, 1989, TsDAVO 342/17/5089: 20.

22 "D. Bartolomeevka, Vetkovskii Region," 1989, NARB 46/14/1261: 132–33.

23 "Indicators of the State of Health," March 10, 1989, NARB 46/14/1264: 32–45.

24 "A. I. Vorobiev to B. E. Shcherbina," June 12, 1989, GAGO 1174/8/2215: 24–27.

25 "A. I. Vorobiev to B. E. Shcherbina," 24–27.

26 "Resolution No. 5-29-8/1," May 15, 1989, GAMO 7/5/3833: 295–306; and "Decision of the Executive Committee of the Province Council of People's Deputies," May 25, 1989, DAZhO 1150/2/3227: 161–67.

27 "Teachers of Komarin to M. S. Gorbachev," October 18, 1988, GAGO 1174/8/2215: 113–15; "To E. I. Chazov," n.d. 1988, GARF 8009/51/4340: 111–14; "Collective of the Korosten' Seamstress Association to K. I. Masyk," September 22, 1989, TsDAVO 342/17/5089: 150–51; and "M. I. Vorotinskii to the Council of Ministers, UkSSR," April 5, 1990, TsDAVO 342/17/5238: 60–67.

28 "Collective MPK-157 to V. S. Venglovskaia," April 14, 1989, GARF 5446/150/1624: 42–44; "On Consideration of the Collective Letters," April 14, 1989, "On Consideration of Collective Letters of Residents of the Village of Povch, Luginskii

Region," April 20, 1989, TsDAVO 342/17/5089: 2, 12–13; "To the Chairman of the Council of Ministers, N. I. Ryshkov," May 16, 1989, GARF 5446/150/1624: 1–3; "Residents of Komarin, Bragin Region, to the Kremlin, Moscow," October 18, 1989, and "Workers of Korma to the Politburo TsK KPSS," 1989, GAGO 1174/8/2215: 103–6, 205–7; "To the People's Deputy UkSSR, A. A. Dron'," April 28, 1990, and "Request of Participants in the Meeting of Malyn Residents," April 29, 1990, TsDAVO 324/17/5328: 30, 98–104.

29  "Dear Comrade K. I. Masyk," January 5, 1990, TsDAVO 342/17/5238: 5–7.

30  "A. A. Grakhovskii to B. E. Shcherbina," March 14, 1989, GAGO 1174/8/2215: 101–2.

31  "Explanatory Note," no later than fall 1989, RGAE; and "Comments and Suggestions for the Concept of a Threshold 'Lifetime Dose,'" September 4, 1989, TsDAVO 342/17/5089: 172–73.

32  "On the Exacerbation of the Situation at Stations of the Southwest Railroad," May 14, 1990, SBU 16/1/1288: 47–48; "On the Radiological Situation in Korosten'," October 19, 1989, TsDAVO 342/17/5089: 159–61; and "On Measures for Medical Service to the Population of Narodychy Region," 1989, TsDAVO, 324/17/5091: 54–59.

33  "On the Radiological Situation," May 18, 1989, "Consideraton of the Collective Letters of Workers of Collective Farm Gorki, Narodychy Region," April 14, 1989, TsDAVO 342/17/5089: 27–28; "V. Maryin to the Secretariat Verkhovnii Council," July 5, 1989, GARF 5446/150/1624: 53; "On Results of Consideration of Telegrams of Residents of Kalinovskii Rural Council, Lugin Region," July 13, 1989, TsDAVO 324/17/5089: 47–48; "On Results of Consideration of Questions Raised in an Appeal of People's Deputies," July 9, 1989, TsDAVO 342/17/5091: 41–43; "On the Collective Letters of Residents of Rudnia Radovel'skaia," October 19, 1989, TsDAVO 342/17/5089: 183; "On Allocation of Additional Funds," December 12, 1989, TsDAVO 342/17/5089: 197–98; "A. M. Serdiuk to V. M. Ponomarenko," May 8, 1990, TsDAVO 342/17/5220: 15; "On the Inclusion of Villages of Ovruch Region [in controlled zones]," February 14, 1990, TsDAVO 324/17/5283: 25–26; and "Veprin, Lesan', and Bakunovichi, Cherykaw Region," Summer 1990, GAGO, 1174/8/2336: 43–47.

34  "On Consideration of Collective Letters of Residents of Village Povch, Luginskii Region," and "No. 3-50/753," June 26, 1989, TsDAVO 342/17/5089: 70; "Collective of the Korosten' Seamstress Association to K. I. Masyk"; "On Including the Whole Malyn Region [in controlled zone]," May 3, 1990, TsDAVO 342/17/5238, 79–80; "On Assessment of Health Indicators of Population of Malyn Region," July 7, 1990, TsDAVO 342/17/5240: 62–72; and "P. P. Shkapich to V. F. Kebich," July 3, 1990, NARB 46/14/1322: 4–6.

35  "On Radiological Circumstances in the City of Korosten'," October 19, 1989, TsDAVO 342/17/5089: 159–61.

36  "Appeal to People of the Land!" 1989, GAOOMO 9/187/214: 14; and "Statement," March 27, 1990, TsDAVO 2605/9/1853: 15–20.

37 Alla Yaroshinskaya, "V zone osobo zhestokogo obmana," *Nedelia* (July 24–30, 1989).

38 "On the State and Measures for Medical Provisions to the Population of Narodychy Region," 1989, TsDAVO 324/17/5091: 54–56; and "V. Maryin to the Secretariat Verkhovnii Council."

39 Marples, *Ukraine under Perestroika,* 50.

40 "On the Situation in Narodychy Region," June 16, 1989, SBU 16/1/1275: 27–28.

41 "Informational Communiqué," 1989, 16/1/1279: 49–52, 86–86.

42 "Informational Communiqué," April 10, 1989, 16/1/1273: 120–21; and "Informational Communiqué," September 25, 1989, SBU 16/1/1279: 83–84.

43 "On Processes Connected with the Construction and Operation of Nuclear Power Plants in the Republic," April 29, 1990, SBU 16/1/1284: 198–200.

44 "On Creation of a State Committee to Investigate the Fact of Mass Graves," October 9, 1990, SBU 16/1/1288: 233.

45 "To People's Deputy, USSR, V. E. Golavnev from Voters," 1989, GAGO 1174/8/2215: 75–76.

46 "Fenomen Chumaka, Kashpirovskogo i Dzhuny: kak sovetskie tseliteli zavladeli soznaniem millionov," StarHit.ru, accessed March 28, 2018, http://www.starhit.ru/eksklusiv/fenomen-chumaka-kashpirovskogo-i-djunyi-kak-sovetskie-tseliteli-zavladeli-soznaniem-millionov-133418/.

47 "Meeting Protocol," July 12, 1989, GAGO 1174/8/2215: 36–38; and "On Relocation of Residents from Villages of Narodychy Region," February 2, 1989, TsDAVO 342/17/5089: 6.

48 "Proposal for a Joint Environmental Program of the Northern Ukraine," November 15, 1990, 996 GPA.

49 "Protocol No. 15," October 31, 1989, GAOOMO 9/187/99: 18; and "Memo," 1990, Arkhiv Kricheva 588/1/173: 8–23.

50 "On Urgent Measures," May 12, 1989, NARB 4R/137/627: 224; "Protocol No. 15"; and "Resolution," May 25, 1990, GAMO 7/5/8940: 25–35.

51 "Directive No. 120-r," February 26, 1988, GAMO 7/5/3776: 23. On Krasnopole Region, see "Resolution," October 17, 1989, GAMO 7/5/3838: 57.

52 "TsK KP BSSR Memo on Liquidation," 1989, NARB 4R/156/627: 126–38; V. G. Evtukha, "Lecture on the 12th Session of the Supreme Soviet, BSSR," 1989, NARB 4R/158/625: 20–61. For a similar case, see "Conference Participants to the Chair of the Supreme Soviet," 1990, NARB 46/14/1322: 63–64.

53 "Memo," February 21, 1990, Arkhiv Kricheva 588/1/181: 11–15.

54 Author interview with Alexey Nesterenko, July 22, 2016, Minsk; and "On Creation of State Service, BSSR," October 19, 1990, NARB 1256/1/302: 255–57.

55 "On the Ecological Pilgrimage," Lozytska personal papers; and author interview with Natalia Lozytska, July 16, 2016, Kyiv.

56 "On the Situation in Narodychy Region," June 16, 1989, SBU 16/1/1275: 27–28;

and "To People's Deputy A. A. Dron'," April 28, 1990, TsDAVO 324/17/5328: 98–104.

57  Author interview with Natalia Lozytska. For KGB report, see "On Processes Connected with the Construction and Operation of Nuclear Power Plants in the Republic."

58  William Krasner, "Baby Tooth Survey—First Results," *Environment* 55, no. 2 (March 2013): 18–24.

59  "On the Situation Emerging in Narodychi Region of Zhytomyr Province," June 1989, GARF 5446/150/1624: 12–40; "Olevsk Regional Council to V. A. Masol," October 8, 1989, TsDAVO 342/17/5089: 178; "Presidium of the Supreme Soviet, UkSSR," February 14, 1990, TsDAVO 1/22/1125: 20–21; and "Appeal," August 29, 1990, NARB 46/14/1322: 54.

60  "Resolution No. 5-29-8/1," May 15, 1989, GAMO 7/5/3833: 295–306; "A General Important Question," May 2, 1989, and "On the Group Inquiry of Deputies," June 6, 1989, TsDAVO 324/17/5089: 29–42, 43; "On Question No. 3," July 24, 1989, TsDAVO 342/17/5089: 124; "V. Voinkov to V. Kebich," November 5, 1990, NARB 46/14/1332: 202; "Resolution of the Party-Economic Collective of Vykhovskii Region," July 26, 1990, and "On Consideration of the Collective Resolution," August 3, 1990, NARB 46/14/1322: 15–18, 11; and "Appeal."

61  Alla Yaroshinskaya, *Bosikom po bitomu steklu*, vol. 1 (Zhytomyr: Ruta, 2010), chap. 5.

62  "To the Presidium of People's Deputies, USSR," June 5, 1989, "V. M. Kavun, V. N. Yamchinskii to N. I Ryzhkov," June 27, 1989, and "To Comrade Shcherbina," June 19, 1989, GARF 5446/150/1624: 11, 51, 54–55.

63  For analysis, see Olga Kuchinskaya, *The Politics of Invisibility: Public Knowledge about Radiation Health Effects after Chernobyl* (Cambridge, MA: MIT Press, 2014), 1605–73 [Kindle]. For ideas on using the disaster to bring in foreign currency, see "Proposal of the Ministry of Health, USSR," July 11, 1989, TsDAVO 342/17/5089: 98–101. For Belarusian requests for foreign aid, see "Accounts Open for Chernobyl Clean-up Aid," August 22, 1989, TASS, GPA 1625.

64  "To the Council of Ministers, USSR," July 26, 1989, GARF 5446/150/1624: 5–8. See also "On the Rationale of Future Residency in Particular Villages," June 1, 1989, TsDAVO 342/17/5089: 76–77.

65  "On the State Project," October 19, 1989, TsDAVO 342/17/5089: 185; and "On the Necessity of Eradicating the Accident Thresholds," November 30, 1989, TsDAVO 324/17/5091: 85–87. See also David R. Marples, *Belarus: From Soviet Rule to Nuclear Catastrophe* (New York: St. Martin's Press, 1996), 45, 89–91.

66  V. K. Vrulevs'kii, *Vladimir Shcherbytsky: pravda i vymysli* (Kyiv: Dovira, 1993): 220, 245; P. Tron'ko, "V. V. Shcherbytsky (1918–1990)," and Iu. I. Shapoval, "V. V. Shcherbytsky: Osoba polityka stred obstavin chasu," *Ukraiins'kyi istorychnyi zhurnal*, no. 1 (2003): 109–17, 118–29.

67  Olha Martynyuk interview with Anatolii Artemenko, December 22, 2016, Kyiv.

68  "Informational Communiqué," January 26, 1987, SBU 16/1/1247: 74–76.

69 "On Protest among Employees of the Academy of Sciences," November 29, 1989, SBU 16/1/1279: 174–76.

70 Author interview with Alla Yaroshinskaya, May 27, 2016, Moscow.

## Send for the Cavalry

1 Pierre Pellerin and Dan Beninson. For reports on their Chernobyl assessments, see Stuart Diamond, "Chernobyl's Toll in Future at Issue," *New York Times*, August 28, 1986; and Judith Miller, "Trying to Quell a Furor, France Forms a Panel on Chernobyl," *New York Times*, May 14, 1986.

2 "M. Titov to Tsk Kompartii BSSR," June 28, 1989, NARB 4R/156/627: 120; and "From Women of Belarus!," no earlier than March 30, 1990, RGAE 4372/67/9743: 130–33. On other requests for experts, see "No. 7061/2," July 18, 1989, TsDAVO 342/17/5089: 114.

3 "V. Valuev, Chair KPB BSSR, to TsK KP Belorussia," June 22, 1989, NARB 4R/156/627: 114.

4 "V. Mar'in to E. V. Kachalovskii," July 14, 1989, TsDAVO 342/17/5089: 110; and "On the Necessity of Eradicating the Accident Thresholds," November 30, 1989, TsDAVO 324/17/5091: 85–87.

5 "Assessment on Visit of the Group of WHO Experts, June 19–25," June 1989, RGAE 4372/67/9743: 437–40. The comments were reprinted in "The International Chernobyl Project: An Overview and Assessment of Radiological Consequences and Evaluation of Protective Measures" (Vienna: IAEA, 1990).

6 As quoted in Yaroshinskaya, *Boll'shaia lozh'*, 209–11.

7 A. V. Stepanenko, "Toward a Concept of Living on Territory Contaminated with Radionuclides," January 27, 1990, RGAE 4372/67/9743: 11–22.

8 "V. Doguzhiev to Gomel' oblispolkom," July 11, 1989, GAGO 1174/8/2215: 29–32.

9 "V. P. Platonov to N. I. Ryzhkov," July 27, 1989, NARB 4R/156/627: 144–45.

10 Yaroshinskaya, *Boll'shaia lozh'*, 209–11.

11 Author telephone interview with Fred Mettler, January 7, 2016.

12 Tchertkoff, *The Crime of Chernobyl*, 71–72.

13 "Resolution of the First All-Union Conference on Radiobiology," Moscow, August 21–27, 1989, RGAE 4372/67/9743: 399–403; "Proposal of the Academy of Sciences, UkSSR toward a Concept for Safe Living in the Regions," September 6, 1989, RGAE 4372/67/9743: 1–3; and "To the Soviet Ministers USSR," July 26, 1989, GARF 5446/150/1624: 5–8. For objections from the Institute of Radiation Medicine on grounds of genetic risk, see "V. G. Andreev to O. O. Bobyliova," August 31, 1989, TsDAVO 342/17/5089: 164–65.

14 Brodie, "Radiation Secrecy and Censorship after Hiroshima and Nagasaki," 842–964.

15 "Protocol of the Meeting of Scientists and Specialists of Ukraine SSR and Belarus SSR," January 16, 1990, Moscow, RGAE 4372/67/9743: 361–66.

16 "To the General Director, UNESCO, Federico Maior," October 4, 1990, UNE-SCO, 361.9(470) SC ENV/596/534.1; Waight to Tarkowski, "Advisory Group to the Soviet Union," January 9, 1990, World Health Organization Archive (WHO) E16-522-6, jacket 1; "Duty Travel Report," P. J. Waight, IAEA Advisory Group 676, Vienna, December 10–15, 1989, WHO E16-522-6, jacket 1; "To the State Expert Commission," February 10, 1990, GARF, 3/59/19: 127; "Note on the Secretary General's Meeting with Mr. Lev Maksimov," March 7, 1990, UN NY, 1046/14/4, acc. 2001/0001; and "Economic and Social Council," July 9, 1990, UNDRO UN NY, S 1046/14/4.

17 Author telephone interview with Valentina Drozd, February 16, 2016.

18 "U. P. Spizhenko to E. I. Chazov," November 30, 1989, TsDAVO 342/17/5091: 85–88.

19 T. I. Gombosi et al., "Anthropogenic Space Weather," *Space Science Reviews* 212, no. 3–4 (November 2017): 985–1039; and R. C. Baker et al., "Magnetic Disturbance from a High-Altitude Nuclear Explosion," *Journal of Geophysical Research* 67, no. 12 (1962): 4927–28.

## Marie Curie's Fingerprint

1 On antinuclear protests, see Marples, *Ukraine under Perestroika*, 115.

2 Author interview with Alexander Kupny, June 14, 2014, Slavutych, Ukraine.

3 Kyle Hill, "Chernobyl's Hot Mess, 'the Elephant's Foot,' Is Still Lethal—Facts So Romantic," *Nautilus*, December 4, 2013, accessed April 29, 2015, http://nautil .us/blog/chernobyls-hot-mess-the-elephants-foot-is-still-lethal.

4 Aleksander Kupny, *Zhivy poka nas pomniat* (Kharkov: Kupny, 2011), 83.

5 Spencer R. Weart, *The Rise of Nuclear Fear* (Cambridge, MA: Harvard University Press, 2012), 81; and Brodie, "Radiation Secrecy and Censorship after Hiroshima and Nagasaki," 847.

6 Weart, *The Rise of Nuclear Fear*, 82.

7 Craig Nelson, "The Energy of a Bright Tomorrow: The Rise of Nuclear Power in Japan," *Origins: Current Events in Historical Perspective* 4, no. 9 (June 2011), accessed April 26, 2015, http://origins.osu.edu/article/energy-bright-tomorrow -rise-nuclear-power-japan.

8 For an eyewitness account of the dangers of the first-strike system, see Daniel Ellsberg, *The Doomsday Machine: Confessions of a Nuclear War Planner* (New York: Bloomsbury, 2017).

9 Angela N. H. Creager, "Radiation, Cancer, and Mutation in the Atomic Age," *Historical Studies in the Natural Sciences* 45, no. 1 (February 2015): 22–23.

10 Barak Kushner, "Gojira as Japan's First Postwar Media Event," in *Godzillas Footsteps: Japanese Pop Culture Icons on the Global Stage*, ed. William M. Tsutsui and Michiko Ito (New York: Macmillan, 2006), 42–50.

11 David Holloway, "The Soviet Union and the Creation of the International

Atomic Energy Agency," and Elisabeth Roehrlich, "The Cold War, the Developing World, and the Creation of the International Atomic Energy Agency (IAEA), 1953–1957," *Cold War History* 16, no. 2 (May 2016): 177–93, 195–212.

12 Carolyn Kopp, "The Origins of the American Scientific Debate over Fallout Hazards," *Social Studies of Science* 9 (1979): 403–22.

13 Martha Smith-Norris, "The Eisenhower Administration and the Nuclear Test Ban Talks, 1958–1960," *Diplomatic History* 27, no. 4 (September 2003): 503–41.

14 Jacob Darwin Hamblin, "Exorcising Ghosts in the Age of Automation: United Nations Experts and Atoms for Peace," *Technology and Culture* 47, no. 4 (October 2006): 734–56; and Toshihiro Higuchi, "Atmospheric Nuclear Weapons Testing and the Debate on Risk Knowledge in Cold War America, 1945–1963," in *Environmental Histories of the Cold War*, ed. J. R. McNeill and Corinna R. Unger (New York: Cambridge University Press, 2010), 301–23.

15 Hamblin, "Exorcising Ghosts."

16 Susan Schuppli, "Dirty Pictures: Toxic Ecologies as Extreme Images," Radioactive Ecologies Conference, Montreal, Canada, March 15, 2015.

17 Usually this work is not recognized as "nuclear." Gabrielle Hecht, *Being Nuclear: Africans and the Global Uranium Trade* (Cambridge, MA: MIT Press, 2012).

18 Eoin O'Carroll, "Marie Curie: Why Her Papers Are Still Radioactive," *Christian Science Monitor*, November 7, 2011.

19 Jan Beyea, "The Scientific Jigsaw Puzzle: Fitting the Pieces of the Low-Level Radiation Debate," *Bulletin of the Atomic Scientists* 68, no. 3 (2012): 13–28.

20 "To the Council of Ministers, SSSR," 1989, TsDAVO 342/17/5089: 111.

## Foreign Experts

1 Waight to V. A. Ivasutin, President, Bryansk Regional Council of Trade Union, July 26, 1990, WHO E16-445-11: 1; "Information on the Realization of the UN Program," 1991, NARB 507/1/5: 131–41; Waight to Tarkowski, "Advisory Group to the Soviet Union," January 9, 1990, WHO E16-522-6, jacket 1; "Briefing for Dr. Nakajima's Meeting with Dr. Chazov, Soviet Ministry of Health," March 1990, WHO E16-445-11, #1; "To the State Expert Commission of Gosplan SSSR on Chernobyl," February 10, 1990, GARF, 3/59/19: 127; and "Notes on the Chef de Cabinet's Meeting with the Permanent Representative of the Ukrainian Soviet Socialist Republic," April 12, 1990; "Notes on the Secretary-General's Meeting with the Deputy Prime Minister Kostiantyn Masyk," April 18, 1990, UN NY, 1046/14/4, acc. 2001/0001; and "Working Plan," October 17, 1990, NARB 507/1/1: 75–87; John Willis to Doug Mulhall, David McTaggart, August 14, 1990, GPA 1625.

2 "The UNESCO Chernobyl Project," June 30, 1990, UNESCO 361.9 (470), UNESCO Archive, Paris; and Michel Hansenne, ILO, Geneva, to Nakajima, January 15, 1991, WHO E16-445-11, #3.

3  "Memorandum of Understanding between the Ministry of Health of the USSR and the World Health Organization on Efforts to Mitigate the Health Consequences of the Chernobyl Accident," April 30, 1986 [*sic*], 1990, WHO E16-445-11, #2. For the range of proposals, see "Burton Bennet to Dr. M. K. Tolba," UNEP, October 22, 1990, UNSCEAR Correspondence Files.

4  Silini to C. Herzog, Director, Division of External Relations, IAEA, October 29, 1986, Correspondence Files, 1986, UNSCEAR Archive.

5  The IAEA Board of Governors minutes for 1989 read: "The experts generally supported a specified lifetime dose of 350 mSv resulting for the Chernobyl accident of the exposed inidividuals in the critical groups in the USSR" in "The Annual Report for 1989," Board of Governors, April 27, 1990, BOG IAEA, Box 33047.

6  From Enrique ter Horst, Assistant Secretary-General, ODG/DIEC, to Virendra Daya, Chef de Cabinet, EOSG, April 16, 1990, United Nations Archive, New York, S-1046 box 14, file 4, acc. 2001/0001. See also on the IAEA's "credibility gap," Bruno Lefèvre, "Mission Report—USSR," July 22–31, 1990, UNESCO, 361.9(470) SC ENV/596/534.1.

7  Author interview with Abel Gonzalez, June 3, 2016, Vienna. "Economic and Social Council Appeals for Co-operation and Aid to Mitigate Consequences of Chernobyl Nuclear Accident," United Nations Database, July 13, 1990, ECOSOC/5254. For a statement by WHO director that IAEA was in charge of the assessment, see *Nuclear Truth*, dir. Vladimir Tchertkoff (Feldat Film, Switzerland, 2004).

8  "Travel Report," May 28, 1990, WHO E16-445-11, 1; "Report of a Special Meeting of the Inter-Agency Committee on Response to Nuclear Accidents (IAC/RNA), May 29, 1990, WHO E16-445-11, #1. The American delegate sought to use this committee as a clearinghouse to approve Chernobyl programs carried out by UN agencies; "Economic and Social Council Appeals for Co-operation."

9  On postponement of a large proposed UNESCO Chernobyl program to follow the IAEA's lead, see "Proposal UNESCO Chernobyl Project," May 10, 1990; B. Andemicael, IAEA, to H. Blix, "Consequences of Chernobyl Accident: ECOSOC Decision," May 18, 1990; ADG/SC to the Director-General, May 17, 1990, UNESCO, 361.9(470) SC ENV/596/534.1, Part I; and ECOSOC/5254 United Nations Database, July 13, 1990.

10  IAEA Board of Governors, 744th meeting, February 27, 1991, International Atomic Energy Agency (IAEA) Archive, Box 33007.

11  "Third Meeting of the Inter-Agency Task Force on Chernobyl," September 19–23, 1991, WHO E16-445-11, no. 5.

12  Itsuzo Shigematsu to Dr. P. Waight, PEP, EHE, WHO, September 25, 1990; and Waight to Shigematsu, October 4, 1990, WHO E16-445-11, #2.

13  "On Questions of Ukraine's Economic Relations with the West," August 10, 1990, SBU 16/1/1288: 130–37.

14  "Informational Notes," May 14, 1987, SBU 16/1/1250: 42–44.

15  "To Comrade V. Kh. Doguzhiev from A. Kondrusev and V. Vozniak," March 5, 1990, TsDAVO 342/17/5220: 10–11.

16  "On Subversion Activity Abroad," February 15, 1990, "On Preparations for the
     Student Strike," February 15, 1990, "On Tendencies in the Development of Neg-
     ative Occurrences," February 2, 1990, "On the Development of the Situation
     in Lviv Province," March 13, 1990, "On Activization of Subversive Activity by
     Radio Free Europe," April 19, 1990, "On the Gathering of Spilka of Independent
     Ukrainian Youth," April 23, 1990, "On Processes Related to Construction and
     Exploitation of NPP," April 29, 1990, SBU 16/1/1284: 83–85, 90–91, 96–97, 128–
     31, 173–74, 176–178, 198–200; "On Circumstances in Kyiv," September 27, 1990,
     "On Measures to Control Events in Kyiv," September 29, 1990, SBU 16/11288:
     215–16, 221–22.

17  "On Problems Connected with Trips of Soviet Scientists Abroad," November 30,
     1990, SBU 16/1/1289: 85–89.

18  "On Attempts by the Protagonist to Use the New Moments and International
     Relations for Subversion," November 5, 1988, SBU 16/1/1266: 215–18.

19  "On International Collaboration," April 29, 1990, SBU 16/1/1274: 194–96. For
     earlier orders to block foreign espionage, see "Assessments by Foreign Special-
     ists of the Circumstances in the Republic," September 19, 1990, SBU 16/1/1288:
     180–84.

20  "On International Collaboration," April 29, 1990, SBU 16/1/1274: 194–96.

21  "On International Collaboration," April 29, 1990, SBU 16/1/1274: 194–96.

## In Search of Catastrophe

 1  "Travel Report," May 28, 1990, WHO E16-445-11: 1.

 2  Anspaugh and Bennett, "Unsolved Problems," July 23, 1990, GAGO,
     1174/8/2346: 103–4.

 3  Author interview with Valentina Drozd, June 23, 2016, Minsk; and Andersen to
     Sawyer, August 12, 1991, GPA 1803.

 4  B. G. Bennett, "Mission Report," no earlier than November 1990, Correspon-
     dence Files, 1990, UNSCEAR Archive.

 5  "Unsolved Problems," GAGO, 1174/8/2336: 103–4.

 6  Author interview with Abel Gonzalez, June 3, 2016, Vienna.

 7  Email correspondence with A. I. Vorobiev, October 8, 2017.

 8  "On Circumstances Emerging in Narodychy Region," June 1989, GARF
     5446/150/1624: 12–40.

 9  B. G. Bennett, "Mission Report," and appendix: "11/3: Institute of Experimental
     Meteorology, Obninsk, Visited by Teams 1 and 2 on 15 August 1990, Report by E.
     Wirth, Fed Rep. of Germany."

10  "The IAEA Project for a Repeat Assessment of the Situation," October 15, 1990,
     NARB 507/1/1: 33–50.

11  The decree to start the registry: "On Establishing an All-Union Scientific Cen-
     ter," September 9, 1986, NARB, 46/14/1261: 87.

12  "Travel Report," May 28, 1990, WHO E16-445-11: 1; "Unsolved Problems"; "List of Questions."

13  *Trud*, July 12, 1990

14  John Willis to Doug Mulhall, David McTaggart, August 14, 1990, GPA 1625.

15  Pavel Vorobiev, "Do i posle Chernobylia," *Nezavisimaia*, April 28, 2006.

16  Email correspondence with A. I. Vorobiev, October 8, 2017.

17  "Report of the Economic and Social Council," October 29, 1990, UN NY S-1046/14/4, acc. 2001/0001.

18  "Soviet Scientists Report," Bennett, "Mission Report," no earlier than November 1990, Correspondence Files, 1990, UNSCEAR Archive: 48.

19  B. G. Bennett, "Background Information for UNEP Representative to the Meeting of the Ministerial Committee for Coordination on Chernobyl," November 17, 1993, Correspondence Files, 1993, UNSCEAR.

20  "On Information," May 24, 1990, TsDAVO, 342/17/5220: 22–23; and L. N. Astakhova et al., "Particularities of the Formation of Thyroid Pathologies among Children Exposed to Radioactivity," Institute of Radiation Medicine, Report No. 91/763E, November 19, 1991, WHO E16-445-11, No. 6; and F. Fry, "Mission Report," no earlier than November 1990, Correspondence Files, 1990, UNSCEAR: 3.

21  "Bennett to Tolba," October 22, 1990, Correspondence Files, 1990, UNSCEAR Archive; "Commentary on the Existing Norms of Radioactivity for Food Products," July 20, 1989, TsDAVO 342/17/5089: 88–90.

22  "On the Directives of SM UkSSR no. 5182/86," May 20, 1989, TsDAVO 342/17/5089: 67–69; "V A. M. Kas'ianenko, 04-r-16 DSP," June 7, 1989, TsDAVO 342/17/5089: 68; "Chernobyl Humanitarian Assistance and Rehabilitation Programme," May 21, 1993, WHO E16-180-4: 11.

23  Author telephone interview with Jacques Repussard, February 28, 2018; and "Andre Bouville, UNSCEAR, to Chester Richmond," Oak Ridge National Laboratory, Correspondence Files, 1987, UNSCEAR.

24  Martha Smith-Norris, *Domination and Resistance: The United States and the Marshall Islands during the Cold War* (Honolulu: University of Hawaii Press, 2016), 83–90.

25  On uncertainties of mathematical models for dose estimations, see Owen Hoffman et al., "The Hanford Thyroid Disease Study," *Health Physics* 92, no. 2 (February 2007): 99–112.

26  "Radioactive Situation and Residential Conditions," Summer 1990, GAGO, 1174/8/2336: 43–45.

27  "Report of the Economic and Social Council," October 29, 1990, UN NY S-1046/14/4, acc. 2001/0001.

28  F. Fry, "Mission Report"; and Report by an International Advisory Committee, *The International Chernobyl Project, Technical Report* (Vienna: IAEA, 1991), 225–26.

29  Report by an International Advisory Committee, *Technical Report*, 236. For comparison, see two-year doses Belarusians provided, V. S. Ulashchik, "Some Medical Aspects of the Consequences of the Accident at Chernobyl, Based on Byelorussian Data," John Willis to Doug Mulhall, David McTaggart, August 14, 1990, GPA 1625.

30  "On Problems Liquidating the Chernobyl Catastrophe," January 22, 1991, SBU 16/1/1292: 59–63.

31  "Cesium 134-137 Doses among Residents of Cherykaw Region in 1986," Arkhiv Kricheva 588/1/160: 7; and V. A. Shevchenko, Institute of General Genetics, Moscow, "Biological and Genetic Consequences of Nuclear Explosions," 996 GPA; "Draft for Chernobyl Report," November 30, 1990, and Bennett, "Mission Report," *Technical Report*, 212–26. On acknowledgment that dose estimates were too conservative, see "Chernobyl: Local Doses and Effects," Document R. 554, Conference Room Papers, 1994, UNSCEAR.

32  "Draft for Chernobyl Report," November 30, 1990, and Bennett, "Mission Report," no earlier than November 1990, UNSCEAR Correspondence Files, 1990; and Report by an International Advisory Committee, *Technical Report*, 212–26.

33  On the production of uncertainty and dose estimates, see Scott Frickel, "Not Here and Everywhere: The Non-Production of Scientific Knowledge," in *Routledge Handbook of Science, Technology and Society* (New York: Routledge, 2014), 263–76; and Scott Frickel, et al., "Undone Science: Charting Social Movement and Civil Society Challenges to Research Agenda Setting," *Science, Technology, and Human Values* 35, no. 4 (July 1, 2010): 444–73.

34  "Progress Report," March–September 1994, Chernobyl Studies Project, Working Group 7.0, DOE, UCRL-ID-110062-94-6, attachment H.

35  Author telephone interview with Fred Mettler, January 7, 2016.

36  Report by an International Advisory Committee, *Technical Report*, 281–84.

37  "Report to the General Assembly," Conference Room Papers, 1995, UNSCEAR.

38  "Working Group, Document R. 541, Epidemiological Studies of Radiation Carcinogenesis," 1994, Conference Room Papers, UNSCEAR.

39  V. P. Platonov and E. F. Konoplia, "Information on Major Findings of Scientific Work Related to Liquidating Consequences of ChAES Accident," April 21, 1989, RGAE 4372/67/9743: 490–571.

40  F. Fry, "Mission Report," 3.

41  "Duty Travel Report," P. J. Waight, IAEA Advisory Group 676, Vienna, December 10–15, 1989, WHO E16-522-6, jacket 1; and F. Fry, "Mission Report."

42  Greenpeace staff received a copy of the Belarusian report: John Willis to Doug Mulhall, David McTaggart, August 14, 1990, GPA 1625.

43  Author telephone interview with Fred Mettler, January 7, 2016; and "International Chernobyl Project Proceedings of an International Conference," Vienna, May 21–24, 1991, 34, 39.

44  On the scale of low-dose epidemiology, see Conference Room Papers, "Working

Group, Document R. 541, Epidemiological Studies of Radiation Carcinogenesis," 1994, UNSCEAR Archive, Vienna.

45  "International Conference Completes Review of Chernobyl Study," May 24, 1991, UN Archives S-1046/16/3.

46  UN Press Release, "Advisory Body Finds No Health Disorders Directly Attributable to Radiation Exposure in Populations Affected by Chernobyl Accident," May 24, 1991, UN Archives S-1046/16/3.

47  International Chernobyl Project Proceedings, 38.

48  "International Conference Completes Review of Chernobyl Study," May 24, 1991, UN Archives S-1046/16/3.

49  *International Chernobyl Project: An Overview* (Vienna: International Atomic Energy Association, 1991), 32.

## Thyroid Cancer: The Canary in the Medical Mine

1   "Press Conference by Ms. Anstee on Chernobyl Pledging Conference, 20 September," September 19, 1991, UNA S-1046/16/3/2001/0002.

2   "Minister P. Kravchenko to V. F. Kebich," April 3, 1991, NARB 507/2/13: 31–37; and "Resolution Adopted by the General Assembly, 45th Session, Agenda Item 12," December 21, 1990, WHO E16-445-11, No. 3.

3   "International Chernobyl Project Proceedings of an International Conference," Vienna, May 21–24, 1991, 23, 34–38, 40.

4   "Medical Aspects of the Chernobyl Accident," *Proceedings*, Kyiv, May 11–13, 1988 (Vienna: IAEA, 1989), 21–24; L. N. Astakhova et al., "Particularities of the Formation of Thyroid Pathologies among Children Exposed to Radioactivity," Institute of Radiation Medicine, Report No. 91/763E, November 19, 1991, WHO E16-445-11: 6.

5   "Medical Aspects of the Chernobyl Accident."

6   Angela Liberatore, *The Management of Uncertainty: Learning from Chernobyl* (Amsterdam: Gordon and Breach, 1999), 3.

7   Toshihiro Higuchi, "Atmospheric Nuclear Weapons Testing and the Debate on Risk Knowledge in Cold War America, 1945–1963," in *Environmental Histories of the Cold War*, ed. J. R. McNeill and Corinna R. Unger (New York: Cambridge University Press, 2010), 301–23. On the proliferation of simplified computer simulations and the simplifications that follow, see J. R. Ravetz, "'Climategate' and the Maturing of Post-Normal Science," *Futures* 43 (2011): 149–57.

8   See testimony of Konoplya, Hotovchits, Bar'yakhtar, and Dushutin, "Proceedings."

9   Report by an International Advisory Committee, *The International Chernobyl Project, Technical Report* (Vienna: IAEA, 1991), 388.

10  Rosen, "Medical Aspects of the Chernobyl Accident," 61.

11  Marples reports the first jump in thyroid cancers in Ukraine in 1986. See Marples, *Belarus: From Soviet Rule to Nuclear Catastrophe*, 104. For an early report of tumors, see "Check of the State of Endocrinology Services," March 23, 1988,

NARB 46/12/1262: 83–85; and "On Deputies' Group Questions," June 9, 1989, TsDAVO 324/17/5089: 38–42. The Ukrainian journal *Zdorov'e* published news of Chernobyl health problems including the growth of thyroid cancer in April 1989. WHO officials learned of the cancers at a 1990 conference, Riaboukine, Travel Report, February 12, 1991, WHO No. 16-441-4; and Rodzilsky to Prilipko, November 30, 1991, WHO E16-445-11: 5.

12  Andrej Lyshchik et al., "Diagnosis of Thyroid Cancer in Children," *Radiology* 235, no. 2 (May 2005).

13  Author interview with Valentina Drozd, June 15, 2016, Minsk.

14  Author interview with Valentina Drozd, June 15, 2016, Minsk. L. N. Astakhova, E. P. Demidchuk, E V Davydov, A.N. Arinchin, S. M. Zelenko, V. M. Drozd, T. D. Poliakova, "Health Status of Belarusian Children and Adolescents Exposed to Radiation as Consequence of the Chernobyl AES accident," Vestnik Academia meditsinskikh nauk, SSSR, Vol 11, no. 11 (1991): 25-27.

15  Author telephone interview with Valentina Drozd, February 12, 2016.

16  They often referred to a small Swedish study of adults treated with iodine-131 that had shown no rise in cancers. Keith Baverstock, "The Recognition of Childhood Thyroid Cancer as a Consequence of the Chernobyl Accident: An Allegorical Tale of Our Time?" *Journal of the Royal Society of Medicine* 100 (2007): 1–3.

17  V. N. Novoselov and V. S. Tolstikov, *Atomnyi sled na Urale* (Chelyabinsk: Rifei, 1997), 127. For Ilyin's far more modest predictions of thyroid cancer, see L. A. Ilyin et al., "Radiocontamination Patterns and Possible Health Consequences of the Accident at the Chernobyl Nuclear Power Station," *Journal of Radiological Protection* 10, no. 3 (1990): 3–29.

18  Smith-Norris, *Domination and Resistance*, 72, 89–93.

19  Barbara Rose Johnston and Holly M. Barker, *Consequential Damages of Nuclear War: The Rongelap Report* (Walnut Creek, CA: Left Coast Press, 2008), 173–91.

20  Wilfred Goding, High Commissioner of the Pacific Islands, UN Trusteeship Council, June 13, 1961, as quoted in Smith-Norris, *Domination and Resistance*, 90.

21  "Minutes NCI Thyroid 131 Assessments Committee," August 24–25, 1987, NCI, RG 43 FY 03 Box 5, part 3; *Cancer in Utah: Report No. 3, 1967–77: Utah Cancer Registry*, September 1979, acc. no. 0331726, Nuclear Testing Archive (NTA); and Joseph L. Lyon et al., "Childhood Leukemias Associated with Fallout from Nuclear Testing," *New England Journal of Medicine*, no. 300 (1979): 397–402.

22  Howard Ball, *Justice Downwind: America's Atomic Testing Program in the 1950s* (New York: Oxford University Press, 1986), 158–72; and Leopold, *Under the Radar*, 169–80.

23  Wise International, "Fallout study mishandled; scientists' past raises questions," Nuclear Monitor Issue, no. 498, September 25, 1998.

24  Author telephone interview with Joseph Lyon, March 7, 2018; and statement of Joseph Lynn Lyon, M.D., M.P.H., Professor, on the National Cancer Institute's Management of Radiation Studies before the Committee on Governmental Affairs, United States Senate, One Hundred Fifth Congress, October 1, 1997.

25  "Testimony of Joseph L. Lyon," October 23, 1981, no. 0067276, NTA.

26  Author telephone interview with Lynn Anspaugh, January 12, 2016; and author email correspondence with Owen Hoffman, March 11, 2018.

27  Kathleen M. Tucker, Robert Alvarez, "Trinity: "The most significant hazard of the entire Manhattan Project" Bulletin of Atomic Scientists, July 15, 2019.

28  Letter from Dr. Richard D. Klausner, "On the National Cancer Institute's Management of Radiation Studies," United States Senate: 58.

29  Author email correspondence with Owen Hoffman. See also O. Hoffman et al., "Dose Assessment Communication," Health Physics (2011): 591–601; and O. Hoffman et al., "A Perspective on Public Concerns about Exposure to Fallout from the Production and Testing of Nuclear Weapons," 736–49.

30  S. Simon, A. Bouville, and C. Land, "Fallout from Nuclear Weapons Tests and Cancer Risks," American Scientist 94, no. 1 (January 2006): 48–57.

31  Rob Hotakainen, Lindsay Wise, Frank Matt, and Samantha Ehlinger, "The Hidden Legacy of 70 Years of Atomic Weaponry," McClatchy Report, December 11, 2015. For a study showing children with better nutrition taking in more radioactive iodine from nuclear tests, see Merril Eisenbud et al., "Iodine-131 Dose from Soviet Nuclear Tests," Science 136, no. 3514 (May 4, 1962): 370–74.

32  Iodine releases of 45 million curies from Chernobyl, 145 million from the National Testing Service. Testimony of F. Owen Hoffman, Ph.D., Oak Ridge, on the National Cancer Institute's Management of Radiation Studies Before the Committee on Governmental Affairs, United States Senate, One Hundred Fifth Congress, September 16, 1998.

33  Author email correspondence with Owen Hoffman, March 11, 2018; Statement of Jan Beyea, Exposure of the American People to Iodine-131 from Nevada Nuclear-Bomb Tests: Review of the National Cancer Institute Report and Public Health Implications, Institute of Medicine and National Research Council (Washington, DC: National Academy Press, 1999). For failed attempts to shift cows to aged feed in Minnesota during high fallout periods, see Kendra Smith-Howard, Pure and Modern Milk: An Environmental History Since 1900 (Oxford: Oxford University Press, 2017), 133–34.

34  David Philipps, "Troops Who Cleaned Up Radioactive Islands Can't Get Medical Care," New York Times, January 28, 2017.

35  See Tom Foulds Collection, University of Washington Special Collections (TFC UWSC); and A. Körblein, "Perinatal Mortality in West Germany Following Atmospheric Nuclear Weapons Tests," Archives of Environmental Health 59, no. 11 (November 2004): 604–9.

36  See, for example, "Testimony of Dr. Rosalie Bertell," U.S. Senate Committee on Veterans' Affairs, April 21, 1998.

37  Department of Energy, Richland Operations Office, "Requests for Assistance Resulting from Chernobyl," June 1, 1986, WH 423500, TFC UWSC.

38  "Chernobyl Implications Report," June 12, 1987, RG 431-01/1358, box 2, National Archives and Records Administration (NARA).

39 "Progress Report," October 1993–January 1994, Chernobyl Studies Project, Working Group 7.0, DOE Opennet, UCRL-ID-110062-94-4.

40 "More Chernobyls Unavoidable," *Los Angeles Times*, June 27, 1987.

41 Quote from "Report of the Economic and Social Council," October 29, 1990, UN NY S-1046/14/4, acc. 2001/0001.

42 Hoffman et. al., "A Perspective on Public Concerns about Exposure to Fallout from the Production and Testing of Nuclear Weapons," 744; Gary J. Hancock et al., "The Release and Persistence of Radioactive Anthropogenic Nuclides," in *A Stratigraphical Basis for the Anthropocene*, ed. C. N. Waters (London: Geological Society, 2014), 265–81; John R. Cooper, Keith Randle, and Ranjeet S. Sokhi, *Radioactive Releases in the Environment: Impact and Assessment* (New York: Wiley, 2003), 17; Hoffman et al., "A Perspective on Public Concerns about Exposure to Fallout from the Production and Testing of Nuclear Weapons," 736–49; Australian Mission to the United Nations, November 26, 1974, and R. H. Wyndham to Dr. S. Sella, November 15, 1973, UN NY S-0446-0106-09; and "Study of the Radiological Situation at the Atolls of Mururoa and Fangataufa," IAEA Board of Governors, Technical Cooperation Report for 1997, April 30, 1998, IAEA BOG, Box 33054. "Report of the United Nations Scientific Committee on the Effect of Atomic Radiation to the General Assembly," 2000, accessed April 29, 2018, http://www.unscear.org/docs/reports/gareport.pdf. On equivalency with Hiroshima bombs, see "General Overview of the Effects of Nuclear Testing," CTBTO, accessed May 29, 2018, https://www.ctbto.org/nuclear-testing/the-effects-of-nuclear-testing/general-overview-of-theeffects-of-nuclear-testing/.

43 Testimony of Owen Hoffman, "National Cancer Institute's Management of Radiation Studies," Hearing before the Permanent Subcommittee on Investigations, U.S. Senate, September 16, 1998, 48.

44 Coline N. Waters et al., "Can Nuclear Weapons Fallout Mark the Beginning of the Anthropocene Epoch?" *Bulletin of the Atomic Scientists* 71, no. 3 (May 2015): 46–59.

45 "On the Work of BSSR Health Ministry, 1986–1989," no earlier than March 1989, NARB 46/14/1260: 1–15.

46 Eckart Conze, Martin Klimke, and Jeremy Varon, *Nuclear Threats, Nuclear Fear and the Cold War of the 1980s* (Cambridge: Cambridge University Press, 2017), and Andrew S. Tompkins, *Better Active Than Radioactive!: Anti-Nuclear Protest in 1970s France and West Germany* (New York: Oxford University Press, 2016): 196–98.

## The Butterfly Effect

1 I. Riaboukhine, "International Program on the Health Effects of the Chernobyl Accident," December 13, 1991, WHO E16-445-11: 6.

2 Matiukhin to Riaboukhine, November 19, 1991, WHO E16-180-4, 6; and Ozolins, "On Meeting at Neuherberg," Munich, January 15, 1991 [*sic*], January 8, 1992, WHO E16-180-4, 10.

3 Astakhova et al., "Particularities of the Formation of Thyroid Pathologies among Children Exposed to Radioactivity from the Chernobyl Accident," 6. The doses in this report are similar to those of the long-term NCI study. See Hatch et al. 2005.

4 Doctors detected cystic nodules in half as many children from the control group as from the case group of Khoiniki. Valentina Drozd et al., "Systematic Ultrasound Screening as a Significant Tool for Early Detection of Thyroid Carcinoma in Belarus," *Journal of Pediatric Endocrinology and Metabolism* 15, no. 7 (2002): 979–84.

5 Author telephone interview with Keith Baverstock, November 3, 2015.

6 For first notifications, see "Duty Travel Report," Dr. P. J. Waight, IAEA Advisory Group 676, Vienna, December 10–15, 1989, WHO E16-522-6, jacket 1; and "WHO IPHECA Task Group, Obninsk, January 7–11, 1990," WHO E16-445-11: 3.

7 Waight, "Meeting at Neuherberg, Munich, January 15, 1991 [*sic* 1992]," and "Travel Report Summary, Waight," January 16, 1992, CEC Meeting on the Reports of an Excess of Thyroid Cancer in Belarus, GSF, Neuherberg, Germany, WHO E16-445-11: 7; and U. Riaboukhine Duty Travel Report, Germany, October 21–25, 1991, WHO E16-445-11: 6.

8 Astakhova et al., "Particularities of the Formation of Thyroid Pathologies among Children Exposed to Radioactivity from the Chernobyl Accident."

9 Waight, "Meeting at Neuherberg, Munich, January 15, 1991 [*sic* 1992]," WHO E16-445-11: 7; and J. Sinnaeve, "Radiation Protection Research to Waight," January 29, 1992, WHO E16-180-4: 10. Waight believed the problem was serious enough to advocate treatment programs in Belarus for thyroid cancer. See "Waight to Manager PEP," January 20, 1992, WHO E16-445-11: 7.

10 "On Reviewing the Implementation of IPHECA (Russian translation)," June 11, 1992, WHO E16-445-11: 9; and "Tarkowski to Napalkov," June 10, 1992, WHO E16-180-4: 10.

11 Author telephone interview with Keith Baverstock, November 3, 2015.

12 Author email correspondence with Keith Baverstock, December 21, 2015. Kreisel, when asked, said he would not discuss personnel or political issues but did add that he thought "a lot of problems could have been avoided had this person [an un-named person disloyal to the WHO] been released from WHO." Author telephone interview with Wilfred Kreisel, July 23, 2018.

13 Baverstock et al., "Thyroid Cancer in Children in Belarus after Chernobyl," and Sinnaeve, CEC, to V. S. Kazakov, Minister of Public Health, Belarus, September 8, 1992, WHO E16-445-11, 11. "Project Document: Thyroid Cancer in Belarus after Chernobyl," January 25, 1993, WHO E16-445-11: 18.

14 Keith Baverstock et al., "Thyroid Cancer after Chernobyl," *Nature* 359 (September 3, 1992).

15 I. Shigematsu and J. W. Thiessen, "Letter to the Editor," *Nature* 359 (October 22, 1992): 680–81; V. Beral and G. Reeves, "Letter to the Editor," *Nature* 359 (October 22, 1992): 680–81; and E. Ron, J. Lubin, and A. B. Scheider, "Thyroid Cancer Incidence," *Nature* 360 (November 12, 1992): 113.

16 Fred A. Mettler et al., "Thyroid Nodules in the Population Living Around Chernobyl," *Journal of the American Medical Association* 268, no. 5 (August 5, 1992): 616–19.

17 Kreisel, Directory Division of Environment Health, to J. E. Asvall, Regional Directory, WHO, Europe, September 25, 1992, WHO E16-445-11: 11.

18 Email correspondence with Keith Baverstock, December 15, 2016.

19 Author telephone interview with Wilfred Kreisel, July 23, 2018.

20 "S. Tarkowski to Napalkov," June 10, 1992, WHO E16-445-11, 9; "Kreisel to Chikvaidze," Department of Humanitarian Affairs, UN NY, September 28, 1993, WHO E16-445-11: 20.

21 Email correspondence with Baverstock, December 15, 2015; Gonzalez to Napalkov, August 10, 1993, WHO E16-445-11: 19; and "Report by the Director-General, IPHECA," November 30, 1992, WHO E16-445-11: 13.

22 As UN officials wrote about various UN thyroid projects, "There appear to be few arrangements for collaboration on these studies." Strengthening of International Cooperation and Coordination of Efforts to Study, Mitigate and Minimize the Consequences of the Chernobyl Disaster, September 28, 1993, UN NY S-1082/35/6/ acc. 2001/0207. See the debate, for example, in Shigenobu Nagataki, ed., *Nagasaki Symposium on Chernobyl: Update and Future* (Amsterdam: Elsevier, 1994).

23 "Rhiaboukine Travel Report to USSR," August 23, 1991, WHO E16-441-4.

24 "Study of Thyroid Cancer and Other Thyroid Diseases Following the Chernobyl Accident (Belarus)," accessed January 15, 2016, http://chernobyl.cancer.gov/ thyroid_belarus.html. On complaints of slow pace, see Anspaugh, "Foreign Trip Report," in "Progress Report," March–September 1994, Chernobyl Studies Project, Working Group 7.0, DOE, UCRL-ID-110062-94-6. For the range of thyroid studies, see "Inventory of International Health Related Activities in the USSR on the Consequences of the Chernobyl Accident" (n.d. 1991), WHO E16-445-11, No. 5; and E. Cardis, B. K. Armstrong, and J. Esteve, International Agency for Research on Cancer, Lyon, "Opinion of the Conduct of Epidemiological Studies of the Consequences of the Chernobyl Accident Within the WHO IPHECA Project," WHO E16-445-11: 7.

25 "Riaboukhine Travel Report Summary to USSR," August 23, 1991, WHO E16-180-4: 5.

26 "Inventory of International Health Related Activities in the USSR on the Consequences of the Chernobyl Accident," August 23, 1991, WHO E16-445-11: 5.

27 On IAEA's failure to return with test results, see author telephone interview with Mona Dreicer, April 13, 2018; and "Martin A. Chepesiuk, MD FRSPC, to Steve Bloom," GPA 1624.

28 Waight, "Travel Report," Paris, March 12–13, 1991, and Obninsk, March 13–16, 1991, WHO E16-445-11: 3; and "Rhiaboukine Travel Report to USSR," August 23, 1991, WHO E16-441-4; "Harry Pettengill, Deputy Assistant Secretary, DOE, to Waight," May 18, 1992, WHO E16-180-4: 10; "Napalkov to Kondrusev, Deputy

Minister of Health, USSR," February 26, 1991, WHO E16-445-11: 3; and "Itsuzo Shigematsu to Nakajima," November 25, 1992, WHO E16-445-11: 13.

29 Urgent Telegram, J. E. Asvall, February 17, 1992, WHO E16-445-11: 7; "Asvall to Napalkov," February 19, 1992, WHO E16-180-4: 10; "S. Tarkowski, DEH to Napalkov, attn. Kreisel," August 27, 1992, WHO E16-445-11: 11; "Kreisel to Tarkowskii, EHE EURO," July 19, 1993, WHO E16-445-11: 19; and H. I. Alleger, Commission of the European Communities, to Kreisel, July 26, 1993, WHO E16-445-11: 19.

30 "Inter-Agency Coordination during Emergecy Situations," September 5, 1990, UNESCO Archive (Paris) 361.9(470) SC ENV/596/534.1.

31 Called the "IPHECA" study, "Memorandum of Understanding Between the Ministry of Health of the USSR and the World Health Organization on Efforts to Mitigate the Health Consequences of the Chernobyl Accident, April 30, 1986 [sic]," 1990, WHO E16-445-11: 2.

32 Author interview with Gonzalez. See also "Waight to Gonzalez," May 17, 1990, WHO E16-445-11: 1. For a long note objecting to the WHO IPHECA study, unsigned, see "Proposed WHO-Soviet International Program on the Health Effects of the Chernobyl Accident (IPHECA)," December 27, 1990, WHO E16-445-11: 3.

33 "Gonzalez to Napalkov," August 10, 1993, WHO E16-445-11: 19. For an earlier statement that the WHO IPHECA program was a continuation of the IAEA study, see "Gonzalez to Kreisel," June 3, 1992, WHO E16-445-11: 9.

34 "U. Riaboukhine Duty Travel Report, Germany," October 21–25, 1991, WHO E16-445-11: 6. For criticism of WHO language that replicated IAEA literature minimizing health impacts, see "Dr. I. Filyushikin, Institute of Biophysics, Moscow, to Kreisel," April 27, 1993, WHO E16-3445-11: 16.

35 "Reviewing the Implementation of IPHECA," June 11, 1992, WHO E16-445-11: 9; and "Napalkov to Gonzalez," September 27, 1993, WHO E16-445-11: 20.

36 "International Chernobyl Project Proceedings," 48; and "B. Weiss to A. Gonzalez," July 2, 1993, WHO E16-180-4: 11; UNESCO to MAB National Committees, November 5, 1991, WHO E16-180-4: 10.

37 Press Release: "IAEA Calls 'Unfounded' Der Spiegel Article," January 28, 1992, UN NY S-1082/27/9/ acc. 2001/0190; and B. G. Bennett, "Background Information for UNEP Representative," November 17, 1993, Correspondence Files, 1993, UNSCEAR Archive.

38 "Bennett to Secretary Hazel O'Leary," May 5, 1994, Correspondence Files, 1994, UNSCEAR Archive.

39 "Strengthening of International Cooperation and Coordination of Efforts to Study, Mitigate, and Minimize the Consequences of the Chernobyl Disaster: Report of the Secretary General," November 13, 1992, Department of Humanitarian Affairs DHA, UN NY s-1082/27/9, ac 2001/0190.

40 Ukrainian Ministry of Health Press Release: "Cooperation in Assessment of Medical Consequences of Chernobyl Catastrophe, January 1993," WHO E16-445-11: 14.

41  B. G. Bennett, "Background Information for UNEP Representative to the Meeting of the Ministerial Committee for Coordination on Chernobyl," November 17, 1993, New York, Correspondence Files, 1993, UNSCEAR Archive. Bennett repeated this sentiment a year later: "Comment on Summary of Funding Requirements for Chernobyl," September 7, 1994, Correspondence Files, 1994, UNSCEAR Archive.

42  "Strengthening of the Coordination of Humanitarian and Disaster Relief Assistance," November 13, 1992, UN NY S-1082/27/9/ acc. 2001/0190.

43  "Strengthening of International Cooperation and Coordination of Efforts to Study, Mitigate and Minimize the Consequences of the Chernobyl Disaster," October 27, 1997, and "Note to the Secretary-General, Chernobyl," October 29, 1997, UN NY S-1092/96/5, acc. 2006/0160.

44  "Note to the File, Chernobyl, Anstee," January 17, 1992, WHO E16-445-11: 7.

45  "Silini to Tolba Telegram," September 26, 1986, and "Silini to Katz," September 12, 1986, Correspondence Files, 1986, UNSCEAR Archive.

46  "Silini to Katz," September 12, 1986, Correspondence Files, 1986, UNSCEAR Archive; and "Silini to Tolba Telegram," September 26, 1986, Correspondence Files, 1986, UNSCEAR Archive.

47  "Bennett to Dr. M. D. Gwynn," UNEP, January 13, 1992, Correspondence Files, 1992, UNSCEAR Archive.

48  V. Arkhipov, "Nuclear Energy, Environment and Public Opinion," January 19, 1990, TsDAVO 2/15/1871, 42–50.

49  "Bennett to Cardis," September 5, 1991, Correspondence Files, 1991, UNSCEAR Archive.

50  "Burton Bennett to Hylton Smith, Scientific Secretary International Commission on Radiological Protection (ICRP)," October 11, 1991, Correspondence Files, 1991, UNSCEAR Archive.

51  Document R. 554, 1994, Conference Room Papers, UNSCEAR.

52  Document R. 556, June 20, 1996, Conference Room Papers, UNSCEAR.

53  Document R. 556, June 20, 1996, Conference Room Papers, UNSCEAR, 29.

54  Among many, see David Michaels, *Doubt Is Their Product: How Industry's Assault on Science Threatens Your Health* (Oxford: Oxford University Press, 2008); Allan M. Brandt, *The Cigarette Century: The Rise, Fall, and Deadly Persistence of the Product That Defined America* (New York: Basic Books, 2007); and Nancy Langston, *Toxic Bodies: Hormone Disruptors and the Legacy of DES* (New Haven, CT: Yale University Press, 2010).

55  Leopold, *Under the Radar*; Brown, *Plutopia*; Johnston and Barker, *Consequential Damages of Nuclear War*; Sarah Alisabeth Fox, *Downwind: A People's History of the Nuclear West* (Lincoln, NE: Bison Books, 2014); Gabrielle Hecht, *Being Nuclear: Africans and the Global Uranium Trade* (Cambridge, MA: MIT Press, 2012); and Jacob Darwin Hamblin, *Poison in the Well: Radioactive Waste in the Oceans at the Dawn of the Nuclear Age* (New Brunswick, NJ : Rutgers Univer-

sity Press, 2008). On the persistence of nuclear secrecy, see Joseph Masco, *The Theater of Operations: National Security Affect from the Cold War to the War on Terror* (Durham, NC: Duke University Press, 2014).

56  "Strengthening of the Coordination of Humanitarian and Disaster Relief Assistance," September 8, 1995, UN NY S-1082/46/5/ acc. 2007/0015; J. Sinnaeve et al., "Collaboration between the Radiation Protection Research Actions of the CEC and the CIS on the Consequences of the Chernobyl Accident," in Shigenobu Nagataki, ed., *Nagasaki Symposium on Chernobyl: Update and Future* (Amsterdam: Elsevier, 1994), 95–114.

57  "N. S. Fes'kov to I. A. Kenik," February 12, 1991, GAGO 1174/8/2445: 15–16.

58  By 1993, in Khoiniki, Belarus, 6 in 1,000 children had thyroid cancer. V. M. Drozd, L. N. Astakhova, O. N. Polyanskaya, V. F. Kobzev, and A. S. Nalivko, "Ultrasonic Diagnostics of Thyroid Pathology in Children and Adolescents Effects by Radionuclides," WHO E16-445-11: 19. See also "Draft Meeting Report, Kyiv, Thyroid Cancer after the Chernobyl Accident," October 23, 1993, Riaboukhine, WHO E16-445-11: 21.

59  Fred H. Mettler, David V. Becker, Bruce W. Wachholz, and Andre C. Bouville, "Chernobyl: 10 Years Later," *Journal of Nuclear Medicine* 37, no. 12 (December 1996): 24.

60  "Anstee to Roland M. Timerbaev," November 1, 1991, WHO E16-180-4: 6; and "Brief for Secretary-General's Meeting," October 15, 1991, UN Archives S-1046/16/3.

61  "International Co-operation in the Elimination of the Consequences of the Chernobyl Nuclear Power Plant Accident," May 24, 1990, UNA S-1046/14/4; "Third Meeting of the Inter-Agency Task Force on Chernobyl," September 19–23, 1991, WHO E16-445-11: 5; and "Briefing Note on the Activities Relating to Chernobyl," June 3, 1993, Department of Humanitarian Affairs DHA, UNA s-1082/35/6/, acc 2002/0207. Japan had previously pledged $20 million in February 1991, which became the basis for funding the IPHECA project. WHO E16-445-11: 3.

62  "Anstee to Napalkov," January 17, 1992, WHO E16-445-11: 7.

63  "Notes of the Secretary-General's Meeting with the Minister of Foreign Affairs of Ukraine," September 22, 1992, United Nations Archive, New York (UN NY) S-1046/14/4, acc. 2001/0001; and "From President Lukashenko to the Secretary-General," October 28, 1996, UN NY S-1082/46/5/ acc. 2007/0015.

64  "Chernobyl: Mission to Russian Federation, Belarus, Ukraine," September 10–16, 1994, UN NY S-1082/46/5/ acc. 2007/0015; "For Information on United Nations, Press Conference Chernobyl," November 30, 1995, UN NY S-1082/46/5/ acc. 2007/0015; "A. M. Zlenko, A. N. Sychev, and S. V. Lavrov to Mr. Boutros Boutros-Ghali," January 9, 1995, UN NY S-1082/46/5, acc. 2007/0015; "Strengthening of the Coordination of Humanitarian and Disaster Relief Assistance," September 8, 1995, UN NY S-1082/46/5/ acc. 2007/0015; "Strengthen-

ing of International Cooperation and Coordination of Efforts to Study, Mitigate and Minimize the Consequences of the Chernobyl Disaster," October 27, 1997, "Press Conference on Funding to Address Effects of Chernobyl Disaster," May 1, 1998, Sergio Vieira de Mello, May 18, 1999, "Note to the Secretary-General," April 23, 2001, UN NY S-1092/96/5, acc. 2006/0160.

65  "Notes of the Secretary-General's meeting with the Minister of Foreign Affairs of Ukraine"; "Meeting of Jan Eliasson and Victor H. Batik," February 25, 1993; and "Meeting with Gennadi Buravkin, Belarus," March 4, 1993, WHO E16-445-11:16.

66  "Strengthening of the Coordination of Humanitarian and Disaster Relief Assistance," September 8, 1995, UN NY S-1082/46/5/ acc. 2007/0015.

67  "Strengthening of International Cooperation and Coordination of Efforts to Study, Mitigate and Minimize the Consequences of the Chernobyl Disaster," October 27, 1997.

68  "Lars-Erik Holm, Chairman of UNSCEAR, to Kofi A. Annan, Secretary-General," June 6, 2000, and Carolyn McAskie, Emergency Relief Coordinator, a.i., aide-mémoire, June 27, 2000, UN NY S-1092/96/5, acc. 2006/0160. For WHO director conceding the IAEA lead in Chernobyl issues, see *Nuclear Truth*, dir. Vladimir Tchertkoff (Feldat Film, Switzerland, 2004).

69  Dillwyn Williams and Keith Baverstock, "Chernobyl and the Future: Too Soon for a Final Diagnosis," *Nature* 440, no. 7087 (April 20, 2006): 993–94.

70  "The Human Consequences of the Chernobyl Accident: A Strategy for Recovering," 2002; and *The Chernobyl Forum: Chernobyl's Legacy, Health, Environmental and Socio-economic Impacts* (UN, 2005).

71  Didre Louvat, Conference Proceedings, *Commemoration of the Chernobyl Disaster: The Human Experience Twenty Years Later*, April 26, 2006 (Washington, DC: 2007), 33.

72  International Chernobyl Project, Proceedings of an International Conference, May 21–24, 1991, Vienna, Austria, 47.

73  Author telephone conversation with Fred Mettler, January 6, 2017.

74  "John Willis to Morten Andersen," April 23, 1991, GPA 1800; "A Group Deputy Question," and "Information to Provide Medical Help to the Pediatric Population," August 6, 1992, TsDAVO, 324/19/33: 8–11.

75  Morten Andersen, "Chernobyl Medical Brief," April 19, 1991, GPA 997.

76  "Andersen to Sawyer," August 12, 1991, GPA 1803.

77  Author telephone interview with Valentina Drozd, February 12, 2016.

78  Susan M. Lindee, *Suffering Made Real: American Science and the Survivors at Hiroshima* (Chicago: University of Chicago Press, 1994), 183.

79  The IAEA technical report mentions various reported cases of cancer, only one of which in Ukraine has been verified, and then this statement: "By the end of 1990, there were 20 verified cases of thyroid cancer in children of UkrSSR. Eleven of these were from non-contaminated settlements." Then later in the same text, "most of the reports of thyroid cancer were anecdotal in nature,"

*Technical Report*, 388. The Summary of the report calls the thyroid cases "hear-say." *The International Chernobyl Project: An Overview and Assessment of Radiological Consequences and Evaluation of Protective Measures* (Vienna: IAEA, 1991), 32–34.

80  Mettler et al., "Thyroid Nodules."

## Looking for a Lost Town

1  V. B. Nesterenko, V. S. Sergienko, and V. F. Shurkhai, "On Radiological Conditions in Veprin, Lesan', Bukunovichi," December 29, 1989, GAGO 1174/8/2336: 41–42.

2  "Conclusions on Radiological Circumstances and Conditions for Living in Veprin, Lesan', and Bukunovichi," (n.d. 1989) GAGO, 1174/8/2336, 43–47.

3  "On Carrying out the Resolution," August 15, 1991, NARB 507/1/5: 108–10; "On Course of Realization," September 26, 1991, NARB 507/1/6: 69–71; and "On Unsatisfactory Execution of Directives," April 1992, GAMO 7/5/4126: 69–71.

4  "Certificate on Analysis of Morbidity," 1991, "Memo," 1990, Arkhiv Kricheva 588/1/173: 35–36, 8–23. "On Inspection of Therapeutic Services of Krasnopole Region," April 18, 1989, NARB 46/12/1262: 8–12; and "Information on the State of Examinations in Regions under Control for 1989," August 1, 1989, NARB 46/12/1262: 45–48. See also V. I. Kulakov et al., "Female Reproductive Function in Areas Affected by Radiation after the Chernobyl Power Station Accident," *Environmental Health Perspectives* 101 (July 1993): 117–23.

5  "Values of Permissible Levels of cs-137," 1989, Arkhiv Kricheva 588/1/181: 5–7. Adults had more at 6,400 bq/kg of cs-137. For outlying high doses for Veprin children (38.9 mSv/3.89 rem for 1990 alone), see "Protocol," June 1990, Arkhiv Kricheva 588/1/184: 2–7.

6  V. N. Sych to E. P. Tikhonenkov, May 24, 1991, GAGO 1174/8/2445: 45–52.

7  "Ministry for Emergencies," July 12, 1999, GAGO 1174/8/3346: 36–41.

8  On estimated doses ranging from 35 to 65 rem, see "Conclusions on the Radiological Situation."

9  Vorobiev, *Do i posle Chernobylia*, 85.

## Greenpeace Red Shadow

1  "On Proposals," October 17, 1990; and "V. Petrovskii to V. Kh. Doluzhiev," July 20, 1990, NARB 46/14/1322: 16–17, 170.

2  "Kreisel to S. Tarkowski, Euro," February 14, 1990, WHO E16-522-6, jacket 1; and "Report of the Seventh Meeting of the Inter-Agency Committee for Response to Nuclear Accidents," Vienna, January 25–26, 1990, UNESCO, 361.9(470) SC ENV/596/534.1, Part I.

3  "Report on Assessment Mission to the Areas Affected by the Chernobyl Disaster, USSR," February 1990, GPA 1819.

4   Author telephone interview with Clare Moisey, February 19, 2018.

5   "Report on Assessment Mission."

6   "Australian Mission to the United Nations," November 26, 1974, UN NY S-0446-0106-09.

7   Paul Lewis, "David McTaggart, a Builder of Greenpeace, Dies at 69," *New York Times*, March 24, 2001; and Frank Zelko, *Make It a Green Peace! The Rise of Countercultural Environmentalism* (New York: Oxford University Press, 2013), 131-160.

8   "Steve Sawyer to David McTaggart," July 1, 1990, GPA 1624.

9   "Spizhenko to Didenko," March 1, 1990, TsDAVO 342/17/5252: 59.

10  "Confidential Report on 'Children of Chernobyl' Trip," April 3, 1990, GPA 1622; and author interview with Andrei Pleskonos, June 1, 2017, Kyiv.

11  "David McTaggart to Steve Sawyer," June 19, 1990, GPA 1624.

12  Doug Mulhall, "Revised Proposal and Steve's Comments," June 30, 1990; and "David McTaggart to Board," June 19, 1990, GPA 1624.

13  "Trip Log," May 1990, GPA 1624.

14  "Summary of Evaluation of Medical Needs to Serve the Children's Dispensary at Kyiv," October 1990, GPA 1804; and author telephone interview with Clare Moisey, February 19, 2018.

15  "To Kostiantyn Masyk from Nikolai Golushko," August 23, 1990, and "To the Council of Ministers, Ukraine SSR, K. I. Masyk," August 22, 1990, SBU 16/1/1288: 152–53, 155–56.

16  "Trip Log," May 1990, GPA 1624.

17  "Trip Log," May 1990, GPA 1624.

18  "Pickaver to Walker," October 10, 1990, GPA 997; and "John Willis to Doug Mulhall and David McTaggart," August 14, 1990, GPA 1625.

19  "David McTaggart to George Soros," 1990, GPA 1624.

20  Steve Sawyer, "Notes from Kyiv Trip," January 20, 1991, GPA 999.

21  "McTaggart to Soros Foundation," June 1990, GPA 1624.

22  "Cunningham to Walker," January 31, 1991, GPA 1799; "Andersen to Walker," February 16, 1991, GPA 1799; "Belcher to Walker," March 15, 1991, and "Mincey to Walker," March 1, 1991, GPA 1799; and "Mincey to Bloom," June 20, 1990, GPA 1624.

23  One physician, Everett Mincey, was a specialist in medical equipment, including radiological tools; "Walker to Andersen," June 18, 1991, GPA 1800.

24  "Anderson to Walker, Sawyer, McTaggart," February 22, 1991, GPA 1799.

25  Author telephone interview with Clare Moisey. On the "limited number of children," see "Mincey to Walker," March 1, 1991, GPA 1799.

26  "First Draft, Report of Meetings in Kyiv," December 11, 1992, GPA 994.

27  Author interview with Andrei Pleskonos, June 1, 2017, Kyiv.

28  "Olga Savran to Judith Walker," April 30, 1991, GPA 1800.

29  "Olga Savran to Judith Walker," April 30, 1991, GPA 1800; and "Andersen to Sawyer," August 12, 1991, GPA 1803.

30  Judith Walker, "A Briefing on the USSR Project," February 11, 1991, 999 GPA.

31  "Morten Andersen Memo to David Squire," August 23, 1991, GPA 1801; "Green-
    peace Kyiv to Yeager," August 26, 1991, GPA 1803; and J. R. Yeager, December 3,
    1991, GPA 1804.

32  "Greenpeace Kyiv to Squire," August 19, 1991, GPA 1803.

33  Author telephone interview with Clare Moisey; and "Morten Andersen re Sci-
    ence Kyiv to Judith Walker," February 25, 1991, GPA 999.

34  "Walker re Ernest McCoy," November 16, 1990, GPA 1819; and "Report of Meet-
    ing with Ricki Richardson," December 14, 1991, GPA 1804.

35  Author telephone interview with Clare Moisey. WHO also was wary of donations
    of equipment and "underexploitation": "Riaboukhine Travel Report Summary
    to USSR," August 23, 1991, WHO E16-180-4: 5.

36  Petryna, for example, argues that Lysenko influenced Soviet radiation medicine
    "by an absence of specific biological description." Petryna, *Life Exposed*, 119–20.

37  Hiroshi Ishikawa, "Radiation Study and the USSR Academy of Sciences in the
    Second Half of the 1950s: Beyond the Lysenkoists' Hegemony," ICCEES Con-
    ference, August 7, 2015, Makuhari, Japan; and Susanne Bauer, "Mutations in
    Soviet Public Health Science: Post-Lysenko Medical Genetics, 1969–1991," *Stud-
    ies in History and Philosophy of Biological and Biomedical Sciences* 47 (September
    2014): 163–72.

38  Ilya Gadjev, "Nature and Nurture: Lamarck's Legacy," *Biological Journal of the
    Linnean Society* 114, no. 1 (January 2015): 242–47.

39  Christopher Sellers, "The Cold War over the Worker's Body: Cross-National
    Clashes over Maximum Allowable Concentrations in the Post–World War II
    Era," *Toxicants, Health and Regulation since 1945*, ed. Soraya Boudia and Nath-
    alie Jas (London: Pickering and Chatto, 2013), 24–45.

40  *Soviet Scientists on the Danger of Nuclear Weapons Tests* (Moscow: Atomizdat,
    1959).

41  Carl F. Cranor, "Reckless Laws, Contaminated People," in *Powerless Science: Sci-
    ence and Politics in a Toxic World*, ed. Soraya Boudia and Nathalie Jas (New York:
    Berghahn, 2014), 197–214.

42  Patrick O. McGowan and Moshe Szyf, "The Epigenetics of Social Adversity in
    Early Life: Implications for Mental Health Outcomes," *Neurobiology of Disease*
    39, no. 1 (July 1, 2010): 66–72.

## The Quiet Ukrainian

1  "Belcher to Walker," February 25, 1991, GPA 1799.

2  "Belcher to Walker," March 15, 1991, and "Donald MacDonald to De Graaf,"
   March 1, 1991, GPA 1799.

3  "Cunningham to Walker," January 28, 1991, GPA 1799.

4  "Greenpeace Kyiv to Squire," August 19, 1991, GPA 1803.

5  Judith Walker, "A Briefing on the USSR Project," February 11, 1991, 999 GPA.

6  "Pickaver to Tykhyy," February 11, 1991, GPA 999.

7  "Sprange to Walker," January 30, 1991, GPA 1799.

8  "Sprange to Walker," February 2, 1991, GPA 1799.

9  "Belcher to Walker," March 26, 1991, GPA 1799.

10  "Belcher to Walker," March 27, 1991, GPA 1799.

11  Steve Sawyer, "Notes from Kyiv Trip," January 20, 1991, GPA 999.

12  Steve Sawyer, "Notes from Kyiv Trip," January 20, 1991, GPA 999.

13  For original plans, see "Protocol of Intentions between Greenpeace and the Shevchenkovskii Regional Council of People's Deputies," June 15, 1990, GPA 1624; "Proposal for a Joint Environmental Monitoring Programme of the Northern Ukraine," November 15, 1990, GPA 996; and "Zindler to Moisey and Mincey," 1990, GPA 999.

14  "Bloom to Lapshin," May 12, 1990, GPA 1622.

15  "Meeting with Scherbak, Gruzin, McTaggart, Sawyer, Walker, Tik [sic]i," January 17, 1991, GPA 1725.

16  "Tykhyy to J. R. Yaeger," December 26, 1991, 1001 GPA.

17  Author interview with Volodymyr Tykhyy, June 1, 2017, Kyiv.

18  Author interview with Andrei Pleskonos, June 1, 2017, Kyiv.

19  "Alan Pickaver to Judith Walker," October 10, 1990, GPA 997.

20  "On Attempts to Create a Military Structure in the Republic," November 24, 1990, SBU 16/1/1289: 66.

21  Author interview with Volodymyr Tykhyy, June 5, 2017, Kyiv.

22  "Certificate No. 8778," October 28, 1986, SBU 68/0463: 24; "Informational Communiqué," March 23, 26, and 30, 1987, SBU 16/1/1249, l. l: 73–77, 87, 95; "Informational Communiqué," October 26, November 11, and December 22, 1988, SBU 16/1/1266: 180–81, 223–24, 305–8; "On Poor Ecological Conditions," February 28, 1990, SBU 16/1/1284: 105–7; "On the Emergency Situation at the Dam," December 4, 1990, SBU 16/1/1289: 94.

23  "The Infolab Program," Tykhyy personal papers; and V. O. Tykhyy, "Review of Greenpeace International Project to Establish a Mobile Laboratory" (n.d.), GPA 999. For comparison with Greenpeace goals, see "Zindler to Moisey and Mincey," 1990, and "Doug Mulhall, re: Dr. Ricky Richardson," October 31, 1990, GPA 1622.

24  Doug Mulhall, "Revised Proposal and Steve's Comments," June 30, 1990, GP 1624. See also "Walker to Gruzin," November 15, 1990, GPA 996; and "Confidential Report on 'Children of Chernobyl' Trip," April 3, 1990, GPA 1622.

25  "Michael Calderbank to Alan Pickaver," July 16, 1991, re: Mobile Lab and Container for GP COC Rad/Tox Project, GPA 1000.

26  "Joint Greenpeace, Greenworld and Renaissance Foundation Radiological Toxicity Monitoring Programme in the Ukraine," Scientific Advisory Group, September 27, 1991, GPA 996; "Hoffman to Pickaver, Monthly Report," January 1991, GPA 999; "Tykhyy to Pickaver," August 16, 1991, GPA 1801.

27  "Science Unit to J. R. Yeager," March 18, 1992, GPA 1725.

28 Tykhyy, "Contamination of Soil, Food and Water by Toxic Chemicals in Ukraine," May 18, 1991, GPA 1000.

29 "Pickaver to Tykhyy," February 11, 1991, GPA 999; and "Tykhyy to Board Members," August 14, 1991, GPA 1001.

30 "Hoffman to Pickaver, Monthly Report," January 1991, 999 GPA; and "Minutes of a Meeting Rad/Tox," March 18, 1992, GP 1725.

31 "Martin Pankratz to Zindler," December 26, 1991, GPA 1001.

32 "Minutes of a Meeting Rad/Tox," May 18, "Service Division Consultation Meeting," July 15, 1992, and "Service Division Consultation Meeting," July 15, 1992, GPA 1725.

33 "Minutes of a Meeting in Kyiv Concerning the Radiological-Toxicological Monitoring Program," Kyiv, September 15, 1992, GPA 1725. GP headquarters decided to keep the lab open: "Alan Pickaver to Steve Sawyer," September 30, 1992, GP 1725.

34 Author interview with Volodymyr Tykhyy, June 5, 2017, Kyiv.

35 Author telephone interview with Iryna Labunska, February 27, 2018.

36 "Case Study of Radioactive Contamination in Zhytomir Ob, Ukraine," January 15, 1993, GPA 1002.

37 Author telephone interview with Olga Savran, January 17, 2018.

38 "McTaggart to Greenpeace Kyiv," August 27, 1991, GPA 1803.

39 "Pickaver to J. R. Yeager," July 16, 1992, GPA 1725.

40 Author interview with Aleksandr Komov, August 2, 2016, Rivne, Ukraine.

41 "Chernobyl Humanitarian Assistance and Rehabilitation Programme," May 21, 1993, WHO E16-180-4: 11.

42 "Chernobyl Humanitarian Assistance and Rehabilitation Programme."

43 "Andersen to Sawyer," August 12, 1991, GPA 1803.

44 "First Draft, Report of Meetings in Kyiv," December 11, 1992, GPA 994.

45 Alexey V. Yablokov, Vassily B. Nesterenko, Alexey V. Nesterenko, and Janette D. Sherman-Nevinger, *Chernobyl: Consequences of the Catastrophe for People and the Environment* (New York: Wiley, 2010).

## The Pietà

1 *Children of Chernobyl*, Yorkshire British Channel 4, 1991.

2 See, for example, James K. McNally, *Children of Chernobyl*, Lethbridge TV, 1990; and Maryann De Leo, *Chernobyl Heart*, Home Box Office, 2003.

3 William Sweet, "Chernobyl's Stressful After-Effects," *IEEE Spectrum* (November 1, 1999); and Elisabeth Rosenthal, "Experts Find Reduced Effects of Chernobyl," *New York Times*, September 6, 2005.

4 J. Lochard, T Schneider, and S. French, *International Chernobyl Project—Input from the Commission of the European Communities to the Evaluation of the Relocation Policy Adopted by the Former Soviet Union* (Luxembourg: Office for Official

Publications of the European Communities, 1992); and author telephone interview with Mona Dreicer, April 13, 2018.

5  Soraya Boudia, "Managing Scientific and Political Uncertainty," in *Powerless Science: Science and Politics in a Toxic World*, ed. Soraya Boudia and Nathalie Jas (New York: Berghahn, 2014), 95–112.

6  "Memo," 1992, NARB 507/1/20: 66–70.

7  "IAEA Project for a Repeat Assessment of the Situation," Moscow, October 15, 1990, NARB 507/1/1: 33; "Chernobyl—Nothing to Celebrate," 1991, GPA 1804.

8  Ilyin et al., "Strategy NKRZ."

9  Nancy Langston, *Sustaining Lake Superior: An Extraordinary Lake in a Changing World* (New Haven, CT: Yale University Press, 2017), 181.

10  "Dr. I. Filyushikin, Institute of Biophysics, Moscow, to Kreisel," April 27, 1993, WHO E16-3445-11: 16.

11  "Protocol No. 3, Meeting of the Collegium State Committee, BSSR," April 22, 1994, NARB 507/1/41, 28–38; Yaroshinskaya, *Bol'shaia lozh'*, 334.

12  Taras Kuzio, *Ukraine: Democratization, Corruption, and the New Russian Imperialism* (Santa Barbara, CA: Praeger Security International, 2015); and Valentin Maslyukov, "A Report from Minsk," *Monthly Review* 50, no. 4 (September 1998): 15–30.

13  Kuchinskaya, *The Politics of Invisibility*, 1606–12 [Kindle].

14  Author interview with Alexey Yablokov, June 5, 2015, St. Petersburg.

15  In *Life Exposed*, Petryna documents the half-hearted attempts to collect health data in the 1990s in Ukraine.

16  "Memo," no earlier than November 1993, GAMO 7/5/4156: 81–94.

17  "On the State of Children's Food," 1992, TsDAVO 324/19/33: 25–28.

18  Tatiana Kasperski, "Nuclear Dreams and Realities in Contemporary Russia and Ukraine," *History and Technology* 31, no. 1 (March 2015): 55–80.

19  Kuchinskaya, *The Politics of Invisibility*, 1539–76 [Kindle].

20  "Report of the Seventh Meeting of the Inter-Agency Committee for Response to Nuclear Accidents," Vienna, January 25–26, 1990, UNESCO, 361.9(470) SC ENV/596/534.1, Part I.

21  Seizin Topçu, "Chernobyl Empowerment?: Exporting 'Participatory Governance' to Contaminated Territories," in *Toxic World: Toxicants, Health and Regulation in the 20th Century*, ed. Soraya Boudia and Nathalie Jas (London: Pickering and Chatto, 2013), 135–58. For medical effects of a reduced clean meal program, see Ian Fairlie, *Torch-2016* (Vienna: Wiener Umweltanwaltsshaft, 2016), 84.

22  Quote from Didier Louvat, Head, Waste Safety Section, IAEA, Conference Proceedings, April 26, 2006, *Commemoration of the Chernobyl Disaster: The Human Experience Twenty Years Later* (Washington, DC: 2007), 25–30.

23  S. Lepicard and G. Dubreuil, "Practical Improvement of the Radiological Quality of Milk Produced by Peasant Farmers in the Territories of Belarus Contaminated by the Chernobyl Accident: The ETHOS Project," *Journal of Environmental Radioactivity* 56, no. 1–2 (2001): 241–53.

24  "Decision," December 29, 1993, GAMO 7/5/4156: 78–94; "Memo," no earlier than November 1993, GAMO 7/5/4156: 81–94.

25  "Memo," 1992, NARB 507/1/20: 66–70; William Sweet, "Chernobyl's Stressful After-Effects," *IEEE Spectrum* (November 1, 1999).

26  "Professor Iuryi Bandazhevskyi: mirnyi atom eto mif," *Tema*, August 30, 2011, accessed July 19, 2015, http://tema.in.ua/article/6677.html.

27  Y. I. Bandazhevsky, "Chronic Cs-137 Incorporation in Children's Organs," *Swiss Medical Weekly* 133 (2003): 488–90; and G. S. Bandazhevskaia et al., "Relationship between Caesium (137Cs) Load, Cardiovascular Symptoms, and Source of Food in Chernobyl Children," *Swiss Medical Weekly* 134 (2004): 725–29.

28  Tchertkoff, *The Crime of Chernobyl*, 201.

29  "Three Sites Singled Out for Nuclear Power Plant in Belarus," No. 2, March 26, 1998, *Interfax*, Belapan Radio, February 4, 1998; and "Lukashenko Says People Will Decide on Nuclear Power Plant," FBIS-SOV-98-035.

30  Author interview with Yuri Bandazhevsky, July 20, 2015, Ivankiv, Ukraine.

31  Author interview with Yuri Bandazhevsky, July 20, 2015, Ivankiv, Ukraine.

32  Iu. I. Bandazhevskii, "*Chernobyl': Ekologiia i zdorov'ia*, vol. 1 (Ivankiv, 2014). For corroborating statistics, see National Research Center for Radiation Medicine, "Thirty Years," 61–65.

33  Dillwyn Williams, "An Unbiased Study of the Consequences of Chernobyl Is Needed," *The Guardian*, January 17, 2010.

34  Eugenia Stepanova et al., "Exposure from the Chernobyl Accident Had Adverse Effects on Erythrocytes, Leukocytes, and Platelets in Children in the Narodichesky Region, Ukraine: A 6-Year Follow-Up Study," *Environmental Health* 7 (2008): 21.

35  M. R. Sheikh Sajjadieh et al., "Effect of Cesium Radioisotope on Humoral Immune Status in Ukrainian Children with Clinical Symptoms of Irritable Bowel Syndrome Related to Chernobyl Disaster," *Toxicology and Industrial Health* 27 (2011): 51–56.

36  Erik R. Svendsen et al., "Cesium 137 Exposure and Spirometry Measures in Ukrainian Children Affected by the Chernobyl Nuclear Accident," *Environmental Health Perspectives* 118, no. 5 (May 2010): 720–27; and Erik R. Svendsen et al., "Reduced Lung Function in Children Associated with Cesium 137 Body Burden," *Annals of the American Thoracic Society* 12, no. 7 (July 2015): 1050–57.

37  Eero Pukkala, "Breast Cancer in Belarus and Ukraine after the Chernobyl Accident," *International Journal of Cancer* 119 (2006): 651–58.

38  The Davis group found an overall "significant increase in leukemia risk with increasing radiation dose." Among the three regions (Ukraine, Belarus, and Russia), it only found a significant effect in Ukraine, where the sample size was the largest. This study matched case and controls on residence, which carries a danger of overmatching because food circulated in local markets where people resided. Overmatching, then, was likely to mean case and controls had similar

exposures from local, contaminated food. Overmatching can lead to underestimation of the effect of interest. Despite this difficulty, an overall significant effect was found. Scott Davis et al., "Childhood Leukaemia in Belarus, Russia, and Ukraine Following the Chernobyl Power Station Accident," *International Journal of Epidemiology* 35 (2006): 386–96. See also A. G. Noshchenko et al., "Radiation-Induced Leukaemia among Children Aged 0–5 Years at the Time of the Chernobyl Accident," *International Journal of Cancer* 127 (2010): 214.

39  A. I. Niagu et al., "Effects of Prenatal Brain Irradiation as a Result of the Chernobyl Accident," *International Journal of Radiation Medicine* 6, no. 1–4 (2004): 91–107.

40  Fairlie, *Torch-2016*, 76; and *20 let posle Chernobyl'skoi katastrofy, natsional'nyi doklad* (Minsk, 2006): 59.

41  K. Spivak et al., "Caries Prevalence, Oral Health Behavior, and Attitudes in Children Residing in Radiation-Contaminated and Non-Contaminated Towns in Ukraine," *Community Dental Oral Epidemiology* 32 (2004): 1–9.

42  Michael Gillies et al. "Mortality from Circulatory Diseases and Other Non-Cancer Outcomes among Nuclear Workers in France, the United Kingdom and the United States (INWORKS)," *Radiation Research* 188, no. 3 (2017): 276–90; and Klervia Leuraud, David B. Richardson, et al., "Ionising Radiation and Risk of Death from Leukaemia and Lymphoma in Radiation-Monitored Workers (INWORKS): An International Cohort Study," *The Lancet. Haematology* 2, no. 7 (July 2015): e276–81.

43  Benedict O'Donnell, "Low-Dose Radiation May Be Linked to Cancer Risk," *Horizon: The EU Research and Innovation Magazine* (May 30, 2016); and Munira Kadhim et al., "Non-Targeted Effects of Ionising Radiation—Implications for Low Dose Risk," *Mutation Research* 752 (2013): 84–98.

### Bare Life

1  On corruption, see "Memo," no earlier than November 1993, GAMO 7/5/4156: 81–94.

2  "Progress Report," March–September 1994, Chernobyl Studies Project, Working Group 7.0, DOE, UCRL-ID-110062-94-6.

3  Author interview with Irina Vlasenko, April 13, 2016, Minsk.

4  In 2016, 82 percent of residents of contaminated zones were under medical observation. "Natsional'nyi doklad Respubliki Belarus," *30 let Chernobyl'skoi avarii* (Minsk, 2016), 16–20.

5  Author interview with Nikolai Kachan, July 28, 2016, Valavsk, Belarus.

6  For a negative appraisal of UN Chernobyl assessments, see European Commission Radiation Protections No. 170, "Recent Scientific Findings and Publications on the Health Effects of Chernobyl" (Luxembourg: Directorate-General for Energy, 2011).

## Conclusion: Berry Picking into the Future

1 In 2015, Ukraine exported 19,000 tons of wild berries, thirty times more than the year before. Email correspondence with Ivgen Kuzin, International Relations Manager of Fruit-Inform, September 11, 2016.

2 Alexis Madrigal, "Chernobyl Exclusion Zone Radioactive Longer Than Expected," *Wired* (December 15, 2009).

3 Author interview with Wladimir Wertelecki, May 5, 2016, Washington, DC.

4 The second highest rates of birth defects of this kind occurred in northern England, near the Windscale Plutonium Plant. Wladimir Wertelecki, "Chernobyl 30 Years Later: Radiation, Pregnancies and Developmental Anomalies in Rivne, Ukraine," *European Journal of Medical Genetics* 60 (2017): 2–11.

5 Author interview with Wladimir Wertelecki, July 12, 2016, Rivne, Ukraine.

6 "From Commission of the European Communities, Com (87), Minutes 894," November 4, 1987, PSP 133, European Union Archive, Florence, Italy; and "Council Regulation (Euratom) 2016/52 of 15 January 2016," accessed May 9, 2018, https://eur-lex.europa.eu/legal-content/EN/TXT/?uri=CELEX%3A32016R00.

7 Khalil Boudjemline, "CBSA's Radiation Detection Program (RADNET)," Technical Reach-Back Workshop, Joint Research Centre, Ispra, Italy, March 28, 2017. Author telephone interview with Khalil Boudjemline, Research Engineer, Canada Border Services Agency, April 21, 2018.

8 *The Rush Limbaugh Show*, November 9, 2016.

9 For arguments to engage in "conspiracies with other forms of life," see Natasha Myers, "From Edenic Apocalypse to Gardens against Eden," in *Infrastructure, Environment and Life in the Anthropocene*, ed. Kregg Hetherington (Raleigh, NC: Duke University Press, forthcoming).

10 Author telephone interview with Sarah Phillips, May 18, 2018.

11 Manfred Dworschak, "The Chernobyl Conundrum: Is Radiation as Bad as We Thought?" *Spiegel Online*, April 26, 2016.

12 National Research Center for Radiation Medicine, "Thirty Years of Chernobyl Catastrophe: Radiological and Health Effects" (National Report of Ukraine: Kyiv, 2016): 7.

13 See analysis by Sonja Schmid, "Nuclear Emergencies and the Masters of Improvisation," *Bulletin of the Atomic Scientists*, April 25, 2016.

14 Sarah D. Phillips, "Contamination of Japan 43—Sociology of Nuclear Disaster," *Nuclear Exhaust*, accessed May 16, 2018, https://nuclearexhaust.wordpress .com/2015/04/30/contamination-of-japan-43-sociology-of-nuclear-disaster -sarah-d-phillips/.

15 "The Legacy of Nuclear Testing | ICAN," accessed May 25, 2018, http://www .icanw.org/the-facts/catastrophic-harm/the-legacy-of-nuclear-testing/.

16 For infants in Utah: 120–420 rads. or 1,200-4,200 mSv. Charles Mays, "Estimated Thyroid Doses and Predicted Cancers in Utah," DOE NV0403156, as reproduced in Andrew Kirk, *Doom Towns: The People and Landscapes of Atomic*

*Testing* (New York: Oxford University Press, 2017), 199–201. For nationwide estimates, see Hoffman et al., "A Perspective on Public Concerns about Exposure to Fallout from the Production and Testing of Nuclear Weapons."

17 Elizabeth Tynan, *Atomic Thunder: The Maralinga Story* (Sydney: New South Publishing, 2016); Nic Maclellan, *Grappling with the Bomb* (Acton: ANU Press, 2017); and B. Danielsson, "Poisoned Pacific: The Legacy of French Nuclear Testing," *Bulletin of the Atomic Scientists* 46, no. 2 (March 1990): 22.

18 Hyeyeun Lim et al., "Trends in Thyroid Cancer Incidence and Mortality in the United States, 1974–2013," *Journal of the American Medical Association* 317, no. 13 (April 4, 2017): 1338–48; B. A. Kilfoy et al. "International Patterns and Trends in Thyroid Cancer Incidence, 1973–2002," *Cancer Causes Control* 20 (2009): 525–31; and F. De Vathaire et al., "Thyroid Cancer Following Nuclear Tests in French Polynesia," *British Journal of Cancer* 103 (2010): 1115–21.

19 Peter Kaatsch, "Epidemiology of Childhood Cancer," *Cancer Treatment Reviews* 36, no. 4 (June 1, 2010): 277–85; National Cancer Institute, "SEER Cancer Statistics Review, 1975–2013," https://seer.cancer.gov/archive/csr/1975_2013/#contents. Upward trends are not a given; in Brazil, for example, they declined. Arnaldo Cézar Couto, "Trends in Childhood Leukemia Mortality over a 25-Year Period," *Jornal de Pediatria (Rio J)* 86, no. 5 (2010): 405–10.

20 Children's Leukemia and Cancer Research Foundation, https://childcancer research.com.au/.

21 Niels Jorgensen, Jaime Mendiola, Hagai Levine, Anderson Martino-Andrade, Irina Mindlis, Shanna H. Swan, et al., "Temporal Trends in Sperm Count: A Systematic Review and Meta-Regression Analysis," *Human Reproduction Update* 23, no. 6 (December 2017): 646–59.

# INDEX